I0644766

Flora's Interpreters

Flora's Interpreters

Popular Botany in Nineteenth-Century America

Ann A. Merrill

THE UNIVERSITY OF GEORGIA PRESS · *Athens*

A Wormsloe
FOUNDATION
PUBLICATION

© 2026 by the University of Georgia Press
Athens, Georgia 30602
www.ugapress.org
All rights reserved
Designed by Louise OFarrell
Set in Adobe Arno by Louise OFarrell
Printed and bound by Friesens
The paper in this book meets the guidelines for permanence and durability of the Committee on Production Guidelines for Book Longevity of the Council on Library Resources.

Use of any part of this book in training for any artificial intelligence (AI), large language model (LLM), machine learning technologies, or similar generative language system without license is expressly prohibited.

Printed in Canada
30 29 28 27 26 C 5 4 3 2 1

EU Authorized Representative
Easy Access System Europe—Mustamäe tee 50, 10621 Tallinn, Estonia, gpsr.requests@easproject.com

Library of Congress Cataloging-in-Publication Data
Names: Merrill, Ann A. author
Title: Flora's interpreters : popular botany in nineteenth-century America / Ann A. Merrill.
Description: Athens : The University of Georgia Press, [2026] | Includes bibliographical references and index.
Identifiers: LCCN 2026001665 | ISBN 9780820374635 hardback
Subjects: LCSH: Botanical literature—United States—History—19th century | Botany—United States—History—19th century | Herbaria—United States—History—19th century | Environmentalism—United States—History—19th century | Botanists—Diaries
Classification: LCC QK14.5 .M47 2026
LC record available at https://lccn.loc.gov/2026001665

For Mom, who loves flowers as much as I do
and in memory of Dad, who counted
all the wildflowers when he hiked

Contents

Acknowledgments

During the many years I dedicated to *Flora's Interpreters*, I frequently read acknowledgments, wondering how anyone ever finished a book. Now I know that it takes the financial, logistical, collegial, and emotional support of many people, and I am honored to thank them here.

I could not have done this project without the resources provided by the Thomson Professorship of Environmental Studies. I am very grateful to Todd Thomson and Melissa McKeithen for their generous gift, which has supported archival research, antiquarian book purchases, and publishing subvention funds. At Davidson College, several deans of faculty have enthusiastically championed the book, including Clark Ross, Wendy Raymond, and Shelley Rigger. Three sabbatical leaves enabled me to write and conduct research away from teaching and administrative duties, including a year funded by the Boswell Family Faculty Fellowship. Rebecca Rodriguez provided expert sleuthing in addition to her day job, as did student assistants Kate Wiseman Reuterswärd and Jeannie Kinnett DeLoach.

Initial archive research was funded by fellowships from the Maine Women Writers Collection and the Winterthur Museum, Garden & Library; a Francois Andre Michaux Fund Library Resident Fellowship from the American Philosophical Society; and an Ernestine Richter Avery Fellowship from the Huntington Library. The staff of these institutions were enormously helpful, particularly Kay Collins (Winterthur), Cally Gurley (MWWC), Dan Lewis (Huntington), and Earle Spamer (APS). A special shout-out to Kasey Grier for creating community among research fellows at Winterthur.

I've loved working with the University of Georgia Press. I thank the very talented design team for exceeding all my expectations. Jon Davies and freelance copyeditor Anne Coulling brought expert eyes to every detail; any remaining errors are, of course, my own. For their encouragement and support, I thank editors past and present Christa Frangiamore, Nancy Grayson, Patrick Allen, Beth Snead, and Nate Holly.

Much gratitude to my colleagues in the English Department at Davidson, especially Shireen Campbell, Randy Ingram, Zoran Kuzmanovich, Elizabeth Mills, Randy Nelson, Alan Michael Parker, and Onita Vaz; extra special thanks to Suzanne Churchill for years of great feedback and to Gail Gibson for being a model mentor. I also thank current and former colleagues in Environmental Studies, including Brad Johnson, Amy Kohout, Chris Paradise, Jen Garcia Peacock, and Matt Samson; and for interdisciplinary friendship, Cindy Hauser and Jaya Jha.

For invitations to share my research in keynote and plenary addresses, I thank Christa Frangiamore (Zahner Conservation Lecture Series, Highlands, N.C.), Richard Hunt (ASLE Symposium on Literature as Mediated Nature, Delaware Valley College), Mary Kuhn (Plant Studies Symposium, University of

Virginia), and Theresa Rodewald (Symposium on Exploring the Poetical Sciences, American Folk Art Museum, New York, N.Y.).

I have been very fortunate to have the exceptional community of the Association for the Study of Literature and Environment (ASLE). For early mentoring and encouragement, I thank John Calderazzo, SueEllen Campbell, Terrell Dixon, and John Tallmadge. For ongoing collegiality, inspiration, and friendship, I am grateful for Joni Adamson, April Anson, Karla Armbruster, Janine DeBaise, Ann Fisher-Wirth, Carmen Flys Junquera, Greta Gaard, John Lane, Mark Long, Ian Marshall, Dan Philippon, Arlene Plevin, Cate Sandilands, and David Taylor. I cherish fellow nineteenth-century Americanist ecocritics Mike Branch, Tina Gianquitto, Rochelle Johnson, Lauren LaFauci, and Jim Warren: this book would absolutely not have been accomplished without y'all.

Several fabulous writing groups have supported me over the years, starting in graduate school, and GW Girls Ellen Garrett, Kim Loudermilk, and Maria Pramaggiore continue to delight and advise. Karen Kilcup generously invited me to join her and graduate students from the University of North Carolina at Greensboro at a pivotal point early in my career. Many thanks to Writer's Bloc members Tina Gianquitto, Tom Hallock, Rochelle Johnson, and Lauren LaFauci. Davidson colleagues Hilary Green and Rose Stremlau were stalwart Zoom writing pals, and I congratulate them on recently completing books of their own.

For keeping my body and soul together, I thank Frank Giancarli, Seth Moliver, and Annie Seier. For decades of laughs, tears, hugs, advice, and adult beverages, I am thoroughly indebted to w(h)iners Suzanne Churchill, Kristi Multhaup, Jeanne Neumann, and Shelley Rigger. Thanks to Amy Diamond for great hikes and willingness to be a general reader. I am blessed with the support of family members Tom, Theresa, and Regina; Beth, Gregg, and Sarah; Jack, Sadie, Jackson, and Oliver; and especially the amazing Grace Merrill, my beloved mom. *Flora's Interpreters* is dedicated to her and my dad Dick Merrill, who died before I finished but inspired me in many ways. My three- and four-legged housemates Rex and Bella make me laugh every single day.

Endless gratitude to the conservationists, botanists, citizen scientists, and many others who dedicate their lives to protecting plants: the remarkable, irreplaceable beings that make life possible.

Flora's Interpreters

FIGURE I.1. Frontispiece, Mason, *A Wreath from the Woods of Carolina* (1859)

Introduction

Flower Books, Popular Botany, and American Environmentalism

Young botanists, as well as those who are older, may derive great pleasure in making excursions into the fields, and upon the hills and mountains, for the purpose of collecting plants. Thus they learn to love every blossom which springs up under their feet; their hearts beat with pleasure when they meet with some little strange flower, which exhibits new traits into the character of the vegetable race. Every murmuring brook shows its banks clad with flowery treasures; the forests and groves exhibit another, but not less beautiful assemblage of plants; and the mountain, the valley, and the sea coast, have all their own peculiar vegetable productions.

—Almira Hart Lincoln Phelps, *Botany for Beginners* (1833)

Flowers cannot be monopolized. The poor have them as well as the rich. It does not require such an education to love and appreciate them, as it would to admire a picture of Turner's, or a statue of Thorwaldsen's. And as they are messengers of affection, tokens of remembrance, and presents of beauty, of universal acceptance, it is pleasant to think that, in them, all men recognize a brief brotherhood.

—Henry Ward Beecher, "A Discourse on Flowers" (1853)

From one end of this vast continent to the other, which embraces every climate from that of the frozen north to the tropics, the earth is carpeted with multitudinous flowers, each coming in its due season, living its pure life, and passing away, unknown, alas, to millions who know not what they miss. From the stunted pine, bending beneath its burdens of almost perennial snows toward the arctic circle, to the graceful palmetto, flourishing beneath an almost vertical sun; from the slender poppy which braves the boreal climate of bleak coasts and dreary barrens trodden only by the Esquimaux, to the delicate Southern orchid daintily clinging to oak or cypress, the continent of North America affords such a variety of plants that not even the botanist can hope to know them all.

—*Wild Flowers of America* (1894)

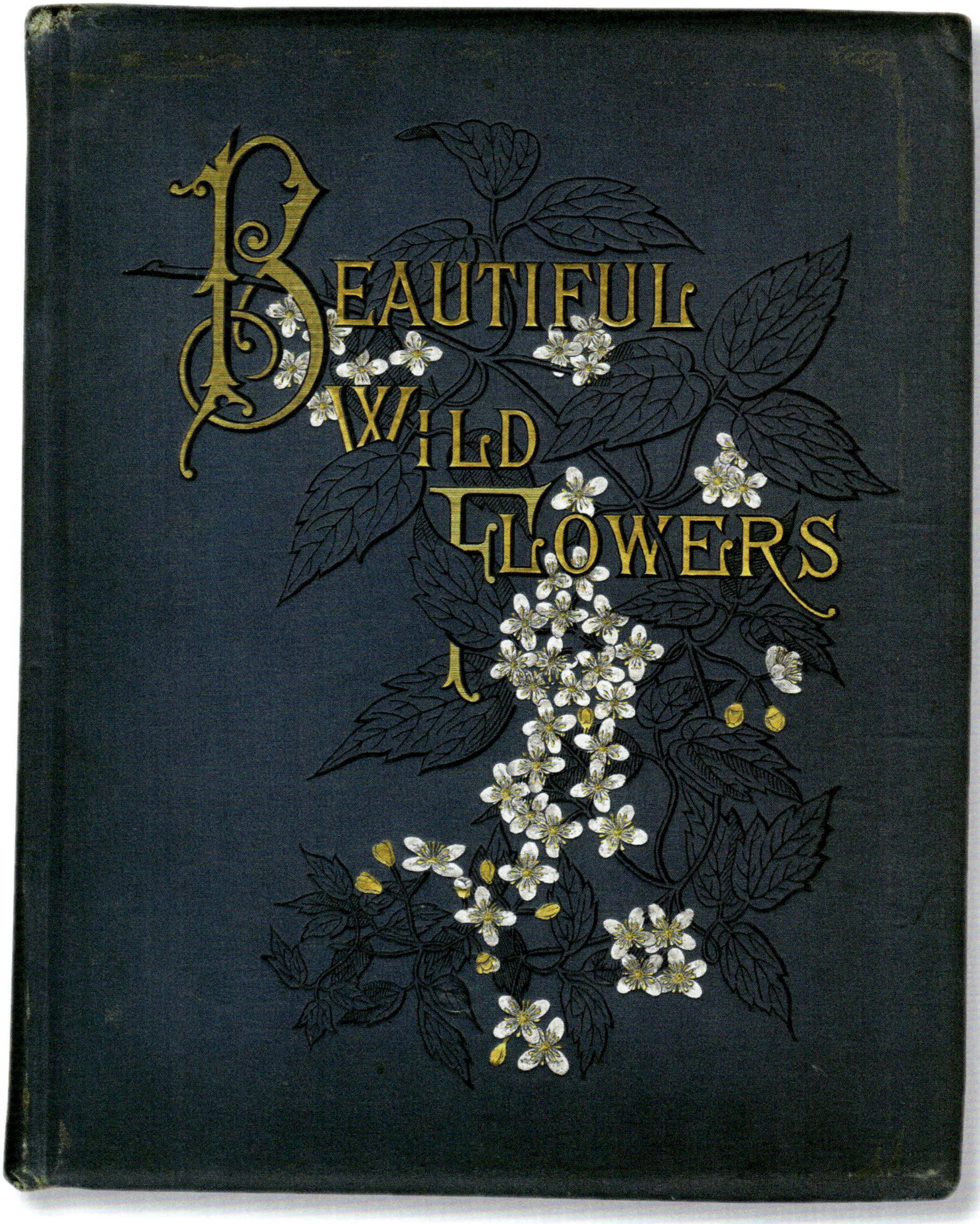

FIGURE I.2. Cover, Sprague and Hervey, *Beautiful Wild Flowers of America* (1882)

IT ALL STARTED WITH AN HERBARIUM. I was taking a tour of the Old Manse in Concord, Massachusetts, where Ralph Waldo Emerson lived from 1834 to 1835 and which the newly wed Sophia and Nathaniel Hawthorne rented from 1842 to 1845; Henry David Thoreau planted them a vegetable garden as a wedding present. As a scholar of nineteenth-century America, I marveled at what those walls had seen and heard. At one point, the tour guide drew our attention to a facsimile page of the herbarium of Sarah Alden Bradford Ripley, a later resident of the house. I had never heard of Ripley nor of an herbarium, but as a lover of plants, I was intrigued by this display of preserved botanical specimens. The next stop on my literary pilgrimage was Amherst, where I visited Emily Dickinson's house; she too kept an herbarium, and it was beautiful. I was hooked.[1]

Twenty years later, I am putting the finishing touches on this book. Along the way, I have spent a lot of time in archives, learning about the many ways that plants were significant to nineteenth-century Americans. I was especially drawn to books, for both their contents and what I encountered in them: pressed flowers, copious marginalia, and fascinating inscriptions. I admired the intricate gold-stamped bindings and the full-color botanical illustrations inside. I studied botany the way someone would in the 1800s, by reading their textbooks. Nonpublished sources were often the most fun to discover—scrapbooks of advertising trade cards, a diary with embroidery, a Bible salesman's canvassing book—all featuring flowers in unusual ways. The more I researched, the more intrigued I became by the intersections of botany, books, and environment. *Flora's Interpreters* is the result.

The epigraphs to this chapter provide some context. The earliest comes from Almira Hart Lincoln Phelps (1793–1884), author of some of the best-selling botany textbooks of the nineteenth century, including *Botany for Beginners*. Phelps lists the many different environments where flowers grow—fields, hills, mountains, brooks, forests, groves, valleys, and seacoast—to entice young botanists to learn and love flowery treasures, assemblages of plants, and peculiar vegetable productions. Phrases heaped upon clauses syntactically reinforce the natural variety she portrays. Phelps was one of many writers in the 1820s and 1830s who described, defended, and celebrated the United States as distinctive, and a large part of that distinction was its natural environments.

Henry Ward Beecher (1813–1887) was a Congregationalist minister, abolitionist, and social reformer; his sister was Harriet Beecher Stowe, author of *Uncle Tom's Cabin* (1851). In his essay "A Discourse on Flowers," he positions flowers as democratic: they cannot be monopolized, are equally available to rich and poor, and do not require the specialized education needed to analyze fine art by Europeans such as the painter J. M. W. Turner and sculptor Bertel Thorvaldsen. Flowers also have the power to signify, as messengers and tokens of beauty while alive and of remembrance when preserved. Beecher's allusion to the language of flowers accords with its popularity in the 1850s, when sentimentalism dominated American literature and culture.

Mary Ann Mason (1802–1881), a native of North Carolina, featured a profusion of her home state's beautiful wildflowers in the decorative title page to *A Wreath from the Woods of Carolina* (figure I.1), a collection of morally didactic stories for children that she wrote and illustrated. Her wreath features botanically accurate renderings of flowers she labeled yellow jessamine, woodbine, blue jessamine, morning glory, honeysuckle, wild crab blossoms, lily of the valley, blue iris, magnolia, violet, meadow lily, passionflower, forget-me-not, trumpet flower, lupine, and wild pinks. Each story is built around one or more of these personified flowers, which also appear in full-page illustrations. The same year that Mason published *A Wreath from the Woods of Carolina*, John Darby published *Botany of the Southern States*; both books continued the work of expanding U.S. botany beyond its New England origins.

The cover of *Beautiful Wild Flowers of America* (figure I.2) depicts virgin's bower or wild clematis (*Clematis virginiana L.*), one of fourteen species featured in this folio-sized volume of full-color plates and detailed information about each flower, including habitat, growing season, uses, history, and folklore. The title specifies that these are *wild*flowers, and as the leaves and stems of the clematis intertwine its letters, they mimic how this climbing vine grows in the wild (figures I.3 and I.4). The title also emphasizes that these are wildflowers of *America*. By the 1880s, the nation included all of the present-day continental United States. When decades of government-sponsored geographical surveys mapped expanding territory, they also detailed its fauna and flora. Commitments to preserving wild lands inspired federal legislation such as S. 392, "An Act to Set Apart a Certain Tract of Land Lying near the Head-waters of the Yellowstone River as a Public Park" (1872), and organizations such as the American Forestry Association (founded 1875) advocated for the protection of native forests.

At the end of the century, *Wild Flowers of America* (1894) included "flowers of every state in the American union" in 288 color plates, from *Abronia turbinata* (transmontane sand verbena), "among the many curious flowers that inhabit the hot plains of the great Southwest," to *Zygadenus glaucus* (mountain death-camas), which "grows in Canada and New England, thence west to Minnesota and northward." The epigraph from this book echoes Phelps's catalog of American environments from sixty years earlier, now including the climates of this "vast continent," from "almost perennial snows" to "almost vertical sun." The stunted pine and graceful palmetto, slender poppy and delicate southern orchid, are but a few of North America's "multitudinous" native species. Even in 1894,

FIGURE I.3 and I.4. Details, cover, Sprague and Hervey, *Beautiful Wild Flowers of America* (1882)

there is "such a variety of plants that not even the botanist can hope to know them all." One year earlier, historian Frederick Jackson Turner presented his essay on "The Significance of the Frontier in American History" at the annual meeting of the American Historical Association. Responding to the 1890 U.S. census report, which declared the frontier "closed," Turner articulated how the frontier, both geographically and conceptually, had shaped American culture and identity: As the boundaries of the United States expanded, "each frontier did indeed furnish a new field of opportunity, a gate of escape from the bondage of the past." He also noted that "the most significant thing about the American frontier is, that it lies at the hither edge of free land."[2] The availability of this "free land"—free for the taking (from its Indigenous inhabitants), free of charge (if you farmed it for five years), and often free of the social constraints of more settled places to the east—also figured prominently in shaping American ideas about its natural environments.

These epigraphs indicate how books from throughout the nineteenth century in the United States used flowers to convey America's distinctive geographical diversity, democracy, and abundance. Each of these authors—Phelps; Beecher; Mason; Sprague and Hervey; and the anonymous compiler of *Wild Flowers of America*—expressed the significance of plants for American culture and identity, reflecting the popularity of botany in nineteenth-century America. Books as diverse as Phelps's botany textbook, Mason's didactic flower fables for children, and Sprague and Hervey's beautiful display all compose the more general category of flower books. The interplay of flower books, popular botany, and American environmentalism is the subject of *Flora's Interpreters*.

Flower Books

In the nineteenth-century United States, the popularity of botany, love of flowers, and new technologies of printing and binding coalesced in the *flower book*, a term I use for the combination of scientific, literary, visual, and material features that characterize the genres I discuss. Genre, like other forms of classification such as plant taxonomy, is both useful and limited. On the one hand, it provides structure and defines similarities, creating order out of chaos. On the other, it can overlook outliers, marginalize departures from the norm, and preclude other ways of organizing the data. The genres I discuss in *Flora's Interpreters*—botany instruction books, herbaria, and flower language dictionaries—all have clear definitions as well as messy overlaps. Botany instruction books included but were not limited to traditional textbooks, as chapter 1 details. The herbarium—a collection of dried, pressed plant species—had both scientific and aesthetic aims. The language of flowers paired a specific plant with an emotional trait, and American authors compiled dictionaries for decoding these relationships. But botany textbooks also included sections on the language of flowers, and flower language dictionaries included botany instruction. The herbarium was artistic and memorial, inspiring plant-based creations well beyond its original taxonomic purpose. Pressed flowers appear in both botany textbooks and flower language dictionaries, material reminders that a person picked that flower, preserved it, and placed it in the book for a reason. As these examples also show, the three terms—botany, plant, and flower—have specific meanings as well as overlapping uses. I use "botany" primarily in scientific and educational contexts—botany textbook, botanical specimen,

FIGURE I.5. Goodale and Goodale, *In Berkshire with the Wild Flowers* (1879–1880)

amateur botany—and "plant" for the whole organism, not just its blooming parts. Flowers had particular cultural applications in nineteenth-century America: the language of flowers, flower missions, flower painting, and flower gardens all resonate in ways that "botany" and "plant" do not in these contexts.[3] I use "flower book" because it encompasses the variety, hybridity, and polyvocality that characterize these books.

These three genres were by no means the only kinds of flower books available in the United States in the nineteenth century. Compilations of poems about flowers, collections of nature essays about flowers, and horticulture and garden guides are also flower books; ornamental bindings were common to all three types (figures I.5–I.7). Others, like Emma C. Embury's *American Wild Flowers in their Native Haunts* (1845; alternately titled *Nature's Gems*), combined in one volume several features of flower books: ornate binding, full-page botanical illustrations, descriptions of plants, and celebration of American natural environments.

The gold-stamped image on the cover of Embury's text (figure I.8) depicts a wooden trellis covered in climbing vines of leaves and blooms, an urn-shaped vase or planter draped with flowers, and birds flying, perching, and nesting in an overall environment of healthy profusion, prefiguring the variety of content inside. The frontispiece (figure I.9) features both a landscape and an array of flowers, a pairing atypical of "the traditional botanical conventions of showing the plant uprooted against an empty background."[4]

FIGURE I.6. Lorimer, *Among the Trees* (1869)

FIGURE I.7. *Vick's Flower and Vegetable Garden* (n.d.)

FIGURE I.8. Cover, Embury, *American Wild Flowers in their Native Haunts* (1845), alternately titled *Nature's Gems*

Embury (1806–1863) was a poet, essayist, fiction writer, and magazine editor, a combination not uncommon for women writers at the time.[5] She introduces *American Wild Flowers in their Native Haunts* by stating, "Every one hears of our towering mountains, our mighty rivers, our dense forests, our ocean-like lakes and our boundless prairies. The grand features of nature are so imposing that we forget the lesser beauties, . . . the blossoms which fringe our rushing streams and enamel our sunny vallies are rarely noted among the characteristics of American scenery." This epic catalog of American landscapes was, as the epigraphs also show, a staple of nineteenth-century descriptions about the nation's diverse topography. Embury asserts the equal importance of American flowers, especially those "in the wilderness,—blossoms sown by the wind, nursed by the shower, peering from their covert on the hill-side, smiling upon us from the cleft of some dark ravine, looking down tenderly from the face of some rugged cliff."[6] The book has twenty full-page illustrations that show a specific American landscape and one or two native wildflowers that grow there. The frontispiece, like Mason's in *A Wreath from the Woods of Carolina*, depicts an assortment of native flower species, including lady's slipper and trillium, which are not covered elsewhere in the book but are still "nature's gems," indicating how precious they are. Each illustration is accompanied by "botanical and local descriptions" written by the book's illustrator, Edwin Whitefield, and a selection of poems and fiction, most of which Embury wrote. Both landscapes and literature compose the "native haunts" of these American wildflowers.

The third plate provides an example of the book's illustrations (figure I.10). The plate reverses what Embury critiqued in her introduction; here, the flowers are the main event: vividly colored, botanically accurate, grand, imposing, and filling most of the page. By contrast, the rugged landscape of Passaic Falls fades into the background, in soft-focus black and white. Whitefield's description of *Houstonia cærulea* (innocence or fairy flax) and *Geranium maculatum* (spotted crane's-bill or crow-foot geranium) notes their geographic ranges and bloom times; he also praises the *Houstonia* as "a general favorite" and the *Geranium maculatum* as "much more worthy of cultivation than many of the exotic species so universally nurtured in our green-houses." The foregrounding of full-color flower specimens against a fainter background also appears in *The Scriptural Album with Floral Illustrations*, the subject of chapter 6. Margaret Welch notes that "the emergence in this period of the . . . representative specimen against its suggested haunt . . . gives the species the attention usually accorded humans."[7]

With flower books as my primary sources, one of the analytical frameworks for *Flora's Interpreters* is print culture studies and history of the book. This approach analyzes books not just for their content but also their production, distribution, circulation, and reception. Social contexts such as literacy rates, access to libraries and affordability of books, gender expectations, socioeconomic class, and racial inequalities are necessary for understanding what people read—and where, when, how, and why. Scholars of American book history analyze publishers such as Ticknor and Fields or Harper Brothers; technologies of book production including innovations such as the steam press, case binding, and chromolithography; the circulation of books and periodicals among networks of African Americans, women, rural communities, and factory workers; sites of reading from schoolroom to parlor to city streets; and more.[8] Their work has been instrumental in shaping my understanding of book culture in the nineteenth-century United States.

FIGURE I.9. Frontispiece, Embury, *American Wild Flowers in their Native Haunts* (1845)

Books are also material objects, and their physical components—paper, cloth, leather, thread, glue—all have meaning. As book history scholar Jonathan Senchyne argues, "The materiality of texts on paper can be as involved in its meaning-making processes as are the words written or printed upon them."[9] How readers interacted with their books can be inferred from material interventions such as inscriptions, marginalia, and the objects that are saved in them. Things I have found in flower books include locks of hair, pieces of ribbon, newspaper clippings, calling cards, a needlepoint bookmark, a prayer card, and, of course, lots of pressed plant specimens—all of which signify. My analysis of the book engages the material features of both its production (bindings, paper, hand-colored illustrations) and its reception (inscriptions, marginalia, pressed flowers).

FIGURE I.10. Houstonia and crowfoot geranium, Passaic Falls, New Jersey, between pages 45 and 46, Embury, *American Wild Flowers in their Native Haunts* (1845)

Almost every flower book I discuss was a physical object I held in my hands. Start with the binding: Is it tight or loose? A binding that has come detached decreases a book's value for antiquarian booksellers but increases the legibility of its production components (figures I.11–I.14). The decorative spine of my copy of Alphonso Wood's *Class-Book of Botany* (1855) has come detached (figure I.12); when I bought the book, it was placed inside the front cover. This book used case binding, a process that facilitated the mass production of books starting in the 1830s; as one source on print culture notes, "These bindings formed cases into which the folded and stitched sheets would be inserted (with glue, rather than sewing, connecting the sheets to the binding). Their decorations would be stamped as a whole by a machine . . . and the material covering the boards was a glazed cloth, rather than paper or leather." Figure I.12 shows the glazed cloth and stamped decoration. In Figures I.13 and I.14, details of the exposed spine show additional layers of the binding process: remnants of the glue that attached the case to the sheets (red arrow), the loosely woven cloth between the case and text block (blue arrows), and the red and white striped cloth headband at the head and foot of the spine (green arrows).[10] I provide this level of detail to illustrate some of the many material features of the book. An even deeper dive would discuss what its paper was made of (cotton, linen, or hemp rags; wood pulp), the source of cloth for its binding (who grew the cotton, picked the cotton, spun the thread, wove the cloth, dyed the cloth), how the text block was constructed, how the glue was made—not to mention, who did this work, what they earned, what machinery was used, how noisy it was, and more. Once you start judging a book by its cover, a whole new set of questions appears about a book's history and materiality.

If the anatomy of a book includes its spine, then inscriptions are perhaps tattoos: inked additions to the body after its initial coming into being. They are written by hand, some quite elaborately, others in impossibly tiny script. Inscriptions almost always include a name; if written by the owner of the book, usually first and last. Gift inscriptions note the giver and recipient, and frequently the occasion. Sometimes the book owner's address is included or, in the case of some botany instruction books, the name of their school. Other inscriptions fall into none of these categories, and you'll see some of those in later chapters. Like tattoos, inscriptions are material texts that signify, and their significance can change over time. Names, dates, places, relationships, and occasions all provide insight into how a book was used and what it meant to the people who used it.

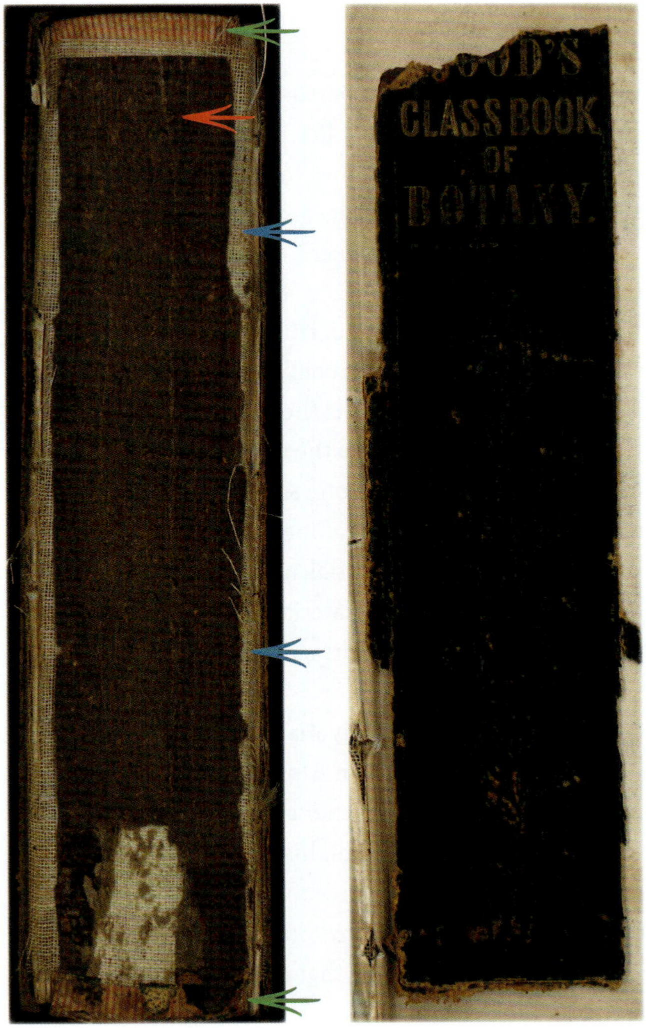

FIGURES I.11–14. Wood, *Class-Book of Botany* (1855). 1.11. (*left*) Bare spine. 1.12. (*right*) Detached spine. 1.13. (*green arrows*) Banding. 1.14. (*blue and red arrows*) Cloth and glue.

Marginalia, the handwritten additions to the margins of a book's pages, are also material texts. Nineteenth-century Americans who interacted with flower books wrote a range of marginalia, from direct commentary about the printed text to nonlinguistic marks such as checks, X's, and dashes. In *Interacting with Print*, the Multigraph Collective notes that "marking practices are determined by various physical considerations, such as the availability of pencils versus pens (or thumbnails), the amount of blank paper in a printed book, and the general proximity of volumes to bodies in the home and on the road." Marginalia are at once material texts and reminders of their makers' own materiality: bodies that existed in a specific place at a particular moment. The insights of material ecocriticism are useful here. As articulated by ecocritic Serenella Iovino, "What is at stake in the 'material turn' is the search for new conceptual models apt to theorize the connections between matter and agency on the one side, and the intertwining of bodies, natures, and meanings on the other side."[11] Both materiality and materialism guide my analysis: the physicality of the book (materiality) *and* the interplay of matter, agency, bodies, natures, and meanings (materialism).

Like inscriptions, marginalia personalize a book and make it unique. My copy of Alphonso Wood's *Class-Book of Botany* is like no other copy: It belonged to Hezekiah W. Marbourg, and his interactions with this book are detailed toward the end of chapter 1. The copy of John Newman's *Illustrated Botany* (1850) I discovered in an archive is so singular that it gets a chapter all its own (chapter 2). Even the meaning of "copy" becomes differently freighted when books are inscribed and annotated: The copy of the poem written in the margin becomes its own unique thing, attached to that page as well as to the person who wrote it, and to whoever reads it later. As material objects, books mediate between their consumers and the physical, actual, material world. Flower books are representations of nature, even as they are made from nature in their paper, cloth, glue, and so forth. The evidence of someone's interaction with a book, in marginalia and inscriptions, is also material (ink or pencil traces on paper, made by a human hand) and—as we might interpret it—a representation of their thoughts and emotions. When I write my name in a book, I am doing several things: claiming it as mine, identifying my ownership to others, maybe even pledging my commitment to reading it carefully and keeping it for a while. When a nineteenth-century American person inscribed a book, were they also claiming it and identifying their ownership to others? We can make reasonable assumptions and conjecture based on additional evidence and other scholars' analyses of similar items. But the knotty complexity of book-*as*-material and book-*mediates*-the-material remains, and in the chapters that follow, I will endeavor to untangle it.

Popular Botany

My emphasis on the popular centers the productions of lesser-known writers and artists that reached wide audiences because of their utility, accessibility, beauty, and recreational or even entertainment value—and how these productions were either explicitly environmental or became environmental because of how people responded to them. I use "popular botany" as another capacious term, one that includes not only the study of botany and the activity of botanizing but also other ways that people interacted with plants. While many Americans engaged directly in agriculture, horticulture, gardening, foraging, forestry, and bioprospecting, I focus on how everyday people's connections to plants were mediated through print and expressed in material culture. Popular botany includes the language of flowers, the topic of chapter 5, and the process of skeletonizing leaves for decoration, discussed in chapter 3; the term "botany" alone excludes these sentimental pastimes and decorative objects. I use "popular" in several of its senses—aimed toward a general audience, garnering widespread interest and approval, and produced by amateurs rather than professionals—and I situate nineteenth-century American popular botany in the intellectual and cultural contexts most relevant to flower books: amateur science, science instruction, civic duty, the culture of sentimentalism, and the nascent conservation movement.

Elizabeth Keeney's *The Botanizers: Amateur Scientists in Nineteenth-Century America* provides this overview of botany's popularity: "After rising to fashion during the 1820s, botany rapidly became the most popular science in America for recreational and pedagogical purposes, and it remained such throughout the century. Tens of thousands of enthusiasts embraced botanizing by collecting, identifying, and preserving specimens. Clubs, correspondence networks, specimen exchanges, and specialized publications arose to meet the demand for botanical culture. Young and old, rural and urban, male and female, joined together in pursuit of the natural history of plants." The activity of "botanizing" that Keeney describes—"collecting, identifying, and preserving specimens"—is central to popular botany, which was at once intellectual, recreational, utilitarian, aestheticized, and even devotional.[12] People wrote about their botanizing in letters, diaries, and memoirs; many amateur botanists shared plant specimens through the mail. Young women wrote to their friends and, when possible, botanized together. In 1810, Sarah Alden Bradford Ripley and Abigail (Abba) Allyn collected plants and worked through Linnaeus's taxonomy to classify them. In an 1845 letter to Abiah Root, fourteen-year-old Emily Dickinson inquired whether her friend had begun making an herbarium yet. Laura Godfrey, who traveled by Pullman car from Bangor, Maine, to San Francisco and back in 1885, noted many flowers in her diary, including in Monterey, California, where "the beach which is a perfect sand bed is beautiful with wild verbena of a lovely pink inclining to lavender."[13]

Just about anyone could botanize: You could do it without special equipment, right outside your door. The writer and naturalist Mary Treat, whose botanical expertise included carnivorous plants, wrote an article for *The Herald of Health* in 1866 on "Botany for Invalids," asserting that "of all sciences it is the most easily comprehended, and at once affords pleasure, and really beneficial effects upon the student, both in mental and physical culture and growth."[14] Botany instruction books encouraged getting outside as a healthy activity that also developed powers of observation. A series of Sunday school books for children described the *Wonders of Vegetation*, and Protestant publishers popularized the botany of the scriptures. Botanizing men, women, and children appeared in fiction and poetry; a childhood spent botanizing was a sure foundation for public service and eminent accomplishments later in life.

The domestic sentimentalism that flourished in mid-nineteenth-century America influenced popular botany, particularly in its valorization of emotion, the gendered experience of women, and the home. Mary DeJong calls sentimentalism "a nearly universal discourse in the nineteenth century," which "work[ed] effectively to bring about much progressive social change, especially for middle-class women," who were its primary producers and consumers. Lisa Stefaniak states that "by the nineteenth century botanical study as a properly feminine pursuit had become strongly linked to prevailing discourses of domesticity," and Mary Kuhn similarly notes that botany "emphasized that home was a place for managing plants" and "promised more orderly, productive, and pleasing environments." Popular botany was by no means exclusively female, however, nor solely sentimental. It was accessible, both locationally and intellectually, and it aligned with American ideals of self-improvement and the pursuit of knowledge.[15]

Informing my analysis of nineteenth-century American popular botany are the frameworks of critical plant studies and plant humanities. Both approaches foreground plants in scholarly analysis and assert their crucial significance to the planet's healthy functioning. Critical plant studies came to the fore in the early 2010s, in part as a parallel to critical animal studies, acknowledging the agency of more-than-human life forms as subjects rather than objects. Also influential were developments in plant sciences, particularly plant neurobiology, that "recast the plant as an intelligent organism capable of coordinated response and signalling behaviours." With scientific evidence for plant agency, sentience, and intelligence, scholars in the humanities and other disciplines developed frameworks for the "philosophical, political, artistic, literary, scientific, historical, economic, intimate, and intersectional dimensions of plant activities in and with different human and more-than-human worlds."[16] Critical plant studies is thus highly interdisciplinary, and more recent developments have expanded the field beyond "Eurowestern plant traditions" to validate and include non-Western approaches, particularly those of Indigenous scholars and practitioners. Prudence Gibson and Catriona Sandilands, working at the intersections of environmental humanities and critical plant studies, attend to historical and literary dimensions, including the "long history of plant involvement in imperialism, colonialism, and commodity capitalism" and "botanical aesthetics and its vegetal limits." Historian Kathleen Cruz Gutierrez allows for the possibility of using "botanical metaphors to draw conclusions or to characterize phenomena" and further asserts that "the constraints of our discipline require attention to textual, material, and oral records—the empirical stuff upon which we assess precision and rigor. That this stuff comes from the hands and mouths of human actors, in my view, is not antithetical to the plant turn."[17] My focus on "the empirical stuff" of flower books engages a nexus of human-book-plant interactions that both reinscribe and resist plant objectification. The insights of critical plant studies remind us that every flower specimen pressed by a person onto paper or into a book was once a living being—one capable of attracting and signifying to humans, as well as signaling and networking with other plants.

Plant humanities is an approach inspired and informed by critical plant studies, with a more applied purpose and specific origin. In September 2018, the Dumbarton Oaks Library and Research Collection in Washington, D.C., received a grant from the Andrew W. Mellon Foundation "to advance Plant Humanities," an "emerging, interdisciplinary field" that "seeks to highlight the enduring significance of plants to human cultures" and, "in dialogue with science," to further "the understanding of plant-human relationships through the methods and perspectives of humanities scholarship, to help address pressing human and environmental challenges in the Anthropocene." Taking advantage of Dumbarton Oaks' strengths in gardens, rare books, and research collections, this initiative foregrounds using primary materials, asking new questions, and telling the vitally important stories of plants from multiple places and time periods. Its Plant Humanities Lab features interdisciplinary narratives and cultural histories of individual species, including those that are edible, medicinal, and horticultural. Grant recipients Yota Batsaki (principal investigator) and Julia Fine (then a postgraduate fellow) explain that "when combined with biology and environmental science, this humanistic approach to plants reveals the processes that helped shape our current cultural, economic, and environmental climate."[18] Informed by plant humanities' conjunction of sciences, arts, and humanities, and critical plant studies' focus on plant agency and human-plant interactions, my analysis of popular botany is likewise interdisciplinary and intersectional.

American Environmentalism

The cultural work of the nineteenth-century American flower book was inherently and often deliberately environmental.[19] I use "environmentalism" as both activism and for what we today recognize as its multiple spaces and concerns, including physical environments of landscapes, topography, ecosystems, habitats, and their more-than-human inhabitants ("the natural world"); issues of resource use, extraction, abundance, and scarcity; realities of species extinction and habitat destruction; knowledge of one's local and larger environments; and conservation and preservation, to name a few. As Rochelle Johnson states in *Passions for Nature*, whether "amassing collections of bird specimens, learning the language of flowers, compiling lists of botanical species, or completing arduous expeditions, many Americans of the nineteenth century undertook considerable efforts to gain knowledge about the physical environment"; the activities Johnson lists all fit my use of "environmentalism." Flower books both reflected and shaped American environmental concerns through a combination of their authors' understandings of the natural world, the impact of those authors' works, how and why readers consumed these books, and how readers in turn left their mark—not only literally upon the page but also more figuratively in how they applied their increased knowledge and awareness. Using present-day ideas about environment in a nineteenth-century context risks anachronism and ahistoricism, but I employ this framework as many environmental humanities scholars before me have done, using the past to shed light on our present.[20]

One important influence on early American environmentalism was natural history: the close observation and detailed recording of natural phenomena. Natural historians, or naturalists, sought a greater understanding of plants, animals, geology, and climate; in the United States, this activity took on nationalist significance. Andrew Lewis, in *A Democracy of Facts: Natural History in the Early Republic*, explains that "ordinary Americans made natural history a part of the nation building process, an exercise as much involved with the creation of national character as it was with plants and animals." One of the most recognized achievements of early American natural history is John James Audubon's *The Birds of America* (1827–1838), featuring accurately detailed watercolors of more than four hundred bird species, available in bound book form and the unbound double elephant folio format of prints measuring thirty-nine by twenty-five inches—today's poster size. Natural history artists and writers published their expertise in mammalogy (mammals), conchology and malacology (shells and the invertebrates living inside them), herpetology (amphibians and reptiles), botany (plants), and, as Audubon did, ornithology (birds). In *The Book of Nature: Natural History in the United States, 1825–1875*, Margaret Welch argues that print was central to conveying this knowledge of America's nature, noting "the overwhelming importance of books in the transmission of natural history practice and discourse before 1875." For naturalists, the book was not only the medium but also their inspiration, as they studied "the book of nature."[21]

Conceptually, the "book of nature" trope held that everything in the material, natural world could be "read" or interpreted for its more abstract—specifically, spiritual—significance. In its nineteenth-century American iteration, it was indebted to the Swedish philosopher Emanuel Swedenborg and his "doctrine of correspondence," which Emerson succinctly articulated in his essay *Nature* (1836) as "Every natural fact is a symbol of some spiritual fact." The "book of nature" was such a powerful metaphor that almost every author who wrote a book about nature in nineteenth-century America invoked it (or so it seems); the writers of botany textbooks, herbarium instructions, and flower language dictionaries certainly did.[22]

Many of the significant movements and events of nineteenth-century America are inextricable from environmentalism, and *Flora's Interpreters* uses these as guideposts. Efforts to articulate a distinctively American identity and culture (exceptionalism), Manifest Destiny and territorial expansion, settler colonialism and the genocide of Indigenous peoples, chattel slavery and plantation agriculture, the Civil War, Reconstruction, urbanization, and industrialization—all had an impact on, and were in turn affected by, the natural environment and peoples' conceptions of it.

The genres and primary materials I discuss are largely overlooked sources for tracing American environmentalism. Botany textbooks, herbaria, and flower language dictionaries all promoted place-based knowledge and science-based approaches to understanding plants. When someone pressed a specimen in a botany instruction book, or noted when and where they found a particular species, this was an act of American environmentalism. So, too, were amateurs' construction of herbaria and less scientific creations involving preserved flowers: Both started with an experience of physical nature in an outdoor environment and often led to a deeper understanding of plants, their habitats, and growing cycles. The language of flowers grounded the emotions of floriography in the physicality of flowers; American compilers of these dictionaries ensured that their readers would also know the science of botany and the wide-ranging literature and lore about plants.

Ecocriticism and environmental humanities are the methodologies informing all of *Flora's Interpreters*, including the contexts of American environmentalism. Ecocriticism was developed in the 1990s by scholars who argued that "environment" was a category just as crucial for literary analysis as gender, race, and class. Ecocritical approaches also recuperated genres such as nature writing; my focus on the flower book as environmental text is informed by ecocriticism's attention to genres beyond the traditional fiction, poetry, and drama. In its early years, ecocriticism engaged primarily Anglo-American literatures and natural environments; in the decades since, its practitioners have expanded to address other Western and non-Western traditions. “Environment” has also been redefined beyond nature and wilderness by environmental justice scholars, postcolonial theorists, and materialist ecocritics, who remind us that the workplace, urban spaces, the planet, and bodies are also environments.[23]

Environmental humanities (EH) is both an outgrowth of and a departure from ecocriticism. As its name indicates, EH is interdisciplinary, including perspectives not just of literary critics but also of other humanistic scholars including historians, philosophers, geographers, anthropologists and political ecologists. More than a sum of individual disciplines, however, EH works to bridge divides between sciences and humanities, including the nature/culture divide, and insists that scientific perspectives alone are necessary but insufficient to solve the world’s increasingly dire environmental problems. Questions of meaning, the impact of cultural assumptions, ways of knowing, and hierarchies of value (particularly those that place humans above all other life forms) inform environmental humanities scholarship and praxis. Like ecocriticism, EH often takes an activist stance, insisting that intellectual work needs to be relevant outside the academy and promote real solutions, not solely theories. *Flora’s Interpreters* is indebted to environmental humanities for its interdisciplinary analyses—with methods deriving from literary, historical, visual, and material culture studies—and its engagement with the natural sciences, particularly botany.

Overview of the Chapters

The six body chapters of *Flora’s Interpreters* follow a dialogic structure of paired genre and “From the Archives” chapters, and each pair connects to an American environmental concern of the nineteenth century that remains relevant today. The genre chapters are broader surveys that span the nineteenth century and analyze examples of the scientific, environmental, literary, visual, and material aspects that characterize these types of flower books. Following each genre chapter is a more focused “From the Archives” chapter, zeroing in on a particular archival discovery that both exemplifies and departs from its genre. The introduction, genre chapters, and conclusion begin with visual and textual epigraphs that introduce and frame the discussion; “From the Archives” chapters open with a single visual epigraph.

Archival research is the backbone of *Flora’s Interpreters*. The published books I discuss are more readily available than other kinds of historical sources such as manuscripts, letters, and diaries. Many nineteenth-century U.S. books have been digitized, and hard copies exist outside special collections, including for sale at often reasonable prices.[24] In my travels through the archives, I have discovered one-of-a-kind books with unique features that typically don’t make it into digitized format, which prioritizes “clean” copies free of marginalia and other additions. By preserving the treasures of the past, archives give researchers access to rare and specialized objects whose significance awaits the interpretive lenses of the scholars who encounter them. Each “From the Archives” chapter explores singular items that expand our understanding of the nineteenth-century U.S. flower book and, in their fascinating combinations of genres and purposes, also give us insights into the people who created them.

Chapter 1, "Botany Instruction Books: Preparing American Citizen Scientists," analyzes how botany textbooks promoted science to a range of learners, from children to adults, and for those both with and without the support of formal education. Paratexts such as prefaces, introductions, and notes to teachers show how this genre additionally advocated for the moral values, healthful exercise, and powers of observation that resulted from studying botany. By describing and classifying native and introduced species of the United States, botany instruction books prepared everyday Americans to become citizen scientists and participate in a larger national project of further identifying the myriad plants of an expanding and developing nation.

Chapter 2, "'You Old Botany': Textbook as Field Guide, Journal, and Commonplace Book," follows the genre chapter on botany instruction books by analyzing a copy of John Newman's textbook *Illustrated Botany* (1850). The book's owner, Mattie Snowden, was a schoolteacher in remote Alpine County, California, in the late 1860s who recorded her botanizing in its margins. But during a several-week field excursion one summer, her journal and herbarium were stolen, and *Illustrated Botany* then served both purposes. The journal entries she pencils into the book discuss plants, poetry, and the people she traveled with, including a Yale professor, a retired Civil War general, and a governor of Nevada.

Chapter 3, "Herbaria: Preserving American Plants," argues that the herbarium—a collection of dried, pressed plant specimens mounted on paper and organized taxonomically—is another kind of flower book, one meant to be read and displayed. Over the course of the nineteenth century, the herbarium evolved from its primary purpose as a scientific tool into more aesthetic and creative formats that retained the aim of learning about plants in as much detail as possible. This chapter analyzes the links between individuals' and the nation's preservation of plants, providing evidence of how the activity of herbarium construction led directly and indirectly to increasing Americans' awareness of the need to protect on a larger scale the species they so painstakingly collected and preserved.

Chapter 4, "'Diary Kept in Flowers': Annie Adams Fields, 'Fragrant Memories,'" follows the genre chapter on herbaria by delving into a diary that resembles an herbarium, with pressed flower specimens on nearly every page. The diary's writer, Annie Fields, was connected to many of the nation's leading and up-and-coming writers through her husband, James T. Fields, editor of *The Atlantic Monthly* and partner of the renowned publisher Ticknor and Fields. In brief, breezy entries from 1862 to 1870, Annie Fields records her visits and conversations; in the years 1862–1865, her entries about the Civil War are short and touchingly to the point. Every flower she preserved had meaning to her, and these "fragrant memories" are yet another iteration of the herbarium's memorial purpose.

Chapter 5, "Flower Language Dictionaries: Cataloging American Abundance," explores the popular phenomenon of the language of flowers, a symbolic system that pairs a particular plant species with a corresponding emotional trait. Nineteenth-century American flower language dictionary authors followed their European predecessors in compiling elaborate collections that not only decoded scores or even hundreds of different flowers but also featured scientific information, literary extracts, beautiful illustrations, and attractive bindings. From their first appearance in the 1820s to the end of the century, in formats from tiny miniatures to folio-sized volumes, American flower language dictionaries conveyed and catalogued the abundance of the nation's plants. But because their copiousness was representational and the resources they extracted were literary, their abundance countered the commodification and utilitarianism that were leading inexorably to the depletion of American flora.

Chapter 6, "'Friendship Is a Holy Thing': *The Scriptural Album with Floral Illustrations*," follows the genre chapter on flower language dictionaries by exploring a token album published in the 1840s, notable for its occasional illustrated pages of biblical scenes with botanical drawings keyed to the language of flowers. Instead of treating just one archival object, this chapter notes the similarities and differences among sixteen individual albums I have researched. Token albums, also called friendship albums, circulated among friends and family who inscribed these books' blank pages with their good wishes, remembrances, occasional sketches, and a lot of verse.

The conclusion, "Critical Plants and Fables of Transformation," begins by analyzing the genre of flower fables and images of personified flowers as agents of transformation and reform in the latter part of the nineteenth century. In both their literary and visual forms, personified plants are often implicitly or explicitly critical, exposing hypocrisy and decrying injustice. The discussion of metaphorically critical plants of the nineteenth century then segues to the literally critical plants of today: endangered, beneficial, and crucial for the ecosystem services they provide. I also consider the afterlife of the nineteenth-century flower book, including in digital formats such as the crowd-sourced plant identification apps Seek and iNaturalist.

A Book About Books

Although the books I discuss are often available in digitized format, the additions to these print materials that are central to my analysis—inscriptions, marginalia, and pressed specimens—originate with an individual, and thus each artifact is unique. Gaining access to these individual artifacts generally requires visiting an archive in person or acquiring them from antiquarian booksellers and online auctions; both approaches require funding and, in the case of going to archives, time away from one's day job.[25] This is the work I have had the privilege to do: to comb through hundreds of individual flower books and discern the patterns that emerge. The analytical frameworks and methods I employ—history of the book, critical plant studies, ecocriticism and environmental humanities, material culture analysis, visual and textual close reading—are all necessary to make informed arguments about the significance of these books. While my coverage of primary sources for each genre is comprehensive, it is by no means exhaustive. Excavating more of these popular sources will yield new insights and counterarguments, identify other important flower book genres, and determine the additional purposes they served.

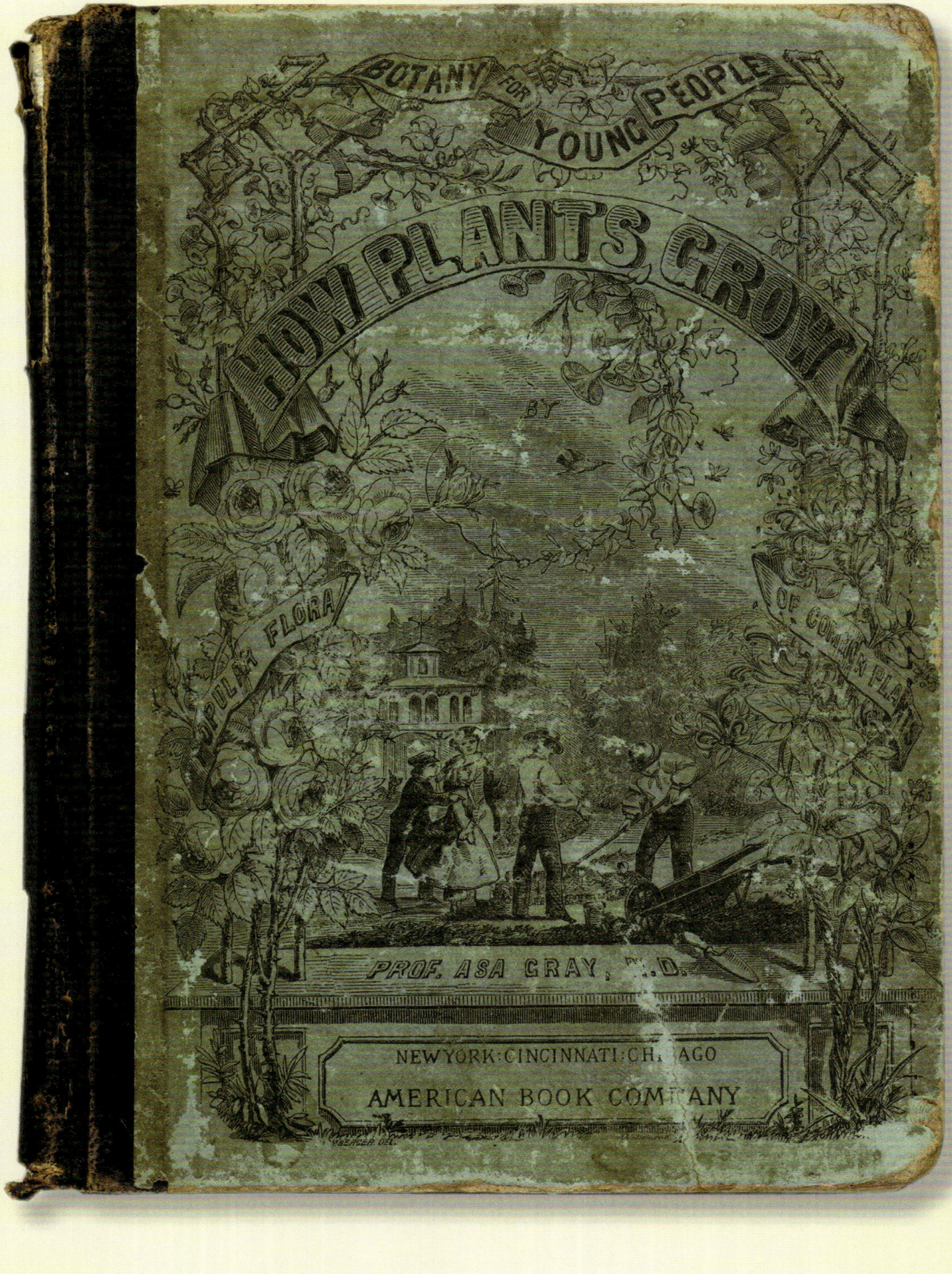

FIGURE 1.1. Cover, Gray, *Botany for Young People*, part 1, *How Plants Grow* (1858)

1

Botany Instruction Books

Preparing American Citizen Scientists

The study of Botany seems peculiarly adapted to females; the objects of its investigation are beautiful and delicate; its pursuit leading to exercise in the open air is conducive to health and cheerfulness. Botany is not a sedentary study which can be acquired in the library; but the objects of the science are scattered over the surface of the earth, along the banks of the winding brooks, on the borders of precipices, the sides of mountains, and the depths of the forest.

—Almira Hart Lincoln Phelps, *Familiar Lectures on Botany* (1829)

No science more effectually combines pleasure with improvement, than Botany. It conducts the student into the fields and forests, amidst the verdure of spring, and the bloom of summer; — to the charming retreats of Nature, in her wild luxuriance, or where she patiently smiles under the improving hand of cultivation. It furnishes him with vigorous exercise, both of body and mind, which is no less salutary than agreeable, and its subjects of investigation are all such as are adapted to please the eye, refine the taste, and improve the heart.

—Alphonso Wood, *A Class-Book of Botany* (1855)

Let this volume, then, fulfil its mission, by scattering flowery truths in the too often waste-places of the world. Let it go to the workshop and the cotton-mill; and the sons and daughters of Toil will find the fable of Aladdin more than realized; for the lighted lamp of science shall unlock a world, rich beyond all human conception, with treasures of immortal life and beauty.

—Frances Green and Joseph Congdon, *Analytical Class-Book of Botany* (1855)

FIGURE 1.2. Cover, Morley, *A Few Familiar Flowers* (1897)

THROUGHOUT THE NINETEENTH CENTURY in the United States, botany instruction books promoted the study of plants to a range of learners: young children, students, and adults. Books for children taught the basics of botany by extolling the wonders of the vegetal world. More advanced textbooks were hundreds of pages long, with detailed illustrations and extensive taxonomies. Through their emphasis on native and introduced species of the United States, botany instruction books laid the foundations for everyday people to do the work of citizen scientists, inviting them to share in a patriotic project of identifying, collecting, preserving, and valuing the myriad members of the plant kingdom.

The images and epigraphs to this chapter show a variety of botany instruction books published between 1829 and 1897. Texts for young learners enticed children with covers showing a profusion of different plants—as is the case with Asa Gray's *Botany for Young People*, part 1, *How Plants Grow* (figure 1.1)—or a single "familiar" blossom, as we see with Margaret Morley's *A Few Familiar Flowers* (figure 1.2). Frances Green and Joseph Congdon specifically invited working-class learners to engage "a world, rich beyond human conception, with treasures of immortal life and beauty." Almira Hart Lincoln Phelps asserted that "the study of Botany seems peculiarly adapted to females," who are presumably, and like the flowers themselves, "beautiful and delicate." As did Phelps, Alphonso Wood emphasized the healthful benefits of botanical field study, whether in the "wild luxuriance" of nature, or in more cultivated places.

These epigraphs also underscore the importance of botany as a science, from Green and Congdon's "lighted lamp of science" to Phelps's widely scattered "objects of . . . science" to Wood's "science [that] effectually combines pleasure with improvement." The study of botany glorified the works of the Creator, taught analytical discipline, and encouraged learners of all ages to get outside and explore their environments: brooks, precipices, mountains, forests, and fields. Botany instruction books became repositories of an ever-expanding catalog of native species; as colonizing settlers moved west, systematic flora (taxonomical compilations of plant species) expanded their range from the Atlantic states to the Midwest to the Pacific coast, and textbook authors announced these ranges on their title pages. Species native to the United States were identified and celebrated as representative of America's exceptional variety, while the cultivated and exotic indicated a nation networked with the world. These books ranged in purpose from fostering moral values to championing the objectivity of scientific inquiry, but all had an implicit or explicit goal of supporting Americans' civic duty to understand the natural world around them, including by careful study of plants.

Books were crucial to the study of botany, both in schools and for independent learners; as Elizabeth Keeney notes, "Schools and textbooks were the media through which many, if not most, botanizers first encountered the science." In the early decades of the nineteenth century, many children still learned at home, taught by parents, older siblings, neighbors, and, for those who had access, tutors. Sally Gregory Kohlstedt explains that in the early nineteenth century, "a basic familiarity with science was introduced *en famille*," and "supplementing and extending relatively private experience was the more public dissemination of ideas through primers, texts, magazines, newspapers, lectures, and museums."[1] More advanced science education, including in botany, was available at academies and female seminaries, but self-taught botanists were more the norm than the exception. The durability of books enabled them to circulate among friends, within generations of families, and in libraries and to be carried into the field. Nineteenth-century books' durability also results in their being available today: Even with loosened bindings and foxed pages full of brown spots, they still provide evidence of how their users interacted with and treasured them.

I use the term "botany instruction book" because it more accurately reflects their range and variety than "textbook." The genre of botany instruction books certainly included conventional textbooks, but it also included Sunday School texts, stories for children, and beautifully illustrated books meant for prominent display. Some of the most widely used textbooks, such as those by Asa Gray (1810–1888) and Almira Hart Lincoln Phelps (1793–1884), remained in print for many decades, had multiple editions, and sold hundreds of thousands of copies.

The study of botany was important in early America for several reasons. Before chemical compounding made drugs widely available in pharmacies, plants were used as medicine, and this knowledge (also known as materia medica) was central to learning and teaching botany, especially at the university level and in medical schools. Home study was also important, as seen in the full title of Daniel J. Cobb's *The Family Adviser; Calculated to Teach the Principles of Botany. Compiled with a Strict Regard to Logick. Containing Directions for Preserving Health, and Curing Diseases. For the Use of*

Families and Private Individuals (1828). Enlightenment science prompted an obsession with classification, and systems like that developed by Carl Linnaeus in his *Systema Naturae* (1735) provided consistent nomenclature to a field otherwise chaotic in the number of common names for the same plant—or the same common name for different species. Studying botany developed discipline of mind, attention to detail, and powers of observation. All of these—medicinal knowledge, classification and identification, intellectual rigor—encouraged citizens to initiate or deepen their understanding of botany.

One botany instruction book lauded Native Americans as the original citizen scientists. Phelps, in an early edition of *Familiar Lectures on Botany* (1836), noted that

> when our forefathers came to this country, they found the natives in possession of much medical knowledge of plants. Having no remedies prepared by scientific skill, the Indians were led, by necessity, to the use of those which nature offered them: and, by experience and observation, they had arrived at many valuable conclusions as to the qualities of plants. Their mode of life, leading them to penetrate the shades of the forest and to climb the mountain precipices, naturally associated them much with the vegetable world. The Indian woman, the patient sharer in these excursions, was led to look for such plants as she might use for the diseases of her family. Each new and curious plant, though not viewed by her with the eye of a botanist, was regarded with scrutinizing attention: the colour, taste, and smell, were carefully remarked, as indications of its properties.

What Phelps describes here are precisely the activities of the citizen scientist: "experience and observation," "scrutinizing attention," "carefully remarked." Her dedication to women's education, seen in the epigraph at the beginning of this chapter, includes noting the contributions of "the Indian woman."[2]

Although amateurs' contributions to science have been acknowledged for centuries, the term "citizen science" first appeared only in the late twentieth century. The U.S. Crowdsourcing and Citizen Science Act of 2016 (15 USC 3724) defines the process as "a form of open collaboration in which individuals or organizations participate voluntarily in the scientific process in various ways." Other sources define the participants in citizen science as members of the general public, amateur or nonprofessional scientists, noninstitutional participants, and those without scientific training. In these present-day definitions, the activities of the citizen scientist include data monitoring, collection, and analysis; hypothesis testing and conducting experiments; and solving real-world problems driven by community concerns.[3] In early America, when people had much less access to information than we do in the twenty-first-century internet age, citizen science was less far-reaching in both its scope and its outcomes.

Studying plants was an important part of natural history. Andrew Lewis notes that "rather than emulate what they perceived as the rigidly hierarchical natural history practices of Europe, American naturalists initially advertised for and welcomed the participation of their fellow citizens. Natural history, they argued, was a tool to investigate, to catalog, to explore, and, ultimately, to know the new nation."[4] This chapter defines botanical citizen science in the nineteenth-century United States as a collective endeavor of locating, identifying, observing, analyzing, collecting, and sharing plant species, specifically as part of everyday people's experience of learning botany from books. These works taught plant science activities to their readers, who in turn left evidence of their work in inscriptions, marginalia, pressed specimens, and other interactions with the material text. I begin by discussing how books for children laid the foundations for citizen science, then show how more advanced learners, especially women, further developed as scientists by using more complex texts. The chapter closes with examples of how two citizen scientists, the writer Henry David Thoreau and the physician Hezekiah W. Marbourg, used their botany textbooks to identify and record their findings both at home and while traveling.

Laying the Foundation of Citizen Science: Teaching Botany to Children

Botany instruction books for children ranged from small, brief pamphlets to copiously illustrated introductions to scientific botany. Some were organized as dialogues or catechisms, some used narrative, and others resembled books for older learners, emphasizing plant taxonomy and physiology. As with books for more advanced learners, the purpose of beginners' books shifts over the course of the nineteenth century, following larger cultural patterns of increasing secularization and greater emphasis on utility.

In the first half of the century, many books for beginners expressed a threefold purpose. Christopher Irving's *A Catechism of Botany* (1826) states the advantages derived from the study of botany: "As an amusement, [it] is both healthful and pleasing, as it attracts us often into the country, and makes us acquainted with the wonderful works of nature: it also enlarges the mind, by implanting new and useful ideas, and fills it with the most exalted admiration of the great Creator of the universe." The twenty-eight engraved illustrations on the title page featuring leaves, flowers, fruits, and seeds show plants as "wonderful works of nature" (figure 1.3). These three aims—healthy outdoor activity, intellectual development, and appreciation of the divine—recur in prefaces in the first half of the nineteenth century. In *Botany for Beginners* (1833), Phelps rhetorically asks her young readers, "Did the great Being who created such a profusion of these beautiful and curious objects, and who also gave to children eyes to see, hearts to love, and understandings to study them, intend they should pass them by with neglect? No, my dear children, it is your duty, as it should be your pleasure to search into the wonders of created nature, to exercise your mental faculties, and to animate your pious feelings in thinking much upon the works of God." Duty and pleasure, mental faculties and pious feelings were alike engaged in the study of botany. The healthy pleasures of an outdoor ramble provided materials for analysis, a process that stimulated the intellect. Studying the wonders of nature would, in turn, lead one to revere the divine. The frontispiece for a later edition of *Botany for Beginners* (figure 1.4) shows an Edenic landscape lush with tropical plant life of the New World, including breadfruit and maize, indicating God's goodness, care, and bounty.[5]

From the early decades of the nineteenth century until after the Civil War, almost all botany instruction books for children made explicit such connections between botany and Christianity, as did books for older learners. What might strike us today as inappropriately religious for science instruction was both familiar and acceptable then. Most children's books were overtly didactic, and connections between faith and intellect (particularly in the study of science) were deemed

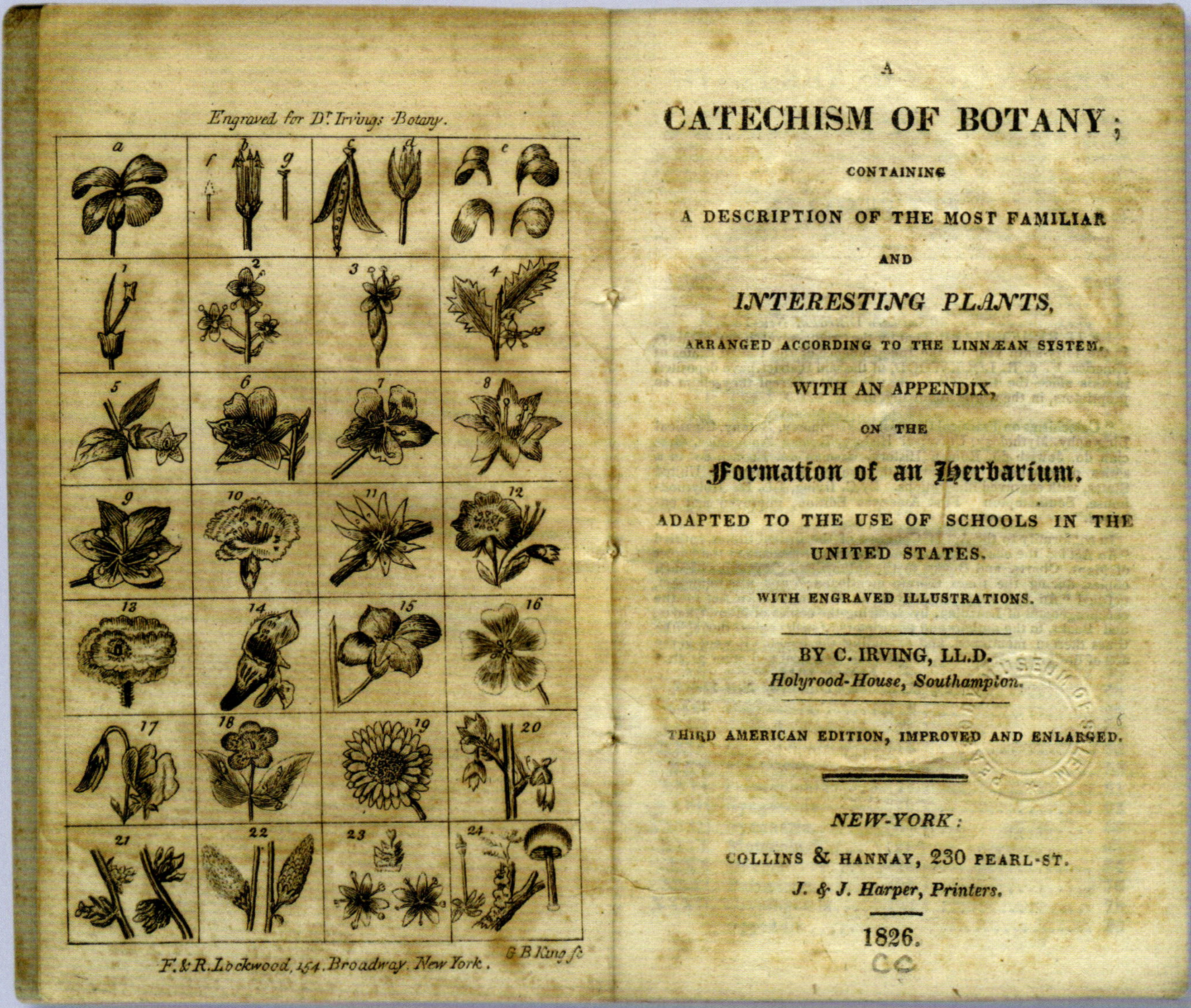

A

CATECHISM OF BOTANY;

CONTAINING

A DESCRIPTION OF THE MOST FAMILIAR

AND

INTERESTING PLANTS,

ARRANGED ACCORDING TO THE LINNÆAN SYSTEM.

WITH AN APPENDIX,

ON THE

Formation of an Herbarium.

ADAPTED TO THE USE OF SCHOOLS IN THE UNITED STATES.

WITH ENGRAVED ILLUSTRATIONS.

BY C. IRVING, LL.D.

Holyrood-House, Southampton.

THIRD AMERICAN EDITION, IMPROVED AND ENLARGED.

NEW-YORK:

COLLINS & HANNAY, 230 PEARL-ST.

J. & J. Harper, Printers.

1826.

FIGURE 1.3. Title page, Irving, *A Catechism of Botany* (1826)

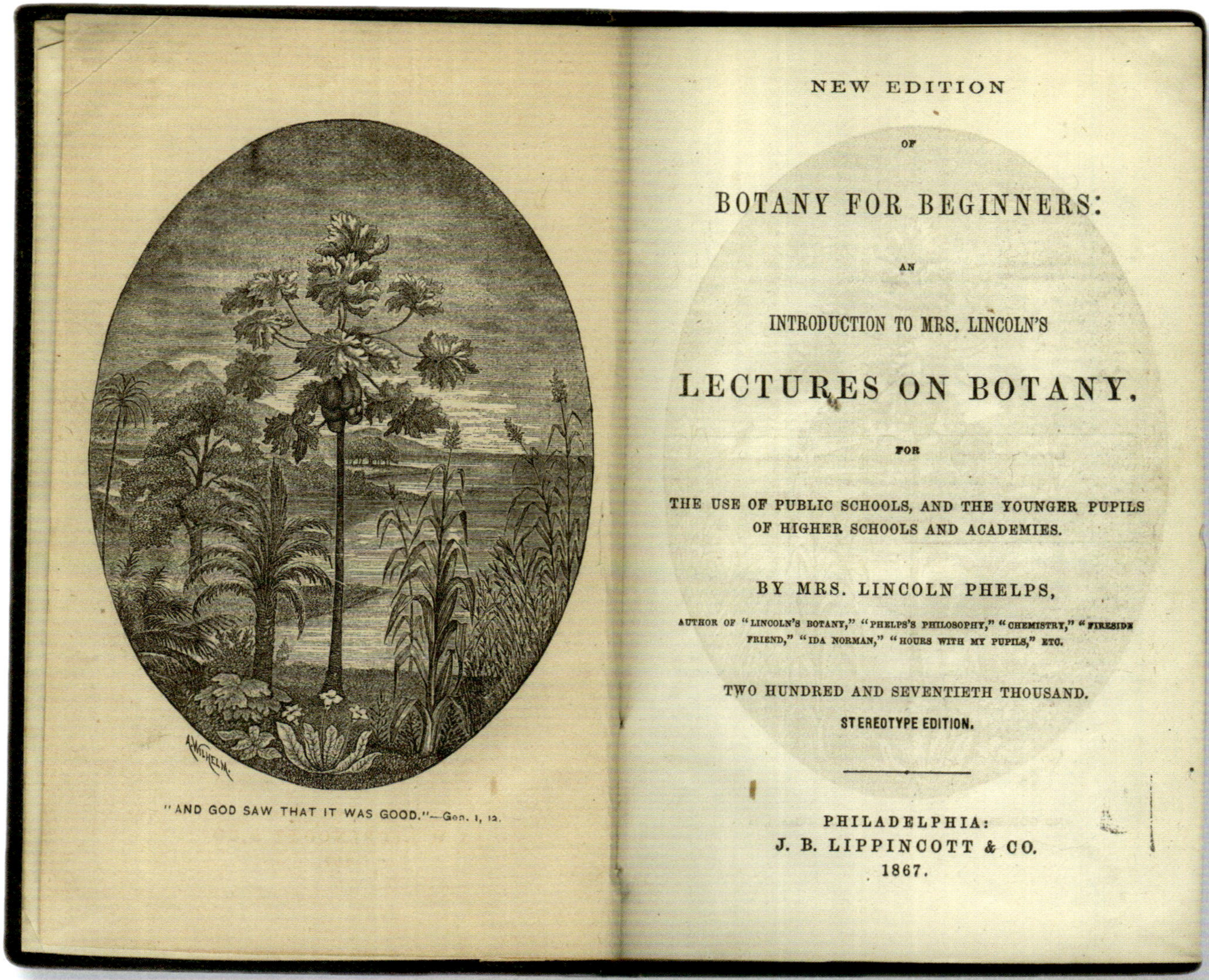

"AND GOD SAW THAT IT WAS GOOD."—Gen. I, 12.

NEW EDITION

OF

BOTANY FOR BEGINNERS:

AN

INTRODUCTION TO MRS. LINCOLN'S

LECTURES ON BOTANY,

FOR

THE USE OF PUBLIC SCHOOLS, AND THE YOUNGER PUPILS OF HIGHER SCHOOLS AND ACADEMIES.

BY MRS. LINCOLN PHELPS,

AUTHOR OF "LINCOLN'S BOTANY," "PHELPS'S PHILOSOPHY," "CHEMISTRY," "FIRESIDE FRIEND," "IDA NORMAN," "HOURS WITH MY PUPILS," ETC.

TWO HUNDRED AND SEVENTIETH THOUSAND.

STEREOTYPE EDITION.

PHILADELPHIA:
J. B. LIPPINCOTT & CO.
1867.

FIGURE 1.4. Title page and frontispiece, Phelps, *Botany for Beginners* (1867)

natural and appropriate, more so than keeping the two strictly separate. In *A Botanical Catechism* (1819), Jane Welsh states that by "viewing the wonders of Nature" children will be led, "though indirectly, to adore with reverence and humility, the Divine *Source* of those beauties" (italics in the original). Phelps's *Botany for Beginners* similarly hopes that "especially for the sake of bringing the young mind to commune with its Maker, by leading it to be familiar with His works, . . . you may learn to read and admire the great volume of nature, of which God is the Author." In his *Introduction to the Study of Botany* (1833), J. L. Comstock identifies the

"Great Author of Nature" in a section on the distribution of seeds, further noting that "the more intimately we become acquainted with the minute parts, or hidden principles of nature, the greater cause do we find for our admiration and astonishment."[6] Young citizen scientists thus began with two kinds of books: the botany instruction book and the "book of nature," and both were essential for learning the minutiae and principles of botany.

Just as these science instruction books often invoked religion, so did religious texts seek to teach science. A series titled *The Wonders of Vegetation* was published by the American Sunday School Union, which was founded in 1824 and "adopted newly mechanized printing processes in order to produce countless copies of small, inexpensive books." Donald Davis and John Mark Tucker explain that "three broad types of material represented the rapidly expanding output: tracts and presentation (award) books, periodicals and annual publications, and textbooks and curriculum materials."[7] Titles in *The Wonders of Vegetation* series included *Various Grasses, The Leaf, The Flower,* and *The Seed.* In *The Wonders of Vegetation: The Seed* (1845), twenty-one of its fifty-one pages (41 percent) mention "God," "the Almighty," "Creator," "He," "Jesus," and similar terms. While the focus throughout is on God, this little book doesn't stint on accurate scientific information, starting right away with terminology such as *plume* and *radicle* in detailed illustrations (figures 1.5 and 1.6). The multiple views in figure 1.5 (bean and germ [1]; spread lobe, plume, and radicle [2]; and the full plant [3]) and in figure 1.6 (globe [a], receptacle [b], and seed carrier [c]) provide a specificity intended to delight the young learner.

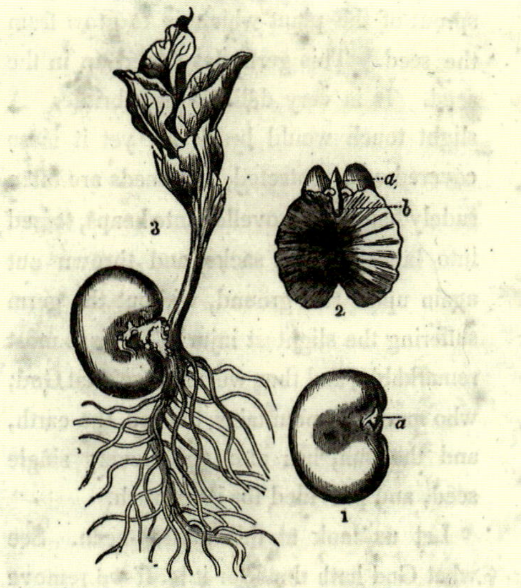

1. The bean, after its soft skin or covering is removed; *a*, the germ.
2. The bean opened, with its lobe spread out: *a*, the young plume; *b*, the young radicle.
3. The young plant, just rising from the ground.

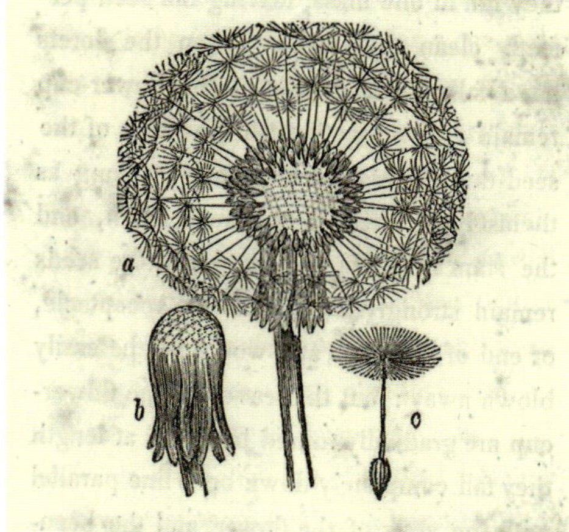

a, Part of a globe of seeds—a "Puff Ball;" *b*, the receptacle turned inside outwards; *c*, a seed with its carrier.

FIGURE 1.5. (*left*) Page 8; FIGURE 1.6. (*above*) Page 38, *The Wonders of Vegetation: The Seed*

Precise illustrations, sophisticated science, and a narrow focus on a single part of the plant all contributed to making the *Wonders of Vegetation* series an important, affordable, and accessible resource for young citizen scientists, while also promoting natural history study as an acceptable activity sanctioned by the Sunday School movement.

Alphonso Wood (1810–1881), author of multiple textbooks for a range of learners, followed his predecessors in extolling the virtues of studying botany as a means for greater appreciation of God and creation. In *First Lessons in Botany* (1851), in a section that strongly recommends using magnifying lenses, Wood links scientific precision to religious admiration: "The more powerful the lens, the more beautiful and wonderful appears the work of the Divine hand in the vegetable structure." Another prolific textbook author, Harvard professor Asa Gray, begins *How Plants Grow,* the first part of his *Botany for Young People,* with a reference to the lilies of the field, a biblical passage much beloved by flower book writers (figure 1.7). The passage continues by emphasizing the young learner's nature literacy: "This book is intended to teach Young People how to begin to read, with pleasure and advantage, one large and easy chapter in the open Book of Nature; namely, that in which the wisdom and goodness of the Creator are plainly written in the VEGETABLE KINGDOM" (capitalization in the original).[8] The pervasive links between studying "the open Book of Nature" and worshipping its Creator underscored the duty of young learners to make such connections. With a firm foundation of observation, analysis, and appreciation, children could become the engaged citizen scientists necessary to build a nation of rigorous thinkers.

Instruction books for children explained the civic responsibilities and rewards of studying botany. Welsh

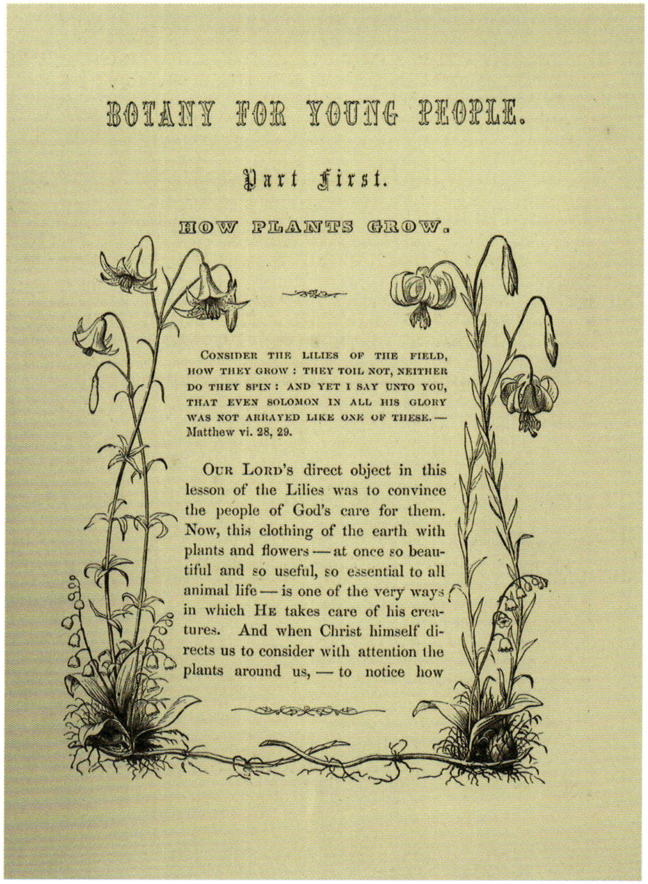
BOTANY FOR YOUNG PEOPLE.

Part First.

HOW PLANTS GROW.

CONSIDER THE LILIES OF THE FIELD, HOW THEY GROW: THEY TOIL NOT, NEITHER DO THEY SPIN: AND YET I SAY UNTO YOU, THAT EVEN SOLOMON IN ALL HIS GLORY WAS NOT ARRAYED LIKE ONE OF THESE.—Matthew vi. 28, 29.

OUR LORD'S direct object in this lesson of the Lilies was to convince the people of God's care for them. Now, this clothing of the earth with plants and flowers — at once so beautiful and so useful, so essential to all animal life — is one of the very ways in which HE takes care of his creatures. And when Christ himself directs us to consider with attention the plants around us, — to notice how

FIGURE 1.7. Page 1, Gray, *Botany for Young People*, part 1, *How Plants Grow* (1858)

asserts her "sanguine hope, that the science of Botany, by being introduced into common Schools, may become a more general study, and prove of increasing utility to this country." For Welsh, that utility included working with actual specimens, whether fresh or dried, and distinguishing between poisonous and nonpoisonous plants.[9] Irving's *A Catechism of Botany* notes native species and their uses: "The potato was carried from this country to Europe"; "The sugar maple is very useful to the inhabitants of this country"; "Indian

FIGURE 1.8. Page 15, Youmans, *The First Book of Botany* (1873)

corn, or maize, is a very valuable American plant"; and, regarding the rubber plant, "The American Indians make boots with it, which water cannot penetrate, and which, when smoked, have the appearance of real leather." In *The Child's Botany* (1828), Samuel Goodrich, a children's book author better known by his pseudonym Peter Parley, lauds the "eminent botanists in this country" ("Dr. Bigelow of Boston," Nuttall, Eaton, and "Dr. Torrey of New York"), showing children that even though America was a young nation, it still produced important scientists. Goodrich assures children that they, too, can contribute to their nation's scientific knowledge: "Much more is known about plants in Europe, than in America, because it is an older country. Therefore I wish to have the children here begin to learn about it, for the Americans ought not to be excelled by the Europeans, since we have many more plants than they have."[10]

After the Civil War, references to God tend to disappear from books for children and beginners. Eliza Youmans's *The First Book of Botany* (1870) has no religious language but does echo antebellum writers' emphases: "Botany, beyond all other subjects, is suited to maintain the mind in direct intercourse with the objects and order of Nature, and to train the observing powers and the mental operations they involve, in a systematic way." The illustration for chapter 1, "The Leaf," shows a teacher and four students engaging both books (the mind) and live plants (the objects of Nature) in an outdoor setting where they have immediate access to the world around them (figure 1.8). In part 2 of Gray's *Botany for Young People*, titled *How Plants Behave* (1872), the fervently religious language of its antebellum companion volume *How Plants Grow* is completely absent. This shift reflects the increasing secularization of American culture and a parallel professionalization of science later in the nineteenth century. Rochelle Johnson notes a third factor, influenced by the work of Charles Darwin and other evolutionists: Scientists "discovered startling information about the age of the earth, the changes brought about by geological processes over long periods of time, and the diversity of life forms throughout the ages."[11] These revolutionary ideas trickled down to botany instruction books for children, along with an emphasis on the tools that could provide deeper knowledge.

FIGURE 1.9. Cover, Herrick, *The Wonders of Plant Life Under the Microscope* (1883)

Sophie Herrick's *The Wonders of Plant Life Under the Microscope* (figure 1.9), written three decades after Alphonso Wood recommended the value of magnifying lenses in *First Lessons in Botany*, asserts that the microscope has "a mission" that prioritizes science: "Not only does it show us the marvellous precision of inorganic nature, and the delicate adjustments of chemical, physical, and vital forces of organic, but it brings us into the very antechamber of that court where life holds its mysterious sway." The binding features a gold-stamped microscope rendered in accurate detail and an ornate font announcing the "Wonders" that await the reader. The black-stamped leaves and flowers in the background signal that "Plant Life" is the foundation of this study: not just plants, but plant *life*, which is active, vital, and marvelously precise. "A Word to the Teacher" in Morley's *A Few Familiar Flowers: How to Love Them at Home or in School* (figure 1.2) recommends keeping up with "the latest researches" and "modern discoveries": thus "the young student has open to him the doors of intelligent interest in what is going on in the scientific world." By the end of the nineteenth century, botany instruction books for children emphasized science and modernity, using microscopes and keeping up with the latest discoveries.[12]

Well-known authors of botany textbooks for young learners, including Comstock, Phelps, Gray, and Wood, also wrote for more advanced students. Their books were often intentional sequels, with authors referencing their more advanced books as the appropriate next step, or vice versa. Comstock, in *The Young Botanist* (1836), states that "soon after the publication of the *Introduction to Botany*, the author perceived the want of a treatise on the same subject of smaller size, and more especially adapted to the situation of those who commence the study without the aid of an instructor." Gray tells the readers of *How Plants Grow* that once they "have thoroughly mastered this little book, they will be well prepared to continue the study in the *Lessons in Botany and Vegetable Physiology*, and in the *Manual of the Botany of the Northern United States*, by the same author."[13] These are savvy marketing strategies; also implicit in this sequence is that children will continue to study botany as they get older.

Systematic Flora as a Resource for the Citizen Scientist: Botany for Older Students

Botany instruction books taught young learners the foundations of citizen science: curiosity, wonder, appreciation, and detailed observation. Books for more advanced learners further developed a specifically *American* citizen scientist, one who would use the ability to identify, categorize, and analyze plants as part of a national effort to catalog native species. New species were continually being discovered by professional botanists and explorers, as well as by dedicated amateurs who developed deep understandings of their local environments. An eighteenth-century source, Humphry Marshall's *Arbustrum Americanum* (1785), asserts that "the Science of Botany . . . merits the attention and encouragement of every patriotic and liberal mind." Lewis states that "the overwhelming majority of early national publications on botany . . . stressed the need for greater study of plants and the importance of new botanical discoveries to the building of the nation."[14]

Nineteenth-century American botany textbooks touted this nationalist project in their prefaces. George Sumner, in *A Compendium of Physiological and Systematic Botany* (1820), stated that his book "was designed as an Introduction to the study of American plants." Amos Eaton, whose *Manual of Botany* was already in its fourth edition in 1824, encouraged students to "search for all they find described in this work, in all parts of our district." There were so many plants yet to be discovered that the "progressive science" of botany "depends on continued research, and the accurate examination of such a multitude of individuals." In the preface to the third edition of her *Familiar Lectures on Botany* (1832), Phelps promoted her book as "a popular *introduction* to the science," with "some of the *most common* indigenous and exotic plants of the various botanical districts of the United States" (emphases in the original). In *An Introduction to the Study of Botany* (1833), Comstock hoped his book would "assist the learner to discover the names of unknown species." By 1848, enough groundwork had been laid that Lewis C. Beck could note that his *Botany of the United States North of Virginia* was indebted to multiple sources: "Torrey and Gray's 'Flora of North America,' Torrey's 'Flora of the State of New York,' and Darlington's 'Flora Cestrica.'" "In determining the geographical range of the species," he explained, "I have derived great assistance from the various local catalogues of plants which have been published within the last ten years." The first edition of Gray's *Manual of the Botany of the Northern United States* also appeared in 1848. Whereas Beck credited published sources, Gray acknowledged the amateur "botanists who have detected and kindly communicated rare or local plants," those "many attentive and zealous correspondents throughout the country; and their daily increasing number."[15] By encouraging readers to find and identify as many plant species as possible, these books ensured that the flora of the United States would continue to be known and studied.

Keeney identifies "three distinct phases" of botany education in the nineteenth-century United States. In the first, from the 1820s to 1840s, "educators saw the worth of botany in the skills and values its study enhanced—observation, love of the Creator, and analytical thinking," as the children's books in the previous section showed; the leading authors in this period were Phelps and Eaton. The second phase, from the 1840s to 1880s, "placed a far greater value on the intrinsic merit of botany as a body of knowledge" and was "epitomized by botanist Asa Gray and educator Alphonso Wood." The third era, "which had its roots in the 1880s but did not reach its apex until well into the twentieth century," split into the Nature-Study movement and the scientific "New Botany."[16] A closer look at Gray's and Wood's best-selling textbooks shows how they each approached further developing the American citizen scientist.

Marshall Sundberg notes that Wood, unlike Gray, "had no formal training in science but learned about plants from lecturers and students in the medical school" at Dartmouth, where he graduated in 1834. His *Class-Book of Botany* was first published in 1845, went through scores of editions, and sold over one hundred thousand copies. The preface to the first edition boldly announces, "That there is need of a new Class-Book of Botany, prepared on the basis of the present advanced state of the science, . . . is manifest to all. . . . Botany should no longer be presented to the learner encumbered with the puerile misconceptions and barren facts of the old school, but as a System of Nature, raised by recent researches to the dignity and rank of a science." The "puerile misconceptions" of the "old school" Wood refers to is the Linnaean system of classification used by Phelps and Eaton. Also known as the artificial system, Linnaean taxonomy was relatively easy to learn—all you had to do was count plant parts—but ultimately obscured any actual similarities or "affinities" among plants, which the natural system (or "System of Nature") made clear. Wood touts another reason why his *Class-Book* is superior: "The multitude of synonyms, by which plants have been heretofore designated, have greatly encumbered the science, and are to be regarded as an evil and only an evil. If we then have a *standard* work on North American Botany, written by North American authors, as we most certainly have, it is absurd for humbler writers to strive to perpetuate this useless jargon of names" (emphasis in the original).[17] One reason for the "multitude of synonyms" for plants was their "discovery" by botanists from disparate places at different times. A European botanist of the early eighteenth century applied one Latin binomial to a genus and species; decades later, a North American botanist applied a different one, and both thought themselves correct and "first." For Wood, the citizen scientist would greatly benefit from being *un*encumbered by a useless multitude of names to learn, thanks to his efforts to standardize them.

Gray's *A Manual of the Botany of the Northern United States* (1848), published just a few years after Wood's *Class-Book*, went through eight major editions, including three after his death. In the preface to the first edition, Gray explains that "this work is designed as a compendious Flora of the Northern portion of the United States, arranged according to the Natural System, for the use of students and of practical botanists," acknowledging both that not everyone learned botany at school and that the flora was of great practical use. Gray also notes that "it was intended to be comprised within the compass of a pocket volume, which might serve as a *vade-mecum* in herborizations, as well as a convenient manual of reference at home." Despite being known as "a herbarium botanist rather than a field botanist," Gray notes the utility of a book that would "go with me" (*vade mecum*) into the field—even if the *Manual*, at over seven hundred pages, was not quite pocket-sized.[18] The emphasis of the *Manual* was taxonomy; "those who desire a more particular acquaintance with the structure, and especially the physiology of plants, may be referred to the author's *Botanical Text-Book* (second edition), or other detailed elementary works."[19]

The title page of the first edition of Wood's *Class-Book* indicated that the flora (a comprehensive list of plants) covered "the Northern United States, particularly New England and New York." Ten years later, the flora in the 1855 edition had expanded to include the northern, middle, and western states. The expansion of the flora reflects the expansion of the nation: between 1845 and 1855, the Oregon Treaty (1846), Treaty of Guadalupe Hidalgo (1848), and Gadsden Purchase (1853) added territory previously claimed by Britain and Mexico to the United States, extending the northwestern border to the forty-ninth parallel, southern border to the Rio Grande River, and western border to the Pacific Ocean. California, Florida, Iowa, Texas, and Wisconsin became states during this decade. In the preface to the 1855 edition, Wood notes that about "nine tenths" of the descriptions for plants in the flora are the result of his "extensive tours . . . in nearly every section of country which this Flora represents." For the remainder, he cites a list of sources, highlighting the work of John Torrey and Gray.[20] Wood's 1855 *Class-Book of Botany* and Gray's 1856 *Manual of Botany* are also significant for how two individuals used them, as will be detailed in the final section of this chapter.

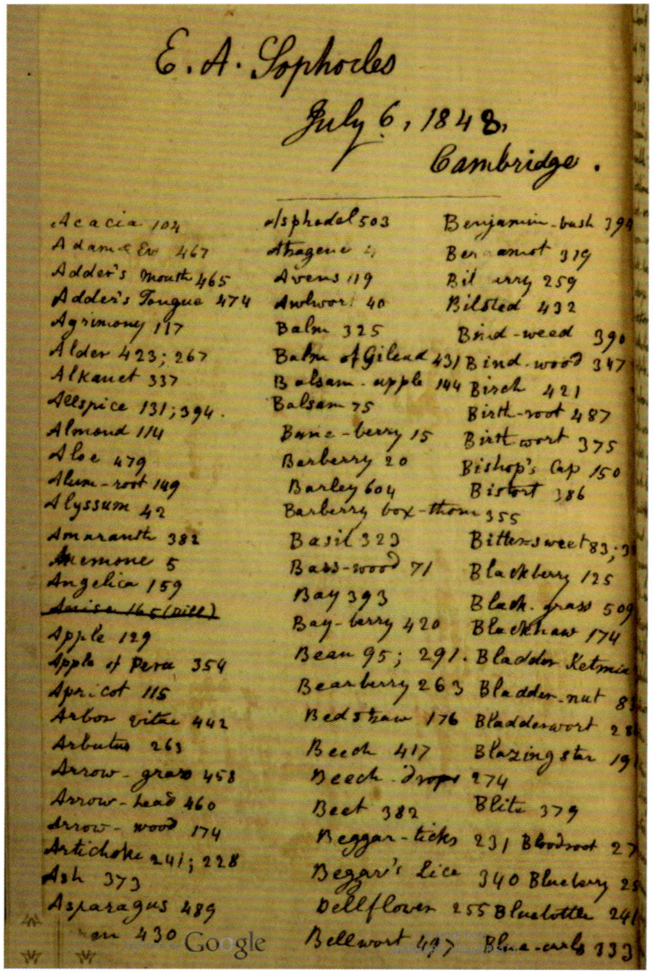

FIGURE 1.10. Inscription and first page of handwritten index by E. A. Sophocles in Gray, *A Manual of the Botany of the Northern United States* (1848), seq. 7, Collection Development Department, Widener Library, Harvard University

One copy of the 1848 edition of Gray's *Manual* belonged to E. A. Sophocles, who inscribed the front and wrote an alphabetical index of the common names of plants Gray includes: from "Acacia" through "Ivy" on four blank pages at the front of the book, and from "Jamestown-weed" to "Yew" on four blank pages at the back (figure 1.10). Evangelinos Apostolides Sophocles emigrated to the United States from Greece as a young man. He was a professor of classics and modern Greek at Harvard from 1842 (when he started as a tutor of Greek) until his death in 1883. His colleagues and students remembered him as a somewhat reclusive and generally irascible figure, who was also very generous and kindhearted.[21] The reasons for Sophocles's handwritten index in Gray are unclear: He made almost no other notes in the text, and the completeness of the list suggests that he did not locate all those species himself. While several sources mention his love for the chickens he raised, none report any botanizing. Perhaps the index was an aid to learning the plants' common names, since their Latin binomials would have been linguistically familiar to him as a professor of classics—or maybe it was simply because Gray did not provide an index of common names in the *Manual*. The top of one page in Sophocles's index shows his trilingual notes on some common New England tree genera and species, in Greek, English, and Latin (figure 1.11). Whatever the reasons for his handwritten index, Sophocles's copy of Gray reveals at the very least an armchair botanist, one whose fascination with languages may have extended to the colloquial terms for his adopted country's flora—adder's mouth, adder's tongue, arrow-grass, arrow-head, arrow-wood, beggar-ticks, and beggar's lice, to name a few from his list. Sophocles and Gray both started their careers at Harvard in 1842; Gray's biographer notes that Sophocles was one of the men who "would be Gray's friends, associates, and sometimes superiors through most of the next thirty years," along with professor of romance languages Henry Wadsworth Longfellow, mathematician Benjamin Peirce, and astronomer William Cranch Bond.[22]

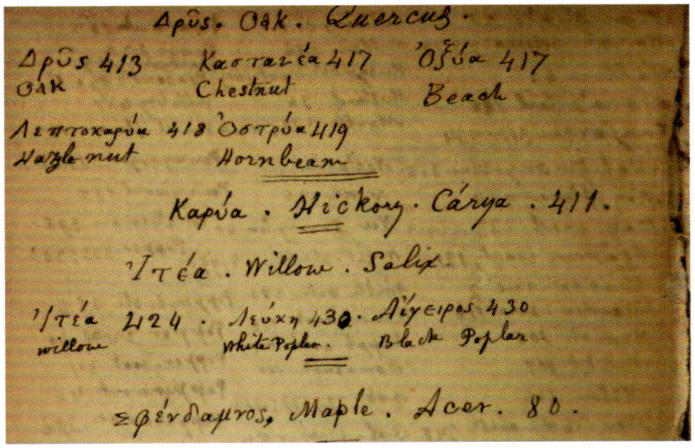

FIGURE 1.11. Detail, E. A. Sophocles's notes in back pages of Gray, *A Manual of the Botany of the Northern United States* (1848), seq. 799, Collection Development Department, Widener Library, Harvard University

Gendered Dimensions of Citizen Science: Teaching Botany to Female Learners

For women wanting to study science in the nineteenth-century United States, the socially acceptable option was botany.[23] Ladies could observe and collect plants while staying close to home, and unlike other natural history subjects, botany did not require the physical stamina or extensive equipment needed to pursue zoology or geology. Echoing Phelps's statement that "the study of Botany seems peculiarly adapted to females," botanist and politician William Darlington asserted that "it is a science peculiarly appropriate to gentle minds. Its cultivation imposes no tax upon the feelings, shocks no sensibility, involves no cruelty."[24] Women's study of botany went beyond such gendered assumptions, however, to become a subject in which they could develop legitimate expertise and contribute to citizen science. Women promoted botany in their roles as teachers, editors, collectors, writers, and illustrators, and these amateur scientists used their botany instruction books as guides.

In a study of science education for women in antebellum America, Deborah Jane Warner states that "books about science directed specifically at women proliferated, as did scientific articles in the general women's magazines, and public scientific lectures attracted large numbers of women. . . . By 1860 the foundations were securely planted for women's involvement in America's scientific enterprise." Nina Baym similarly argues that "a small but assertive and influential group of nineteenth-century women of letters used their access to print to argue energetically that women ought to know the sciences."[25]

Periodicals such as the highly popular *Godey's Lady's Book* promoted women's participation in citizen science through botany. From 1837 until 1877, it was edited by Sarah Josepha Hale, who "persistently advocated science study for women," according to Baym. During this time, *Godey's* regularly included articles about the importance of botany for educating women. In July 1841, an article likely written by Hale noted the potential for women's contributions: "If memoranda were made of the places where such wild flowers are found, the latitude, with the common name, and whether they grow singly or in groups, profusely or sparsely, with the time of flowering, ladies might add something to the history of our Flora worthy of remembrance, and particularly so, would they make themselves acquainted with, and note their botanical characteristics."[26] The activity suggested here—noting the precise location, growing characteristics, and bloom time—showed women how they could contribute as citizen scientists.

Amos Eaton, cofounder of the Rensselaer School (now Rensselaer Polytechnic Institute), was an early proponent of science education for women, including in his botany instruction books. In *Botanical Exercises* (1820), a section for teachers on collecting plants states that "you can generally persuade ladies to go out by turn in small parties and collect plants in the nearest open fields." Even though the passage continues, "and young gentlemen will search the woods, swamps, and the more distant and less accessible [*sic*] places," he has validated women's botanical fieldwork. Eaton's *A Manual of Botany, for the Northern and Middle States of America,* first published in 1817, went through eight editions and was widely used, including by women students. In fact, by the third edition (1822), Eaton closed his preface by remarking, "The recommendation of the study of botany to the attention of ladies, subscribed by the late Governor Strong of Massachusetts, and others, which was published in the second edition, is unnecessary at this day; for I believe more than half the botanists in New-England and New-York are ladies."[27]

One owner of the fourth edition of Eaton's *Manual of Botany* (1824) provided interesting evidence of her activities as a developing botanist. Eliza Roberts Nye of Littleton, Massachusetts, proudly inscribed her copy at least five times: on the front endpaper, front flyleaf, second flyleaf, and back flyleaf (figures 1.12 and 1.13).

FIGURE 1.12. Inscription, front flyleaf, Nye's copy of Eaton, *A Manual of Botany* (1824)

FIGURE 1.13. Inscription, front flyleaf (verso), Nye's copy of Eaton, *A Manual of Botany* (1824)

The sparse genealogical record for Eliza Nye does not mention her education, but her copy of Eaton's *Manual* suggests that she used it for botanizing: it contains over fifty pressed plants and scores of pencil markings, most of them the letter "x" or "H" in the margin by particular species.[28] After nearly two hundred years, most of the dried leaves and flowers are mostly unrecognizable, making it difficult to determine whether the pressed specimens correspond to any of

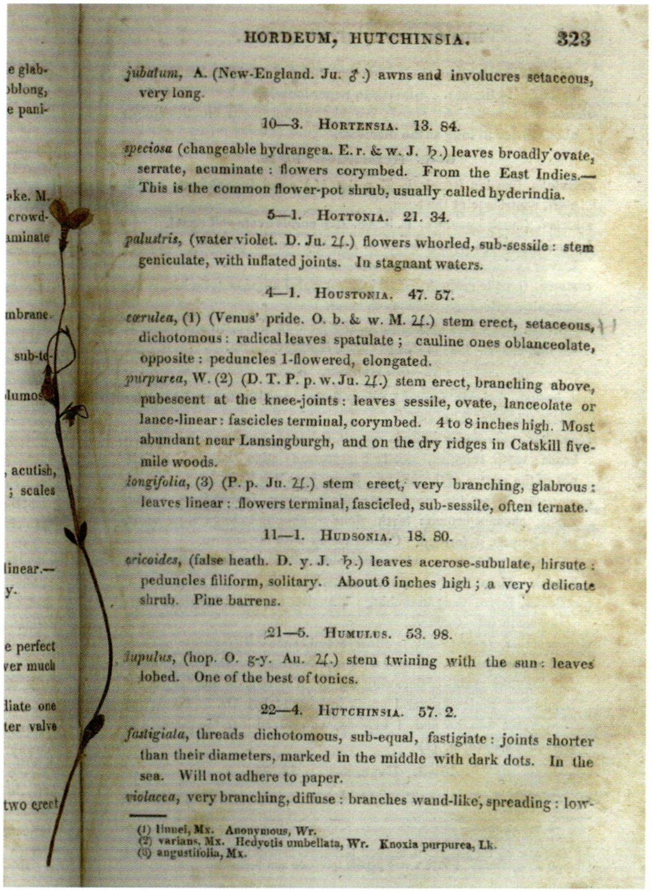

HORDEUM, HUTCHINSIA. 323

jubatum, A. (New-England. Ju. ♂.) awns and involucres setaceous, very long.

10—3. HORTENSIA. 13. 84.

speciosa (changeable hydrangea. E. r. & w. J. ♄.) leaves broadly ovate, serrate, acuminate : flowers corymbed. From the East Indies.—This is the common flower-pot shrub, usually called hyderindia.

5—1. HOTTONIA. 21. 34.

palustris, (water violet. D. Ju. ♃.) flowers whorled, sub-sessile : stem geniculate, with inflated joints. In stagnant waters.

4—1. HOUSTONIA. 47. 57.

cærulea, (1) (Venus' pride. O. b. & w. M. ♃.) stem erect, setaceous, dichotomous : radical leaves spatulate ; cauline ones oblanceolate, opposite : peduncles 1-flowered, elongated.

purpurea, W. (2) (D. T. P. p. w. Ju. ♃.) stem erect, branching above, pubescent at the knee-joints : leaves sessile, ovate, lanceolate or lance-linear : fascicles terminal, corymbed. 4 to 8 inches high. Most abundant near Lansingburgh, and on the dry ridges in Catskill five-mile woods.

longifolia, (3) (P. p. Ju. ♃.) stem erect, very branching, glabrous : leaves linear : flowers terminal, fascicled, sub-sessile, often ternate.

11—1. HUDSONIA. 18. 80.

ericoides, (false heath. D. y. J. ♄.) leaves acerose-subulate, hirsute : peduncles filiform, solitary. About 6 inches high ; a very delicate shrub. Pine barrens.

21—5. HUMULUS. 53. 98.

lupulus, (hop. O. g-y. Au. ♃.) stem twining with the sun : leaves lobed. One of the best of tonics.

22—4. HUTCHINSIA. 57. 2.

fastigiata, threads dichotomous, sub-equal, fastigiate : joints shorter than their diameters, marked in the middle with dark dots. In the sea. Will not adhere to paper.

violacea, very branching, diffuse : branches wand-like, spreading : low-

(1) linnei, Mx. Anonymous, Wr.
(2) varians, Mx. Hedyotis umbellata, Wr. Knoxia purpurea, Lk.
(3) angustifolia, Mx.

FIGURE 1.14. Pressed specimen, page 323, Nye's copy of Eaton, *A Manual of Botany* (1824) the species on a particular page. For example, given its small size and the shape and placement of its leaves, the pressed plant in figure 1.14 could be the *Houstonia cærulea* mentioned on that page. However, it's unlikely that the specimen in figure 1.15 is any kind of *Hypnum*, which is a genus of mosses. The plant in figure 1.16 somewhat resembles a kind of *Aira*, one of the genera appearing on these facing pages; for comparison, the drawing in figure 1.17 shows a detail of *Aira caryophyllea.*

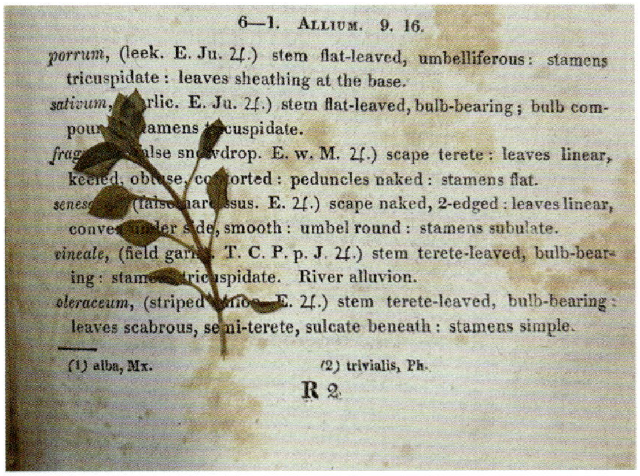

6—1. ALLIUM. 9. 16.

porrum, (leek. E. Ju. ♃.) stem flat-leaved, umbelliferous: stamens tricuspidate: leaves sheathing at the base.
sativum, rlic. E. Ju. ♃.) stem flat-leaved, bulb-bearing; bulb compou amens cuspidate.
frag lse sn drop. E. w. M. ♃.) scape terete: leaves linear, keeled, obt se, co orted: peduncles naked: stamens flat.
senes (false nar ssus. E. ♃.) scape naked, 2-edged: leaves linear, conve er s de, smooth: umbel round: stamens subulate.
vineale, (field gar T. C. P. p. J. ♃.) stem terete-leaved, bulb-bearing: stam ric uspidate. River alluvion.
oleraceum, (striped E. ♃.) stem terete-leaved, bulb-bearing: leaves scabrous, se ni-terete, sulcate beneath: stamens simple.

(1) alba, Mx. (2) trivialis, Ph.

R 2

FIGURE 1.16. Pressed specimen, page 197, Nye's copy of Eaton, *A Manual of Botany* (1824)

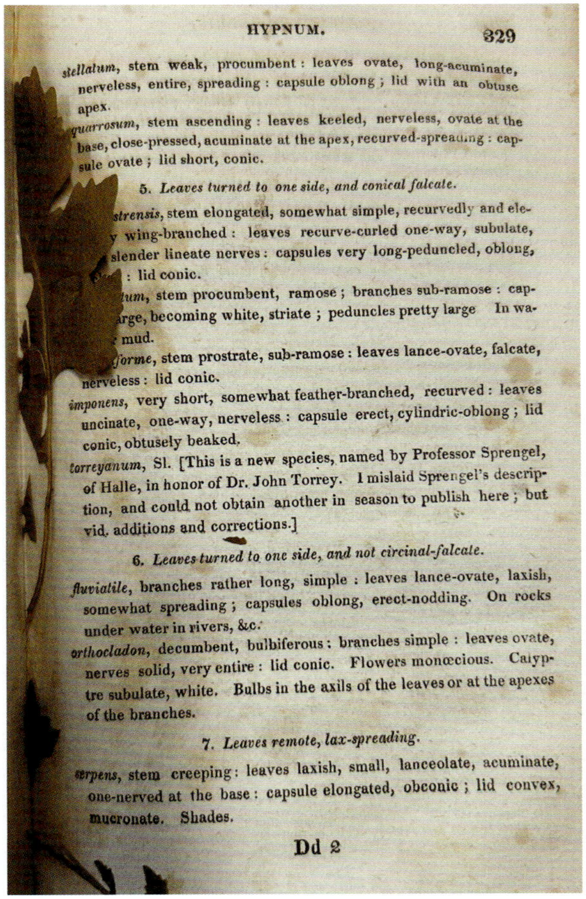

HYPNUM. 329

stellatum, stem weak, procumbent: leaves ovate, long-acuminate, nerveless, entire, spreading: capsule oblong; lid with an obtuse apex.
quarrosum, stem ascending: leaves keeled, nerveless, ovate at the base, close-pressed, acuminate at the apex, recurved-spreading: capsule ovate; lid short, conic.

5. *Leaves turned to one side, and conical falcate.*

strensis, stem elongated, somewhat simple, recurvedly and ele- y wing-branched: leaves recurve-curled one-way, subulate, slender lineate nerves: capsules very long-peduncled, oblong, : lid conic.
tum, stem procumbent, ramose; branches sub-ramose: cap- rge, becoming white, striate; peduncles pretty large In wa- mud.
forme, stem prostrate, sub-ramose: leaves lance-ovate, falcate, nerveless: lid conic.
imponens, very short, somewhat feather-branched, recurved: leaves uncinate, one-way, nerveless: capsule erect, cylindric-oblong; lid conic, obtusely beaked.
torreyanum, Sl. [This is a new species, named by Professor Sprengel, of Halle, in honor of Dr. John Torrey. I mislaid Sprengel's description, and could not obtain another in season to publish here; but vid. additions and corrections.]

6. *Leaves turned to one side, and not circinal-falcate.*

fluviatile, branches rather long, simple: leaves lance-ovate, laxish, somewhat spreading; capsules oblong, erect-nodding. On rocks under water in rivers, &c.
orthocladon, decumbent, bulbiferous: branches simple: leaves ovate, nerves solid, very entire: lid conic. Flowers monœcious. Calyptre subulate, white. Bulbs in the axils of the leaves or at the apexes of the branches.

7. *Leaves remote, lax-spreading.*

serpens, stem creeping: leaves laxish, small, lanceolate, acuminate, one-nerved at the base: capsule elongated, obconic; lid convex, mucronate. Shades.

Dd 2

FIGURE 1.15. Pressed specimen, page 329, Nye's copy of Eaton, *A Manual of Botany* (1824)

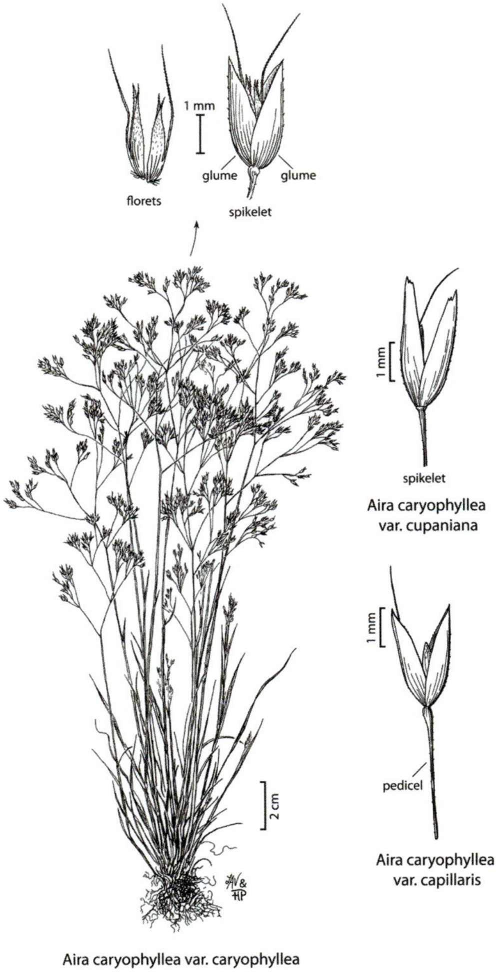

FIGURE 1.17. Detail, *A. caryophyllea,* from *Flora of North America,* illustrators Linda Ann Vorobik and Hana Pazdírková, © Utah State University, used with permission

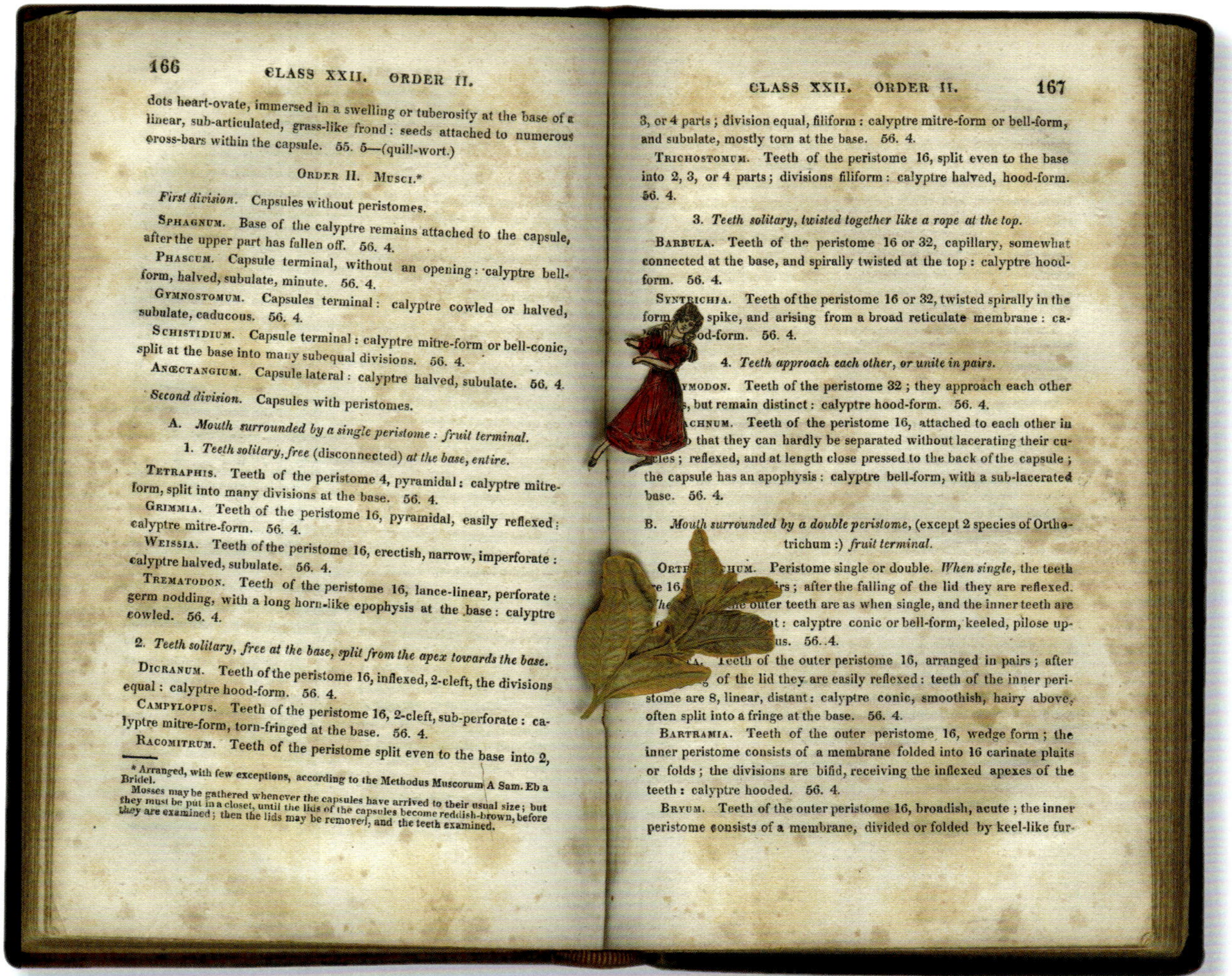

FIGURE 1.18. Print cutout and plant specimen, page 167, Nye's copy of Eaton, *A Manual of Botany* (1824)

Regardless of whether the specimens correspond to the pages where they appear, the presence of more than fifty pressed plants shows an active botanist at work. Nye was fifteen or sixteen in 1827, the date she inscribed on the second flyleaf. A possible glimpse into her youth appears in figure 1.18, which includes a hand-colored cutout of a girl in a red dress. The combination of plant specimen and print specimen indicates that Nye (or another reader) used the *Manual of Botany* for both botanizing and keeping mementos from other print sources.

Some young women attended a female seminary or academy for their postprimary education. In the first half of the nineteenth century, there were several hundred of these schools offering advanced curricula for women, including in the sciences, and that number increased steadily in the second half.[29] Two young women at different academies inscribed the same copy of Green and Congdon's *Analytical Class-Book of Botany* (figures 1.19 and 1.20). Emma L. Stearns (b. 1833) of Ashburnham, Massachusetts, attended New Ipswich Appleton Academy in New Hampshire from fall term 1854 to spring term 1856. Sarah F. Lee (b. 1838) of New Ipswich, New Hampshire, attended Bradford Academy in Massachusetts in 1858 and 1859 and inscribed *Analytical Class-Book of Botany* on April 10, 1858. Given when she attended Bradford Academy, Sarah Lee likely wrote most of the marginalia in this book. The writer repeatedly included the abbreviation "Ex." (perhaps for "Example" or "Exam") and the dates May 14, 19, 21, and 26, 1858, and marked the end of a chapter with "La Fin" or "Finis" (figures 1.21 and 1.22). The inscriptions by two young women who noted their schools raise interesting questions. How did the book get from one to the other, given that they were at schools in different towns? Perhaps the young women became friends in New Ipswich, where Stearns attended school and Lee resided; they were relatively close in age—five years apart—and both well educated. Textbooks cost money, and there were fees for attending these schools; sharing resources such as *Analytical Class-Book of Botany* could have helped defray expenses. Although Stearns and Lee learned from Green and Congdon's book at female academies, not at the "the workshop and the cotton-mill" the authors mention in their preface, these young women may well have "attain[ed] a competent knowledge of this charming science, with perhaps the will to pursue it further, and ma[d]e themselves thorough and accomplished scholars," as the authors also hoped.[30]

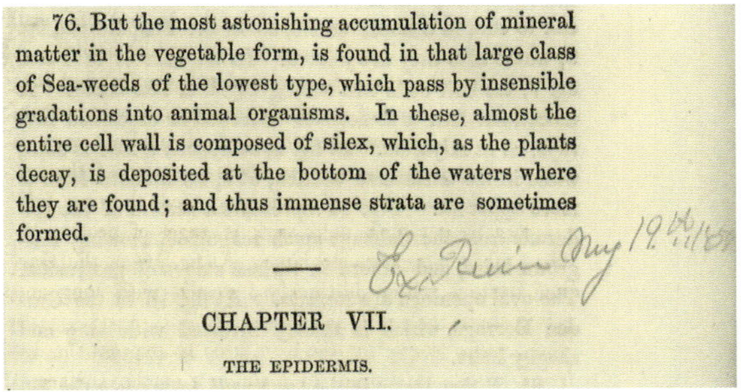

76. But the most astonishing accumulation of mineral matter in the vegetable form, is found in that large class of Sea-weeds of the lowest type, which pass by insensible gradations into animal organisms. In these, almost the entire cell wall is composed of silex, which, as the plants decay, is deposited at the bottom of the waters where they are found; and thus immense strata are sometimes formed.

CHAPTER VII.

THE EPIDERMIS.

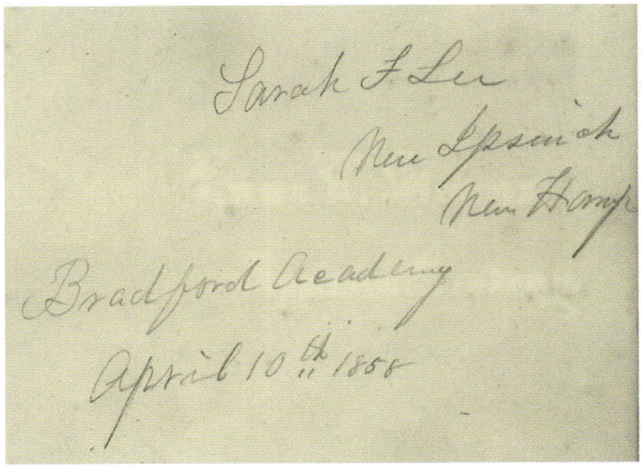

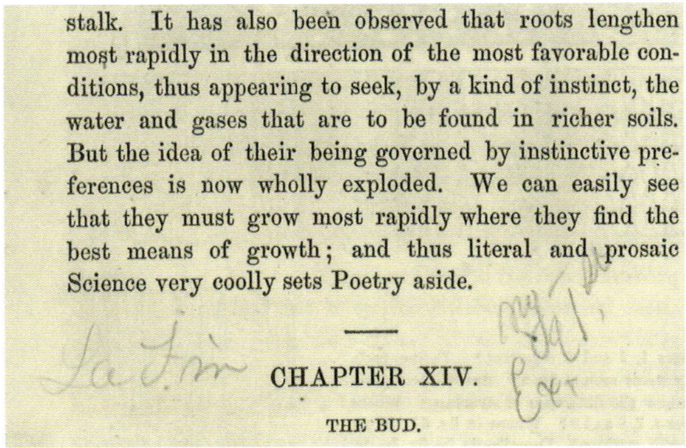

stalk. It has also been observed that roots lengthen most rapidly in the direction of the most favorable conditions, thus appearing to seek, by a kind of instinct, the water and gases that are to be found in richer soils. But the idea of their being governed by instinctive preferences is now wholly exploded. We can easily see that they must grow most rapidly where they find the best means of growth; and thus literal and prosaic Science very coolly sets Poetry aside.

La Fin

CHAPTER XIV.

THE BUD.

FIGURE 1.19. (*top left*) Emma L. Stearns inscription; FIGURE 1.20. (*bottom left*) Sarah F. Lee inscription; FIGURE 1.21. (*top right*) Marginalia, page 11; FIGURE 1.22. (*bottom right*) Marginalia, page 30. Green and Congdon, *Analytical Class-Book of Botany* (1855)

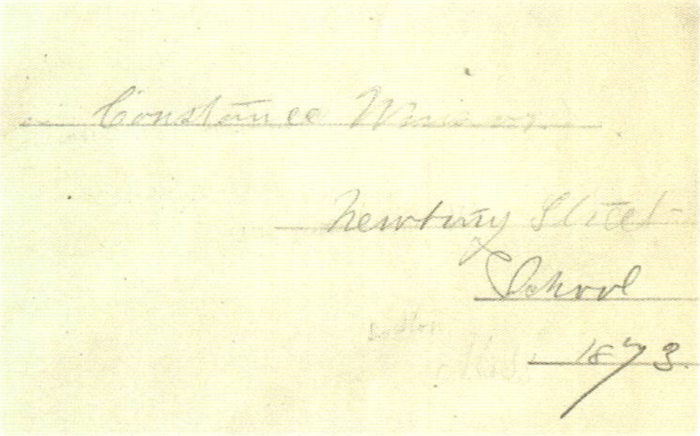

FIGURE 1.23. Constance Winsor inscription in Youmans, *The First Book of Botany* (1870), Text Y8, seq. 1, Arnold Arboretum Library, Harvard University

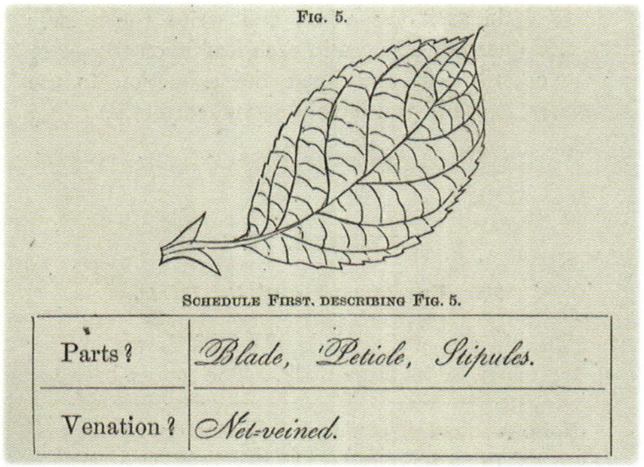
FIG. 5.

SCHEDULE FIRST, DESCRIBING FIG. 5.

Parts ?	*Blade, Petiole, Stipules.*
Venation ?	*Net-veined.*

FIGURE 1.24. Schedule First, Youmans, *The First Book of Botany* (1873)

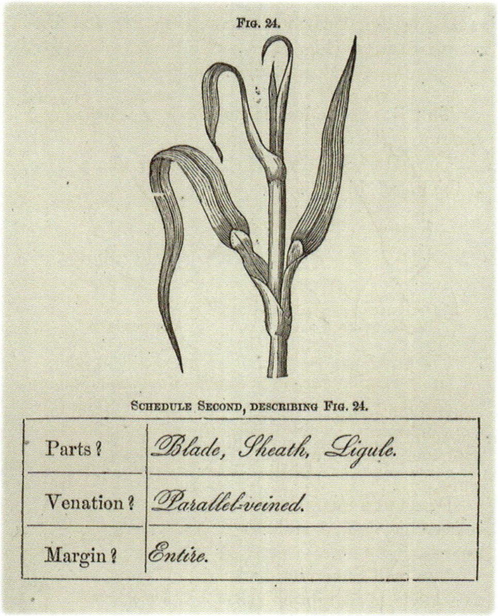
FIG. 24.

SCHEDULE SECOND, DESCRIBING FIG. 24.

Parts ?	*Blade, Sheath, Ligule.*
Venation ?	*Parallel-veined.*
Margin ?	*Entire.*

FIGURE 1.25. Schedule Second, Youmans, *The First Book of Botany* (1873)

Constance Winsor was a few years younger than Emma Stearns and Sarah Lee when she inscribed her copy of Eliza Youmans's *The First Book of Botany* with her name, school, and year (figure 1.23). Winsor was born in 1860 and wrote in her copy of Youmans's *First Book of Botany* as a thirteen-year-old. Youmans's *First Book* was aimed at beginners and designed to "train the observing powers . . . in a systematic way." It uses a method of "schedules" first developed by John Stevens Henslow, an English botanist, priest, and geologist who taught Charles Darwin. Youmans provides thirty-one schedules that guide students through the structures of the leaf, stem, inflorescence, and flower (figures 1.24 and 1.25). In her "Suggestions to Teachers," Youmans notes that although she provides models of the schedules, "the pupil will make them for himself with pencil and paper," which Winsor did. One such example remains in her copy of *The First Book of Botany*, including a pressed plant specimen; the corresponding schedule in Youmans is provided for comparison (figures 1.26 and 1.27). Winsor correctly notes the number of leaflets (nine), repeats the parts from Youmans's schedule, and inserts a representative plant specimen into a cut at the top of the page. She signs her name, the date, and again her school. A circular for the Newbury Street School from June 1874—a year after Winsor completed Schedule Nine—notes the inclusion of the natural sciences in its curriculum.[31] In her later studies in botany, she used Gray's *Manual of Botany* (1875). Winsor married James Noyes in 1890, gave birth to her daughter Penelope in 1892, and died in 1895, "aged not quite 35 years." Penelope Noyes grew up in Cambridge, Massachusetts. In her memoir, she reminisced that with her father, "the Botanic Gardens at the corner of Garden and Linnaean Streets were a favorite place on a

Sunday afternoon"; perhaps her father had gone there with her botany enthusiast mother too.[32]

These examples show young women from the 1820s through the 1870s using their botany textbooks as study guides, specimen repositories, and records of their scientific engagement. While inscriptions like those of Emma Stearns and Sarah Lee are only a small piece of evidence, the inclusion of their schools' names and Emma Stearns's elaborate calligraphy suggest pride in their education. When additional material evidence is present, such as Eliza Nye's pressed specimens and Constance Winsor's schedule, we get a glimpse into young women's botany education at work.

Botanizing at Home and Away: Henry David Thoreau and Hezekiah W. Marbourg

Nineteenth-century American citizen scientists were, as has been shown throughout this chapter, ordinary people who pursued botany for the love of it, while also working as teachers, administrators, editors, clergy, and more. Henry David Thoreau (1817–1862) was an accomplished naturalist who wrote about plants in both his published works and unpublished journals, letters, and field notes; he used his copy of Gray's *Manual of Botany* to record his findings in the last few years of his life. The physician Hezekiah W. Marbourg (1833–1889) recorded his botanizing in Wood's *Class-Book of Botany* only occasionally but over decades, writing brief notes in the margins and pressing specimens.

Thoreau is best known today as the author of *Walden* (1854), a memoir and nature writing classic about his two-year sojourn at Walden Pond outside Concord, Massachusetts, and "Resistance to Civil Government" (1849), an essay about citizens' conscientious duty to

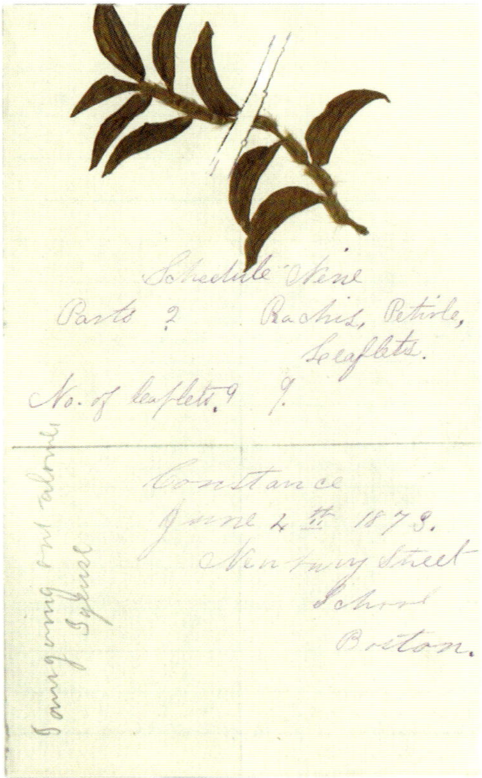

FIGURE 1.26. "Schedule Nine" by Constance Winsor in Youmans, *The First Book of Botany* (1870), Text Y8, seq. 3, Arnold Arboretum Library, Harvard University

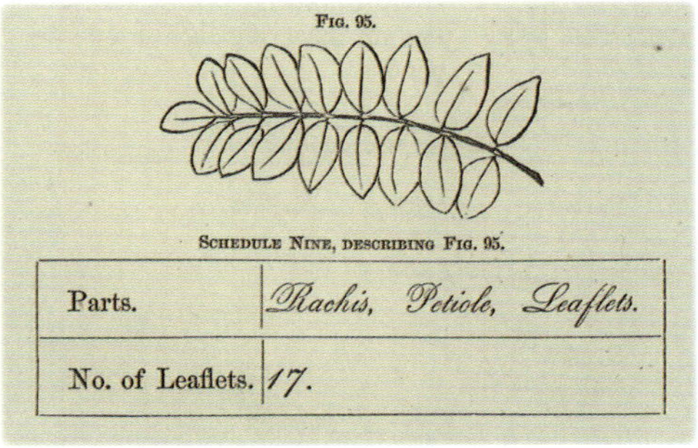

FIGURE 1.27. "Schedule Nine" in Youmans, *The First Book of Botany* (1873)

oppose unjust and overreaching governments. Thoreau began studying botany at the Concord Academy and continued studying natural history at Harvard, including attending lectures on botany. Robert Sattelmeyer, in *Thoreau's Reading*, lists works by botanists John Bartram, William Bartram, Amos Eaton, Asa Gray, Pehr Kalm, John Lindley, Carl Linnaeus, André Michaux, Thomas Nuttall, Almira Phelps, John Torrey, and Alphonso Wood.[33]

Between the publication of *A Week on the Concord and Merrimack Rivers* (1849) and *Walden* (1854), several events may account for Thoreau's increased botanical knowledge. As Ray Angelo speculates, "What more directly crystallized Thoreau's botanical inclinations . . . was the publication in 1848 of the first edition of Asa Gray's *Manual of Botany*," and "the earliest specimens in his organized herbarium" date from 1850. Bradley Dean states that in 1850, Thoreau "began cultivating an interest in science, particularly botany" and "started carrying a botanical guide with him on his afternoon walks." Sattelmeyer writes that Thoreau used "certain basic texts, manuals of botany like Bigelow's and Gray's, that were essential tools for learning taxonomy."[34] Dean also posits "A Possible Source for Thoreau's Increased Interest in Natural Phenomena in 1851": the distribution in 1851 by the Smithsonian Institution of a "Registry of Periodical Phenomena." As Dean explains, "The Smithsonian Institute sent [this] to scientists across the country" and "invited 'all persons who may have it in their power, to record their observations . . . and to transmit them to the Institution.' The circular lists 127 species of plants, using in most cases both common and Latin names, and asks observers to mark opposite each species its date of flowering. The Smithsonian list bears a striking resemblance to the phenological lists Thoreau began assembling at that time." The Smithsonian "Registry" is an early example of crowd-sourcing citizen science, and Thoreau's phenological observations—noting the first appearance and bloom times of local plant species—are being used today as data for tracking climate change in Concord, Massachusetts.[35]

Thoreau botanized widely in and around Concord throughout the 1850s. In 1861, hoping to improve his tuberculosis in a different climate, he embarked on a journey to Minnesota from May 11 to July 11. He and his traveling companion, Horace Mann Jr., botanized along the way; Thoreau recorded his activities in a field notebook and took specimens for his herbarium. In her detailed study *Westward I Go Free: Tracing Thoreau's Last Journey*, Corinne Hosfeld Smith notes that he "had his plant press with him, as well as his compass, microscope, spy glass, and one or two botany manuals. More than likely he consulted one of his own editions of Asa Gray's *Manual of the Botany of the Northern United States*." Thoreau had two editions of Gray's *Manual*, the first edition of 1848 and the second edition of 1856, suggesting that he used sources that were as current as possible.[36]

Thoreau did indeed use Gray on his western journey, as seen in his copy of the 1856 edition of the *Manual of Botany*, which is now part of the Berg collection at the New York Public Library. On the inside right flyleaf, just below his inscribed name, is written:

(910)

Sep 1–6 I have collected a little over 900 flowers—not counting sedges grasses &c.

This number aligns with the more than nine hundred specimens in Thoreau's herbarium. His annotations generally consist of a single place name or place and date, starting in 1858 and ending with September 17, 1861, less than eight months before his death. Many dates and places correspond to his Minnesota journey: Some are written simply as "Minnesota"; others are more specific, such as Goat Island (perhaps near Niagara Falls, or in Minnesota), Dunlieth, St. Anthony, Lake Calhoun, Minnehaha, and Mackinaw. There are also New England locations nearer his home in Concord: Brattleboro, Monadnock, Beverley, Salem Harbor, Wayland, and Acushnet. Some of his marginalia compare descriptions in Gray's text to those of other botanists. For example, on page 18, next to the entry on "MENISPERMUM, L. MOONSEED," where the description reads "Flowers white," "white" is underlined twice, with this note in the right margin: "Wood says 'yellow' Beck says 'greenish yellow' so do others." On page 191, in reference to "A.[ster] Radula," Thoreau wrote, "Plate in Torrey's N. Y. Bot." Thoreau was thorough in his botanizing, using books by Wood, Beck, Torrey, and other experts in his field research.

In another botanical record, Thoreau titled eleven loose pages of phenological data "The Flowering of Plants, Accidentally Observed in '51, with Considerable Care in '52; The Spring of '51 Being 10 Days and More Earlier than That of '52. The Names Those Used by Gray." James P. Warren, in his discussion of "Thoreau's botanical turn," notes that "Gray's *Manual of Botany* gradually takes over as Thoreau's authoritative guide for identifying plants."[37] Thoreau's annotations in Gray's 1856 *Manual of Botany* are a valuable addition to his journal and field notes for these later years of his life. The specific dates—always month and day, sometimes also year—are part of Thoreau's phenological legacy, and unlike his records of Concord, the data from his Minnesota trip have yet to be used for climate change analysis. His marginalia show how books such as Gray's *Manual of Botany* also functioned as field guides, a format that appeared only later in the nineteenth century.

Around the same time that Thoreau was botanizing near Concord and on his journey west, Hezekiah Marbourg recorded his findings near home and farther away. Marbourg inscribed his copy of Wood's *Class-Book of Botany* (1855) twice on the inside front cover (figure 1.28). Hezekiah Wilson Marbourg was born in 1833 in Indiana County, Pennsylvania. He graduated from Pennsylvania (now Gettysburg) College and received an M.D. from Jefferson Medical College in Philadelphia in 1858. He spent most of his career as a physician in Johnstown, Pennsylvania. During the Civil War, he was stationed at Roanoke Island, North Carolina, as a contract surgeon for the U.S. Army. After the war, he continued to practice medicine until his death in 1889 in the massive Johnstown Flood.[38]

Marbourg's copy of Wood's *Class-Book* includes notes next to thirty-three species, usually with a place

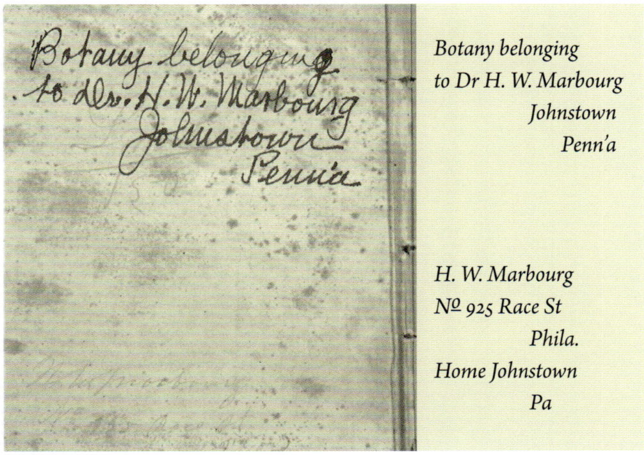

FIGURE 1.28. Inscriptions by H. W. Marbourg in Wood, *A Class-Book of Botany* (1855)

and date, and occasionally a brief comment such as "beautiful" or "pretty." Seventeen pressed specimens remain in the book. The earliest margin note is dated April 25, 1857, and the latest May 2, 1888. He botanized in Johnstown and Philadelphia in 1857, 1858, and 1860, when he was a medical student and physician beginning his practice. He didn't go far afield: The locations he recorded in Philadelphia include Logan Square and Franklin Square, and in Johnstown, the Cement Quarry, a factory, Cambria [Iron] Furnaces, an "old tunnel," and "Eddy Owen's yard market." These urban and industrial sites suggest a botanizer with an eye for plants in places others might overlook—for example, he found the tiny *Hepatica triloba* (liverwort) at the "Sement Quarry" on April 25, 1857 (figure 1.29).

His marginalia resume in 1865, for a little over a week from April 27 to May 5, starting in Norfolk, Virginia, and ending at Roanoke Island. Table 1.1 summarizes his botanizing during this time. These dates and places indicate that he traveled some long distances in this

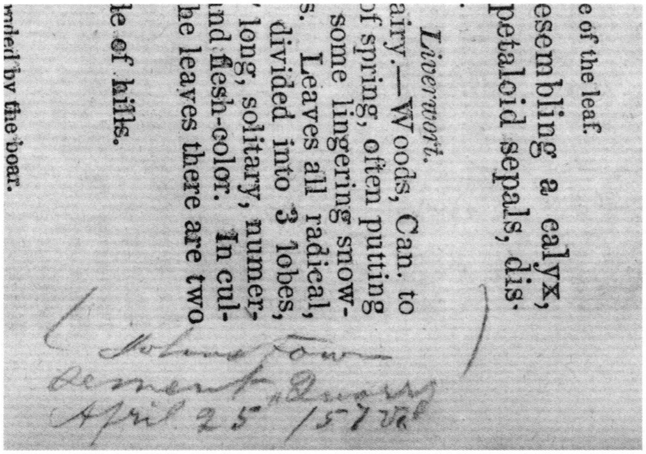

FIGURE 1.29. Marbourg's marginalium, page 141, Wood, *A Class-Book of Botany* (1855)

short time span. On today's roads, Norfolk to New Bern is 162 miles; New Bern to Goldsboro, 60; Goldsboro to Raleigh, 55; Raleigh to Morehead City, 150; Morehead City to New Bern, 35; and New Bern back to Roanoke Island, 170 miles. On April 9 Confederate General Robert E. Lee surrendered to Union General Ulysses S. Grant at Appomattox, effectively ending the war. A few weeks later during his botanizing, Marbourg might have been on leave or circulating among different medical facilities; all the locations where he botanized had hospitals.[39] A week after returning to Roanoke Island, he did a bit more botanizing in the area, finding and pressing a piece of star grass near Beaufort, North Carolina, on May 12, 1865 (figure 1.30). Marbourg completed his service in June.

Maura Flannery notes that "American plant collectors were often in the military. . . . Botanists saw these missions—even the war—as opportunities for collecting in new areas." Although Marbourg made notes for only a few days in spring 1865, he recorded more plants then than in any other time period. As a Northerner in the South, he encountered species whose ranges did not extend to his home state of Pennsylvania. Some he found "beautiful," such as the water lily and umbrella tree; the carnivorous side-saddle plant was "very peculiar," by which he might have meant "strange" or simply "singular" (figures 1.31 and 1.32). Lauren LaFauci argues compellingly for the "mutually constitutive relationship between 'the peculiar institution' [of slavery] and the prevailing societal and scientific conceptions of wetlands as 'peculiar' places."[40] Given that the side-saddle plant (*Sarracenia purpurea*) grows in boggy, swampy environments, Marbourg could be using the contemporary discourse about these "peculiar" aspects of Southern nature.

TABLE 1.1. H. W. Marbourg's 1865 botanizing in North Carolina, as noted in Wood, A *Class-Book of Botany* (1855)

PAGE	SPECIES NOTED: *LATIN BINOMIAL*	SPECIES NOTED: COMMON NAME	LOCATION (IN HWM'S SPELLING)—COMMENT	DATE
235	*Laburnum vulgarum*	golden-chain	Norfolk, Va.	April 27
374	*Kalmia glauca*	swamp laurel	between New Bern and Goldsborough, N.C.	April 29
150	*Magnolia umbrella*	umbrella tree	Goldsborough—"beautiful"	April 29
542	*Iris prismatica*	Boston iris	between Goldsboro & Raughley, N.C.	April 29
463	*Chionanthus virginica*	Virginian fringe-tree	Raleigh, N.C.	April 30
214	*Æsculus hippocastanum*	horse chestnut	Morehead City, N.C.	May 1
418	*Salvia lyrata*	wild or meadow sage	New Bern	May 3
154	*Nymphæa odorata*	water lily	Rhoanoke Island—"beautiful"	May 5
155	*Sarracenia purpurea*	sidesaddle flower	Rhoanoke Island—"very peculiar"	May 5
299	*Lonicera sempervirens*	trumpet honeysuckle	Rhoanoke Island	May 5

Marbourg returned to Johnstown after the war. His first wife, Wilhelmina Rachel Wright, died in 1870; in 1872, he married Esther Wright Nippes, Wilhelmina's sister and the widow of Charles Nippes, who was killed at the Battle of Gettysburg in July 1863. Both Wilhelmina R. W. Marbourg and Esther L. W. Nippes graduated from the Women's Medical College of Pennsylvania as married women and then practiced as physicians, all of which was noteworthy at this time. The 1870 census "counted 525 trained women doctors in America—more than in the rest of the world combined," but "only 137 women were enrolled in regular medical schools, and most of these were in separate women's medical colleges."[41] Records from the Women's Medical College of Pennsylvania indicate that Wilhelmina Marbourg wrote her 1869 thesis on a botanical topic, "*Cimicifuga racemosa*" or black cohosh, "a plant/herb which has been traditionally exploited and extensively used by women," including for "complications associated with childbirth." Esther Nippes wrote hers in 1872 on the gynecological topic "Vasico vaginal fistula," which was "a catastrophic complication of childbirth among 19th century American women." Its history is notorious today because of an Alabama physician who conducted experiments on enslaved women to develop a procedure for repairing this kind of fistula.[42] H. W. Marbourg and Esther Nippes Marbourg established a joint medical practice in Johnstown,

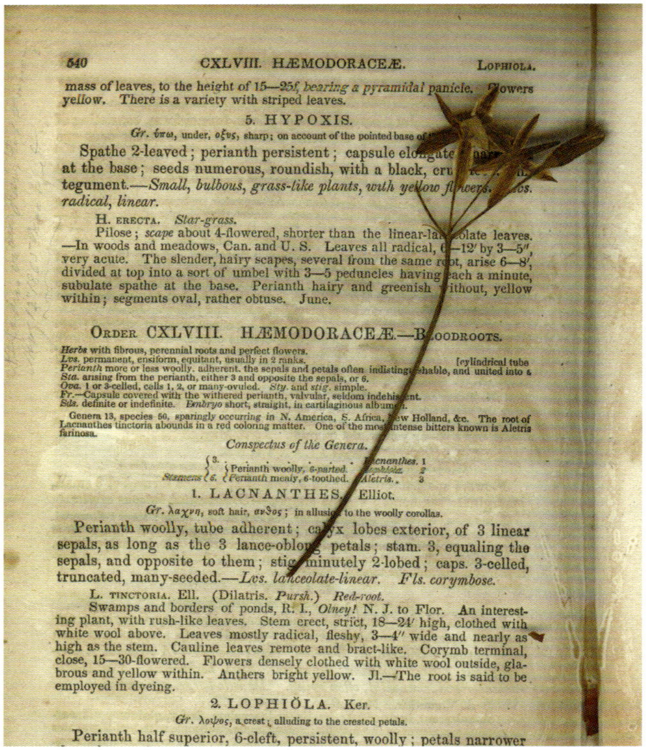
540 CXLVIII. HÆMODORACEÆ. LOPHIOLA.

mass of leaves, to the height of 15—25f, *bearing a pyramidal panicle.* Flowers yellow. There is a variety with striped leaves.

5. HYPOXIS.

Spathe 2-leaved; perianth persistent; capsule elongated ... at the base; seeds numerous, roundish, with a black, cru... tegument.—*Small, bulbous, grass-like plants, with yellow flowers.* ... *radical, linear.*

H. ERECTA. *Star-grass.* Pilose; *scape* about 4-flowered, shorter than the linear-lanceolate leaves.—In woods and meadows, Can. and U. S. Leaves all radical, 6—12' by 3—5", very acute. The slender, hairy scapes, several from the same root, arise 6—8', divided at top into a sort of umbel with 3—5 peduncles having each a minute, subulate spathe at the base. Perianth hairy and greenish without, yellow within; segments oval, rather obtuse. June.

ORDER CXLVIII. HÆMODORACEÆ.—BLOODROOTS.

Genera 13, species 50, sparingly occurring in N. America, S. Africa, New Holland, &c. The root of Lacnanthes tinctoria abounds in a red coloring matter. One of the most intense bitters known is Aletris farinosa.

Conspectus of the Genera.

1. LACNANTHES. Elliot.

Perianth woolly, tube adherent; calyx lobes exterior, of 3 linear sepals, as long as the 3 lance-oblong petals; stam. 3, equaling the sepals, and opposite to them; stig. minutely 2-lobed; caps. 3-celled, truncated, many-seeded.—*Lvs. lanceolate-linear. Fls. corymbose.*

L. TINCTORIA. Ell. (Dilatris. *Pursh.*) *Red-root.* Swamps and borders of ponds, R. I., *Olney!* N. J. to Flor. An interesting plant, with rush-like leaves. Stem erect, strict, 18—24' high, clothed with white wool above. Leaves mostly radical, fleshy, 3—4" wide and nearly as high as the stem. Cauline leaves remote and bract-like. Corymb terminal, close, 15—30-flowered. Flowers densely clothed with white wool outside, glabrous and yellow within. Anthers bright yellow. Jl.—The root is said to be employed in dyeing.

2. LOPHIOLA. Ker.

Perianth half superior, 6-cleft, persistent, woolly; petals narrower

FIGURE 1.30. Pressed specimen and marginalia, page 540, Wood, *A Class-Book of Botany* (1855)

and in the 1870s and 1880s Marbourg still noted some plants in his *Class-Book*. The last marginalium appears next to the entry on the smiling wake-robin (*Trillium erythrocarpum*), which he found on May 2, 1888, at the Cambria Furnaces, part of Johnstown's largest steel and iron works and one of the industrial sites where Marbourg had botanized thirty years earlier.

Pull a thread and see what unravels: I bought this copy of *Class-Book of Botany* because it was affordable; it was in "fair" condition and the decorative spine was detached, as shown in figures I.11–I.14 of the introduction. When the book arrived, I did an internet search for "H. W. Marbourg" and discovered several biographical sources, because records exist for Marbourg's college and medical school education, army service, and medical practice. He was a significant enough citizen of Johnstown, Pennsylvania, to be included in its county history. Marbourg's second wife's first husband was killed in battle, which was also documented. Because his wives graduated from medical school, we have records of them, too, beyond those of their husbands. All these people's lives were directly or indirectly linked to significant events of the later nineteenth century—the Civil War, medical education for women, the practice of medicine during the war, and the Johnstown Flood of 1889—as well as more common ones, such as the marriage of Wilhelmina Marbourg's widowed husband to her widowed sister. A couple of inscriptions, several dozen margin notes, and some pressed specimens are the material evidence that tells part of their story.

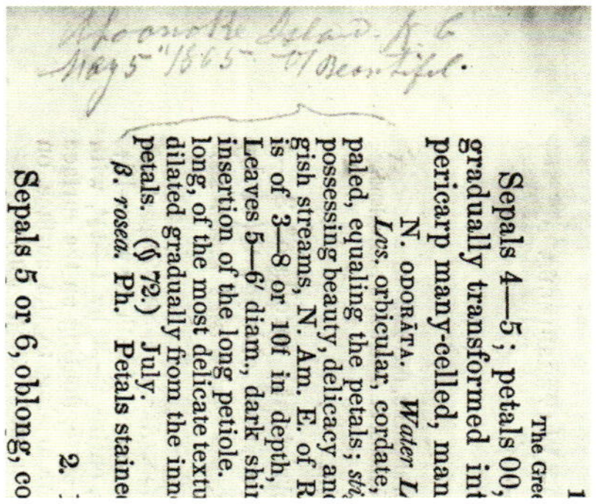

FIGURE 1.31. Marbourg note, page 154, Wood, *A Class-Book of Botany* (1855)

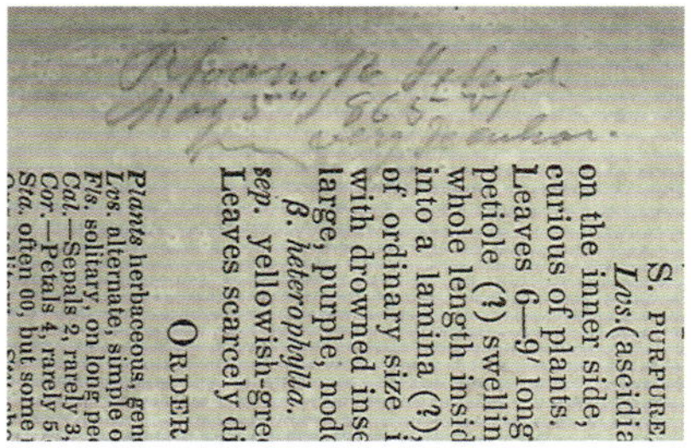

FIGURE 1.32. Marbourg note, page 155, Wood, A *Class-Book of Botany* (1855)

A wide range of people were amateur botanists in the nineteenth century, and as Keeney notes, most of them "never wrote anything, or made a cent from botany, or joined an institution, or subscribed to a botanically inclined periodical." But some of them inscribed their botany instruction books and left other clues about their botanizing, which in turn indicates how they became the citizen scientists that nineteenth-century American botany instruction book authors hoped for. Their efforts and accomplishments were generally modest, local, and personal. Much of what amateur botanists learned and discovered was experiential, from conversations, walks in the woods, schoolroom lessons, and encounters with other botanizers. Even materials that are written down, such as letters, diaries, inscriptions, marginalia, and field notebooks, may or may not be preserved in ways that enable us to access them today. Thoreau's copy of Gray's *Manual of Botany* is available at the New York Public Library because someone knew its value and donated it to the Berg Collection. My access to Marbourg's copy of Wood's *Class-Book of Botany*, on the other hand, was a fluke.[43]

The next chapter focuses on a single botany textbook, John Newman's *Illustrated Botany*, that belonged to someone who left few traces in official sources. However, what she documented between the covers of a single book provides a fascinating and often provocative record of one young woman's experience of amateur botany and much more, shortly after the Civil War in the sparsely populated Sierra Nevada mountains of northeastern California. This botany textbook became her *vade mecum* in ways that Gray likely would not have imagined when he expressed that hope for the first edition of his *Manual of Botany*.

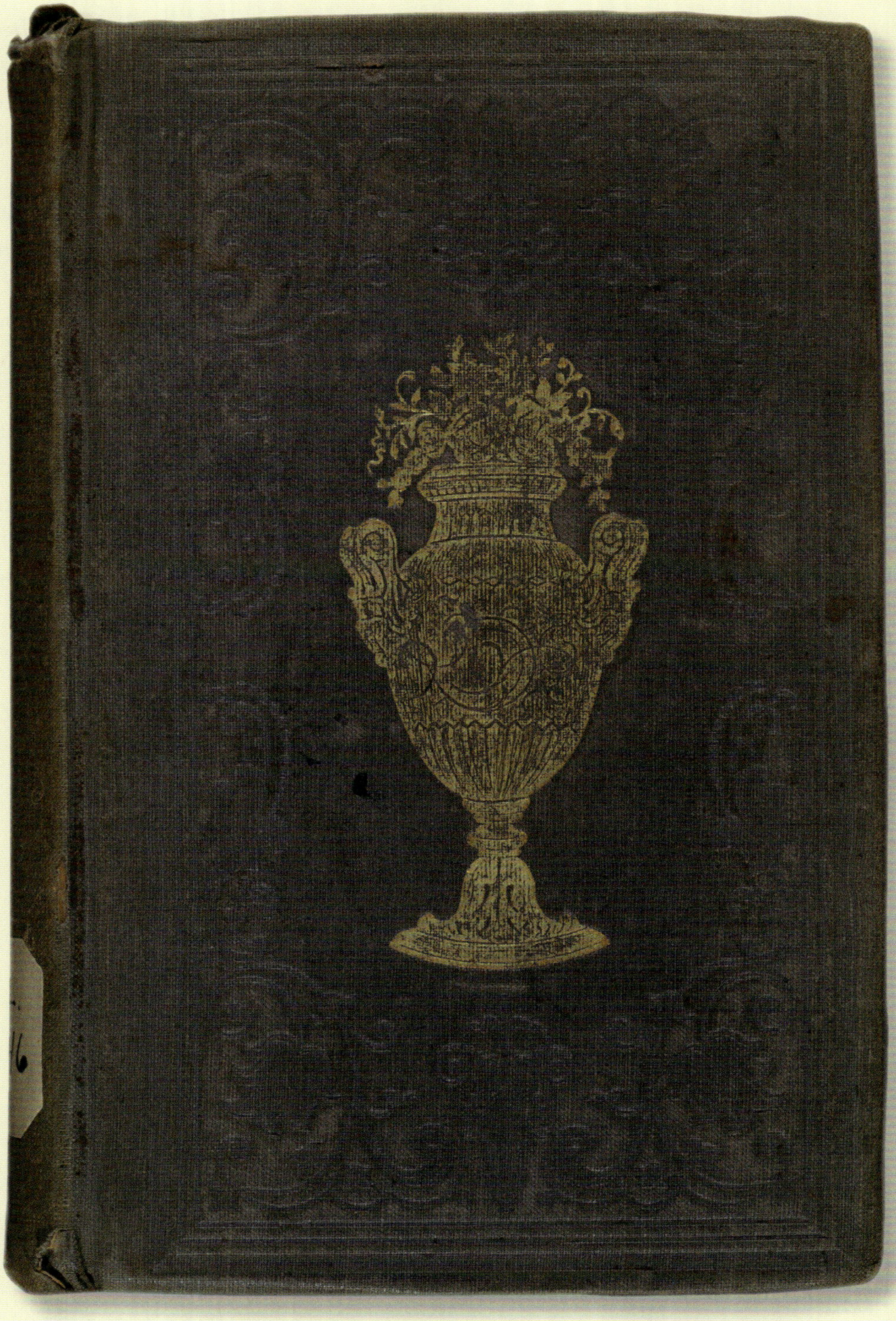

Figure 2.1. Cover, M. S. Snowden's copy of Newman, *Illustrated Botany* (1850), GEN P. L. N46, Arnold Arboretum Archives, Harvard University

2

FROM THE ARCHIVES

"You Old Botany"

Textbook as Field Guide, Journal, and Commonplace Book

As I discuss throughout *Flora's Interpreters,* nineteenth-century U.S. flower books served multiple purposes, and any one text often fits several different genres. The focus of this chapter, M. S. Snowden's copy of John Newman's *Illustrated Botany* (1850), provides a fascinating example of a botany instruction book that also became a field guide, commonplace book, specimen repository, and even the owner's journal.[1] Copious marginalia in a distinctive voice model the dialogical framework of the textbook and indicate an actively engaged reader and citizen scientist. Botanical and literary references join diary entries about well-known figures in California history, ultimately confirming the combination of high and low, scientific and popular, and educational and emotional discourses that characterizes popular botany in nineteenth-century America.[2]

The chapter begins with an overview of the botanical instruction features of *Illustrated Botany,* then discusses Snowden's wide-ranging marginalia on botany, her excursions with prominent people, and frequent literary references. Her copy of *Illustrated Botany* captures in a single volume various interactions of owner and book and reveals the multidisciplinary knowledge of an enthusiastic amateur botanist.

About *Illustrated Botany*

The title page of *Illustrated Botany* announces its intentions and important features, including floral dictionary, glossary of scientific terms, engraved illustrations, and the author's status as a well-published writer:

ILLUSTRATED
BOTANY,
CONTAINING
A FLORAL DICTIONARY,
AND A
GLOSSARY OF SCIENTIFIC TERMS.
ILLUSTRATED WITH
MORE THAN ONE HUNDRED ENGRAVINGS.
BY JOHN B. NEWMAN, M.D.,
AUTHOR OF VARIOUS WORKS ON THE
NATURAL SCIENCES.

These features are typical of other nineteenth-century American botany instruction books. The publisher's preface targets amateurs, stating that the book is "intended for those who have no previous knowledge of the subject." This edition of *Illustrated Botany* was published in 1850, at the century's midpoint, when many kinds of popular botany books flourished, especially those for a general audience. The preface further states that "the aim has been, not only to make it simple enough to be understood without other instruction, but also, by means of ample illustration, in the way of facts and anecdotes, to keep up and gratify curiosity to the end."[3] The main part of the book (chapters 1 to 12), titled "Botany for Young People," is a series of dialogues in which Emily instructs her younger sister Laura, combining "facts" about botany with a wide variety of "anecdotes" that focus on plant taxonomy, structure, and physiology. The text's more literary touches include frequent verses inspired by and personifying the twenty-four classes of Linnaean classification, from "Song of the Monandrian Tribes" to "Song of the Cryptogamian Tribes," as well as excerpts from poets both named (such as Erasmus Darwin, Felicia Hemans, Robert Southey, William Wordsworth, and John Keats) and unnamed.

The book also includes a synopsis of classes and orders; examples for "exercising the learner," alphabetically organized from *Amygdalus* to *Zea*; an illustrated section on medical plants; a floral dictionary, arranged alphabetically from acacia (platonic love) to yew (sorrow); a glossary of technical terms, from acerose to whorls; and additional illustrations, showing mostly leaf shapes and other physiological details (figures 2.2 and 2.3).

John B. Newman was a physician who briefly served as president of Harrodsburg Female College in Kentucky in the early 1850s. He wrote on a variety of subjects, including *The Early Peopling of America, and Its Discovery Before the Time of Columbus* (1847), *The Home Doctor, a Family Manual* (1847), *Fascination, or the Philosophy of Charming* (1848), and *The Natural History of Man: Showing His Three Aspects of Plant, Beast, and Angel* (1852). He even wrote a book-length narrative poem, *Wa-wa-wanda, a Legend of Old Orange* (1860), in the challenging meter of trochaic tetrameter, which Longfellow used for "The Song of Hiawatha" (1855). Newman published other botany books, including *Boudoir Botany* (1846), *Beauties of Flora, and Outline of Botany* (1848), and *The Ladies' Flora* (1854), all of which combine the basics of botany with beautiful illustrations and interesting anecdotes (figures 2.4 and 2.5).

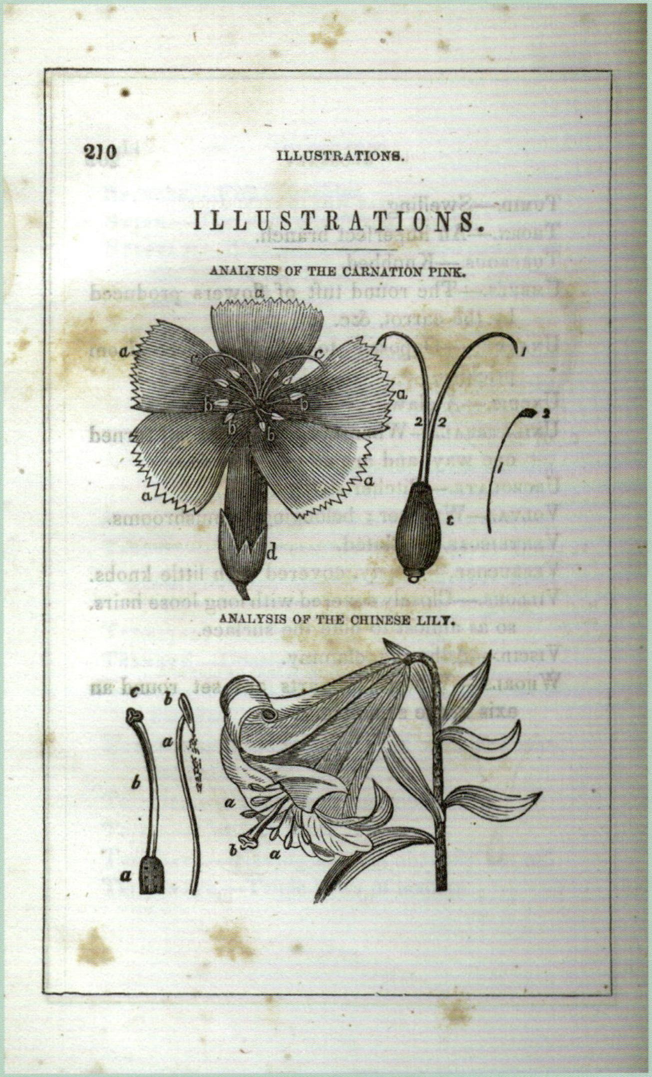

FIGURE 2.2. "Illustrations," page 210, Newman, *Illustrated Botany* (1850)

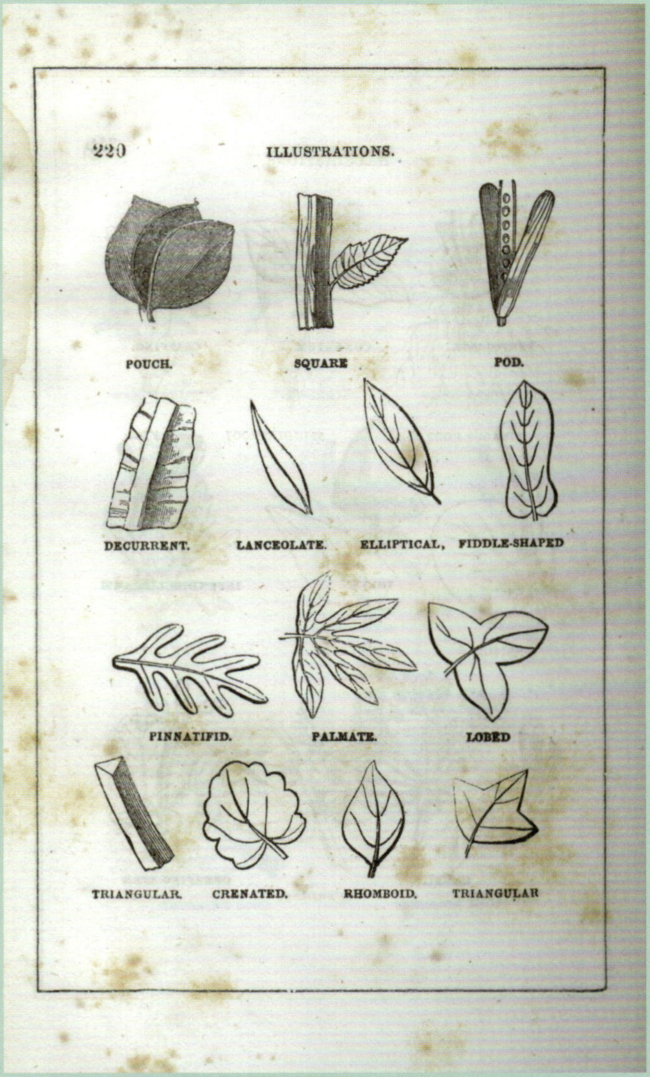

FIGURE 2.3. "Illustrations," page 220, Newman, *Illustrated Botany* (1850)

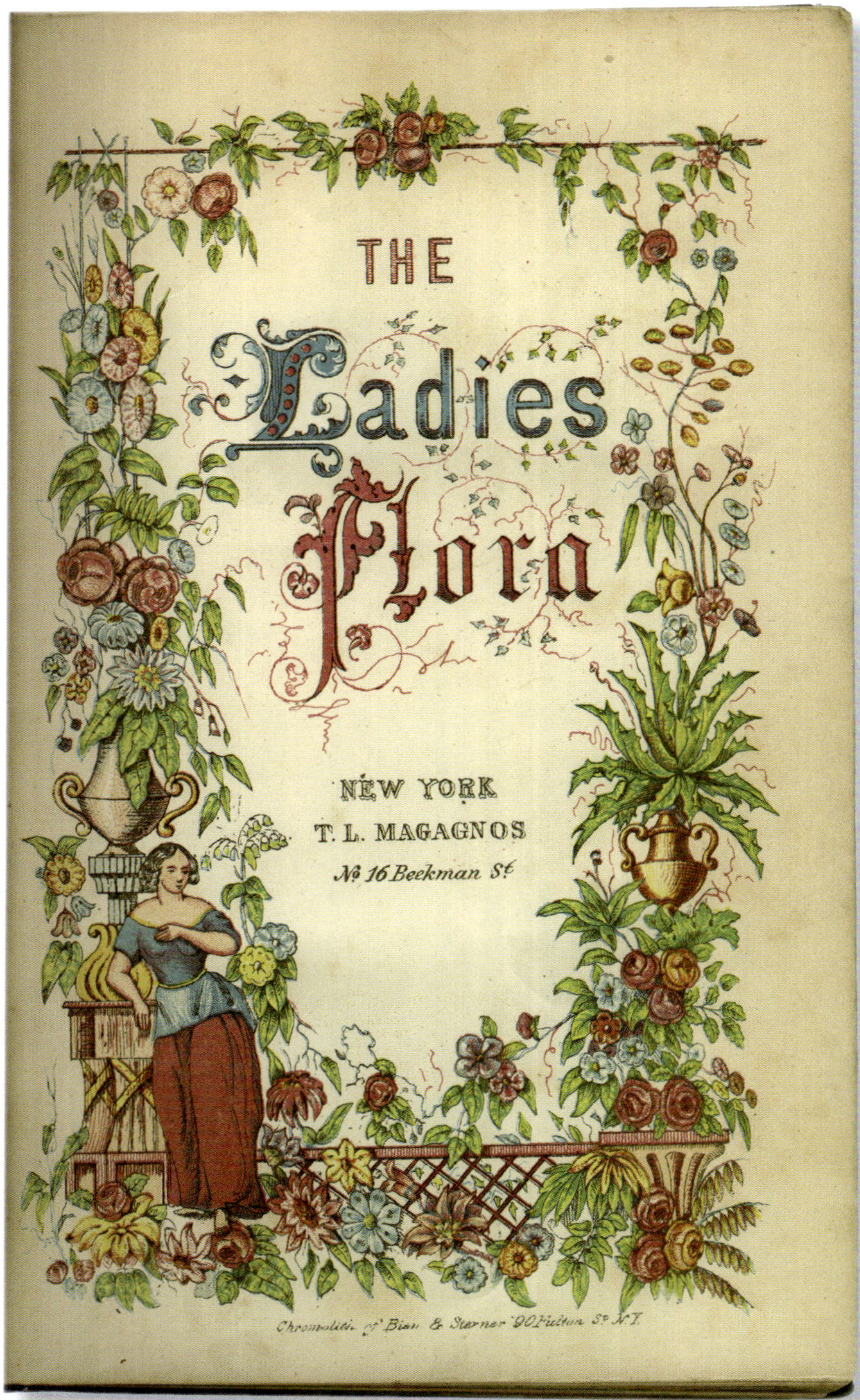

FIGURE 2.4. Title page, Newman, *The Ladies Flora* (n.d.)

FIGURE 2.5. "Celosia," page 98, Newman, *Boudoir Botany* (1847)

FIGURE 2.6. Inscription by M. S. Snowden in Newman, *Illustrated Botany* (1850), GEN P. L. N46, Arnold Arboretum Archives, Harvard University

Introducing M. S. Snowden

Snowden's copy of *Illustrated Botany* is inscribed on the inside front cover with "M. S. Snowden May 10th" (figure 2.6). The penciled inscription is faint, and if there was a year that followed the May 10 date, it has been smudged beyond recognition. Snowden also inscribed the front free endpaper (recto):

> *"Not to know at large of things*
> *remote*
> *from use, obscure and subtle,*
> *but to know*
> *that which before us lies in*
> *daily life,*
> *is the prime wisdom."*

Note: In my transcription of Snowden's marginalia, I retain her original line breaks. The quotation is from book 8, lines 191–194, of John Milton's epic poem *Paradise Lost* (1667), suggesting not only Snowden's reading of literary classics but also a yearning for wisdom found in the matter(s) of daily life.

As with so many everyday nineteenth-century American amateur botanists and flower book owners, details about M. S. Snowden are few. The handwritten annotations in *Illustrated Botany* provide some clues, as they mention well-known people, identifiable places, and many dates by month and day, although only one year, 1868. While Snowden generally identified people solely by first names or initials—Bose, C. C., Mr. Lewis C., Jack, Johnny, and Captain N—others are prominent individuals, such as "Gov. Blaisdell," Henry Blasdel, governor of Nevada; "Gen. Rosecranz," William Rosecrans, a former Civil War general who was active in the mining industry; and "Prof. Silliman," Benjamin

Silliman Jr., a Yale professor of chemistry and a mining consultant. Snowden noted places such as Hope Valley, Scandinavian Canyon, Hot Springs, and the Calaveras Grove of Big Trees, all of which are in or near Alpine County, California, in the northeastern part of the state that borders Nevada. This mountainous county lies in the central Sierra Nevada, and a silver rush in the late 1850s and early 1860s prompted not only a wave of white settler colonizers to the area but also the need "for an organized county government." In 1864, Alpine was one of California's newest counties, created out of "parts of the adjoining counties of Amador, Tuolumne, El Dorado, Calaveras, and Mono."[4]

"M. S. Snowden" was most likely Mattie S. Snowden, who taught school in Alpine County in the 1860s. Historian Karen Dustman reports, "She was a teacher in Alpine from 1865–1866, age 21, born in Delaware (though in 1866–67 it says born in Maryland), according to the Alpine County Superintendent of Schools reports. She lived at Silver Mountain in 1867 and had a 'narrow escape' returning from a ball in Woodfords [a nearby town], when the two-horse wagon in which she and others were riding plunged into the river about a mile above Mt. Bullion."[5] The first time I encountered Snowden's annotations in *Illustrated Botany*, I had no idea who this person was, but I was fairly sure it was a woman. Newman's botany book was targeted to a female audience, with sisters Laura and Emily as the main characters of its instructional narrative. The tone of some of Snowden's snarkier remarks echoes discourse patterns in popular nineteenth-century American fiction by and about women. One of the experts she encountered is "*horrid*, to dissipate all one's illusions" (emphasis in the original). In Maria Susanna Cummins's bestselling novel *The Lamplighter* (1854), the protagonist, Gerty, calls antagonist Nan Grant "a horrid, wicked woman"; later in the novel, Gerty herself is called a "horrid little creature" by a passerby who jostles little Gerty on the street. These literary examples of "horrid" to describe despicable people are echoed in Snowden's marginalia. When Snowden's journal was stolen, she fumed, "The 'chap' who has it will find no compliments for himself in it. 'Twould be far more *flattering* to his egotism to be ignorant of its contents" (emphasis in the original). In another bestseller of the time, Fanny Fern's *Ruth Hall* (1855), Ruth's new publisher, Mr. Walter, lambastes her previous publisher, Lescom: "'Well, upon my word,' exclaimed Mr. Walter, when he had finished Mr. Lescom's letter, 'if this is not the coolest piece of egotism and impudence that I ever saw.'"[6] Again, the specific term and tone are replicated in Snowden's writing. She read widely in fiction and other genres, as the section on her literary marginalia will show.

Snowden's Botanical Marginalia

The previous chapter on botany instruction books details how owners interacted with their books: For example, Hezekiah Marbourg used Wood's *Class-Book of Botany* as a field guide and repository for plants he preserved during the Civil War, and Eliza Nye placed multiple pressed specimens in her copy of Eaton's *Manual of Botany*. When Snowden botanized, she used Newman's *Illustrated Botany* as her guide. Compared with *Class-Book of Botany* or *Manual of Botany*, *Illustrated Botany* takes a more literary approach to botany instruction, combining dialogue, narrative, and verse. In one of the book's early lessons, an annotation gives a glimpse into Snowden's character—or perhaps sense of humor. On page 14, the text notes about a less capable botany student, "She does not learn her subject thoroughly as she proceeds. Persons of her disposition are always poor scholars, and will inevitably remain so till cured of such wretched habits. It is not because the science is so difficult, but merely because they are too indolent." A small line at "they are too indolent" connects to Snowden's penciled comment, written sideways in the left margin: "*I am afraid this is my case too.*" However, her frequent, learned, and informed annotations suggest that she was far from indolent.

Snowden might have initially learned botany in school, or she might have been self-taught. Many of her botanical annotations, almost all of them linked to the printed text, reveal activities of an engaged plant enthusiast: Some discuss scientific details, while others connect to specific places or activities. The nineteenth-century botanist's typical training in taxonomic classification can be seen in notes such as this one, which appears next to text discussing the *Compositæ* order:

> *To this order belongs*
> *the climbing Hemp—it*
> *is found around brooks—*
> *and running rills—*

The annotation combines both scientific and literary discourses: a note about the climbing hemp's taxonomy, as well as the poetically rhyming "found around" and alliterative "running rills."

Snowden most frequently remarked on plants she encountered in and around Alpine County. Marginalia note flowers such as lupine ("a common variety grows in Alpine") and lilies ("those of southern Cal. are of much larger proportions"); also noted are species listed in *Illustrated Botany* that were not found locally, such as tulip ("none indigenous to Alpine"). Some marginalia directly indicate Snowden's botanizing, as in an undated note commenting, "I was very busy collecting treasures," and another on July 4, "Today gathered a great deal of what several in the party thought was the Cardinal flower." Her knowledge of botany can be seen in the use of scientific terms such as "indigenous" (about the tulip) and "endogens," which appears in this handwritten note: "Except Palmetto one or two greens at the south & some green briers all the endogens of this country are herbs." Neither "indigenous" nor "endogens" is listed in *Illustrated Botany*'s glossary of technical terms, suggesting that Snowden likely had additional botany instruction from other sources.[7]

As did many nineteenth-century American amateur botanists, Snowden sent plant specimens to other enthusiasts. In the July 4 annotation about the cardinal flower, she wrote:

> *Can art ever reproduce such*
> *coloring. the flaming petals*
> *are certainly not equaled in any*
> *other that I have met with in this*
> *region save the Snow plant:*
> *which stands alone—(I sent by mail*
> *yesterday one to Mr. H. of H)—who was so*
> *desirous of seeing this wonder of nature*
> *he said by my description of it he*
> *was prepared to worship it but I*
> *fear it will be so withered that*
> *he can form no idea of its coloring.*

Both the cardinal flower (*Lobelia cardinalis*) and snow plant (*Sarcodes sanguinea*) have bright red coloring (figures 2.7 and 2.8). Other species of the *Lobelia* genus,

FIGURE 2.7. Cardinal flower, Sprague and Hervey, *Flowers of the Field and Forest* (1882)

such as *Lobelia inflata* (Indian tobacco) and *Lobelia siphilitica* (great lobelia), were common to the eastern United States, which could be why Snowden sent a specimen of the more western *L. cardinalis* to Mr. H.

Snowden also engaged with the medical plants chapter of *Illustrated Botany*. In reference to text about boneset (*Eupatorium perfoliatum*), Snowden wrote in the lower right corner, "In Md. it is used in stead of Quinine" (figure 2.9). "Md." could be Maryland or perhaps medicine, and during the Civil War, boneset was used as a quinine substitute. On a page that discusses stramonium, she wrote:

> *The Captain needed a supply today as during our tramp he had an attack - but it did not prove severe after smoking a few leaves.*

Datura stramonium (commonly known as jimsonweed), while highly toxic and hallucinogenic in larger doses, is effective in treating asthma in smaller ones.[8] Here is yet another way that botany instruction books were used: as a guide to—and here, also, as a record of—medical treatment.

Snowden's interest in the practical uses of plants extended to her notes about the area's Indigenous peoples. On a page discussing "Sage—Salvia," Snowden noted, "Indians gather it in quantities." While she doesn't say what they used it for, it may have been as a cold remedy or respiratory aid. In an entry dated June 21, at the bottom of a page that discusses raspberry, Snowden wrote, "Gathered quantities of these (thimble berries the Indians call them)." Next to text that discusses soapwort (*Saponaria officinalis*), she commented, "The soapwort of the Sierra's—which the Indians use for cleansing purposes."[9] The "Indians" that Snowden references would likely have been the Washoe, or Wašišiw. According to Dustman, the local white inhabitants "respected the Indians' superior knowledge of the natural environment." Snowden's interest in Indigenous peoples' botanical knowledge might have resulted from interactions with Wašišiw people during her time in Alpine County. In later decades of the nineteenth century, activist white women increasingly turned their attention to Indigenous peoples, and Snowden might have known of this nationwide work.[10]

FIGURE 2.8. Snow plant, Thayer, *Wild Flowers of the Pacific Coast* (1887)

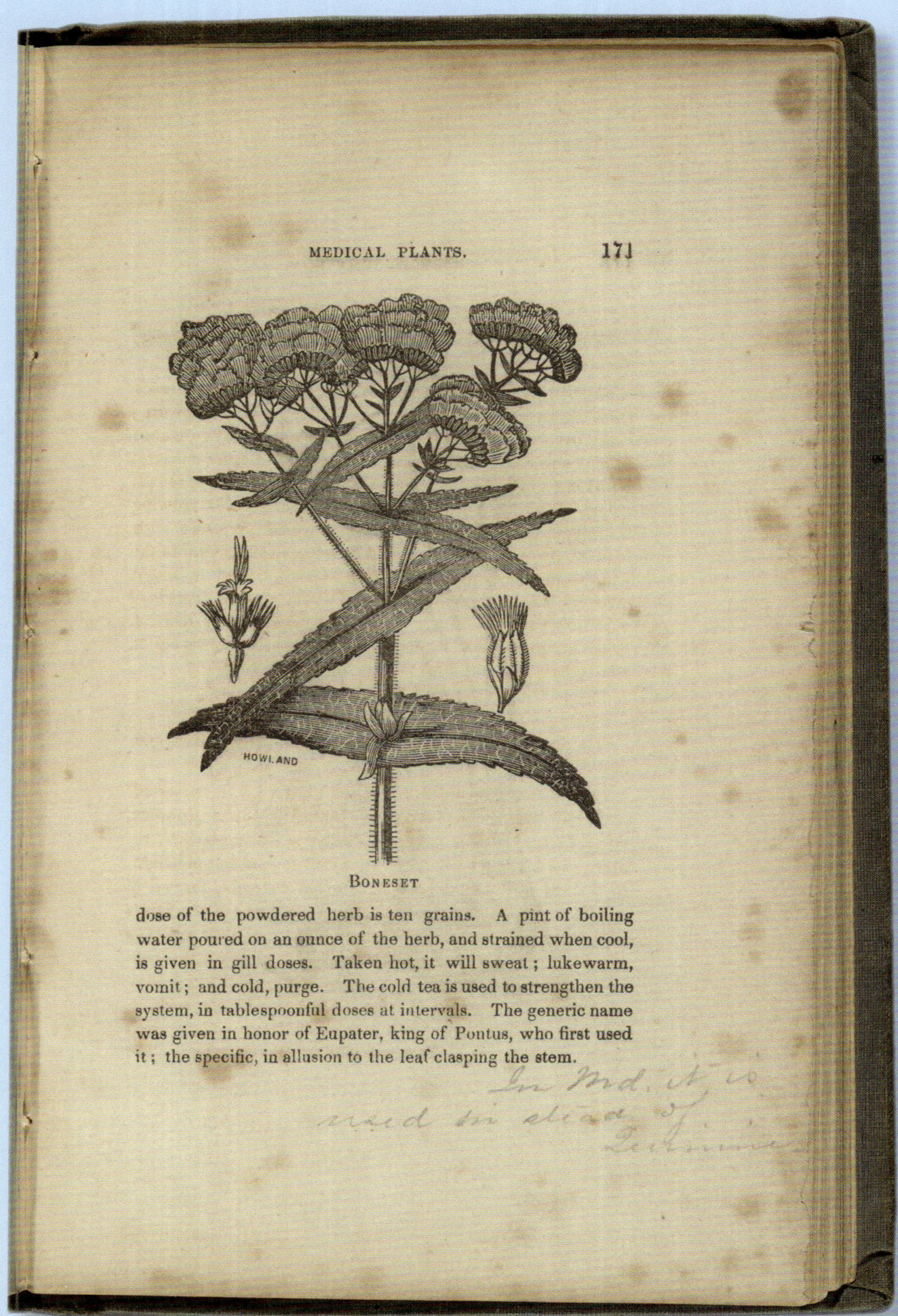

BONESET

dose of the powdered herb is ten grains. A pint of boiling water poured on an ounce of the herb, and strained when cool, is given in gill doses. Taken hot, it will sweat; lukewarm, vomit; and cold, purge. The cold tea is used to strengthen the system, in tablespoonful doses at intervals. The generic name was given in honor of Eupater, king of Pontus, who first used it; the specific, in allusion to the leaf clasping the stem.

In Md. it is
used in stead of

FIGURE 2.9. Boneset, page 171, Snowden's copy of Newman, *Illustrated Botany* (1850), GEN P. L. N46, Arnold Arboretum Archives, Harvard University

Illustrated Botany Becomes Snowden's Journal

What makes this copy of *Illustrated Botany* especially noteworthy is that it becomes Snowden's journal (figure 2.10).

> *My journal & herbarium were taken*
> *from the rock in my absence for two*
> *minutes. It is too mean—I am certain*
> *of one thing—that the "chap" who has it*
> *will find no compliments for himself*
> *in it—'twould be far more flattering to his*
> *egotism, to be ignorant of its contents.*
> (Emphasis in the original)

As a botanizer, Snowden brought her herbarium into the field, and she wrote about her excursions in a journal. When these two important records were taken, she despairingly turned to the blank spaces in her *Illustrated Botany* for both purposes. Her longer journal entries were often dated and described people, places, and activities.

One of the people she wrote about most frequently was "Prof. Silliman." Benjamin Silliman Jr. (1816–1885) was professor of chemistry at Yale, as was his father, Benjamin Silliman Sr. (1779–1864). Silliman Jr. was instrumental in developing the nascent petroleum industry in the United States and was active in mining in the western states, including as a consultant. In 1867, he "investigated various mineral deposits in Calaveras County," located just west of Alpine County. While Silliman was not known as a botanist, he would have studied botany as part of his science education at the secondary and university levels.[11]

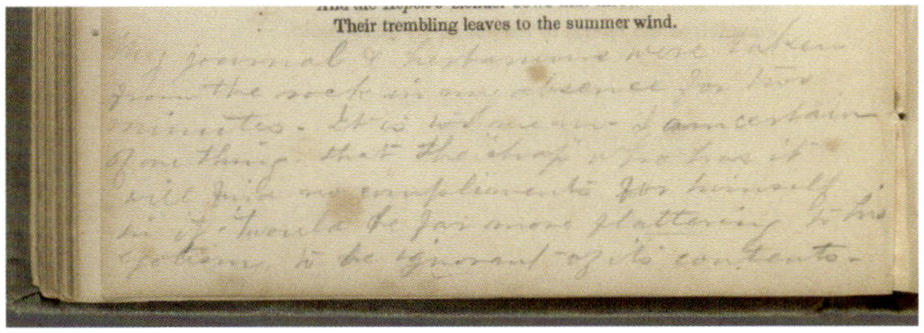

FIGURE 2.10. Snowden's marginalia, page 96, in Newman, *Illustrated Botany* (1850), GEN P. L. N46, Arnold Arboretum Archives, Harvard University

Silliman's botanical knowledge is mentioned in one of Snowden's first annotations:

> *Prof. Silliman doubts this is the flora of*
> *Alpine CO but as he was known to make*
> *so many blunders in the geology of the Co—*
> *(by the*
> *practical*
> *miners)*
> *we did*
> *not trust*
> *his opinion*
> *implicitly.*

Judging from a pencil mark in the margin, "the flora of Alpine CO" that Silliman was doubting may have been types of poisonous plants, as the text here discusses "a rule framed by botanists, which says, that all plants with five stamens and one pistil, with a dull colored lurid corolla, and of a nauseous sickly smell, are always poisonous." Botany instruction books frequently begin with a section on poisonous plants: At a time when medical first responders were almost always family members or neighbors, knowing which plants were toxic was lifesaving. Snowden also notes Silliman's "many blunders in the geology of the Co[unty]." One of his more notorious geological errors took place in Ojai, California, in late 1864. His positive and optimistic reports of oil deposits there were questioned by investors, and, explains George E. Webb, "rumors of salted petroleum samples and negative comments concerning Silliman's scientific integrity also surfaced."[12] The discovery of significant silver ore seams in Alpine County likely also attracted his attention.

Snowden's journal entries asserted her botanical knowledge and recorded her disagreements with Silliman and other noteworthy people. On June 1, she wrote:

> *Gen. Rosecranz who has been our guest for several days was out to-day in the cañon. This he said is imaginative. I replied you would not say so had you examined the foliage crossing the Isth. along the Chagres r He and Prof. S could botanize together and geologize for they each one have a theory of their own for everything.*

Snowden links Rosecrans's opinion to the book's text by bracketing this nearby passage: "In temperate climate the leaves of trees are apart from each other and light, so as to allow the sun to shine on the flowers and trunk, but under a torrid zone they are broad, thick, and firm."[13] While Rosecrans disputed the idea, saying "This is imaginative," Snowden countered with her firsthand experience of having "examined the foliage crossing the Isth[mus of Panama] along the Chagres r[iver]." For people traveling from the Eastern Seaboard of the United States to California in the 1860s, before the transcontinental railroad was completed in 1869, the typical route was to take a steamship from an eastern port to Panama, travel overland across the isthmus, and resume ship travel from there to a West Coast port, usually San Francisco.[14]

In another example, Snowden mentioned Silliman's botanical knowledge but did not defer to it. The text on page 87 mentions Arethusa (a genus of orchids), and Snowden wrote,

> *Prof. S. claims that this is*
> *limited to the eastern border along the coast of N. E. &*
> *down [?]*
> *but it certainly has a facsimile here and is found in wet*
> *springy places. It is Shelley who has written so*
> *beautifully of*
> *it. "Arethusa arose*
> *from her couch of snows" & c.*

"Here" is presumably near Alpine County—far from the New England coast—and Snowden is familiar enough with local flora and ecosystems to know that *Arethusa* is "found in wet springy places." "Shelley" is Percy Bysshe Shelley, an English Romantic poet, whose "Arethusa" is an eponymous poem about a water nymph. At the bottom of this page, Snowden wrote, "Arethusa was named in honor of a nymph of Diana," and in this one brief annotation, she demonstrated knowledge of botany, poetry, and classical myth.[15] Notes such as this one indicate that this amateur woman botanist was no shrinking violet: She had knowledge, skills, and confidence, and no blundering professor or bloviating general was going to tell her otherwise.

Snowden's age, gender, and class status as a teacher in a remote California town starkly contrast those of the older, male, well-known, and well-off Professor Silliman and General Rosecrans. Her familiarity with botany and direct observations of local flora were superior to Silliman's ill-informed claims, and she allied herself with the "practical" knowledge of the Western miners against the "theory" of the Eastern professor.[16] The decades after the Civil War—during which many women assumed leadership roles in their male relatives' absence—heralded new opportunities and cultural shifts that countered the gendered proscriptions of antebellum domestic sentimentalism. Mattie Snowden's journal entries suggest one woman's experience of these changes, inspiring her self-assured words and assertive actions.

Snowden's Literary Marginalia

Snowden frequently quoted literature in her marginalia, most often poetry, and always in reference to the printed text of *Illustrated Botany*. Snowden's familiarity with both British and American poets would have been typical of readers in the mid-nineteenth century, a time when poetry was ubiquitous and people could recite lines as easily as we do song lyrics and advertising jingles today. Poems circulated widely in both print and manuscript.[17] Rufus Griswold's bestselling anthology *The Poets and Poetry of America* (1842) featured portraits of Richard Henry Dana, William Cullen Bryant, Fitz-Greene Halleck, Charles James Sprague, and Henry Wadsworth Longfellow on its title page and over one hundred other poets in its pages (figure 2.11). Gift books of poetry such as John Greenleaf Whittier's *Jack in the Pulpit* proliferated in the nineteenth century (figure 2.12). Single-author books as well as anthologies were cherished, and poems appeared in a wide range of newspapers and magazines. Copyright protection for non-U.S. authors did not exist until 1891, and even American authors were frequently reprinted, with or without their permission. Everyday Americans quoted poems in their letters, wrote lines of verse in their friends' keepsake albums, and kept handwritten compilations of poetry, known as commonplace books.

FIGURE 2.11. Title page, Griswold, *The Poets and Poetry of America* (1842), New York Public Library Digital Gallery, via Wikimedia Commons

Snowden's literary marginalia connect the book's botanical information to texts familiar to her. Where *Illustrated Botany* mentions the arbutus tree, Snowden noted its taxonomy—as she did with the climbing hemp—then quoted Longfellow:

> *Does not our Arbutus or may flower*
> *belong to the heath family?*
> *Longfellow calls it*
> *"Children lost in the woods, and*
> *covered with leaves in their slum*
> *ber" (where John Alden crossed the*
> *brook by the ford).*

The quotation is from "The Courtship of Miles Standish" (1858), a long narrative poem by Longfellow about the love triangle of Miles Standish, John Alden, and Priscilla Mullins, Puritans who sailed on the *Mayflower* and settled Plymouth Colony in the early seventeenth century. This section of the poem personifies mayflowers:

> So through the Plymouth woods John Alden went on his errand;
> Crossing the brook at the ford, where it brawled over pebble and shallow,
> Gathering still, as he went, the May-flowers blooming around him,
> Fragrant, filling the air with a strange and wonderful sweetness,
> Children lost in the woods, and covered with leaves in their slumber.[18]

Snowden later mentions the protagonist of "The Song of Hiawatha" (1855), another long narrative poem by Longfellow, where Newman's text discusses *Sagittaria*. Snowden wrote "arrow-heads" (the common name for *Sagittaria*) in the left margin and added, sideways along the right margin, "Could they have been the kind Hiawatha was searching for?" Longfellow was an internationally known literary figure and the most popular American poet of the nineteenth century. "The Courtship of Miles Standish" and "The Song of Hiawatha" were widely reprinted and immensely popular for decades following their initial publication in the 1850s.

FIGURE 2.12. Cover, Whittier, *Jack in the Pulpit* (1884)

Snowden also quoted less well-known poetry (quotation marks are hers):

"Bright berries in the gloomy green!
To every soul that lives
Some bright hopes glow
their griefs between
When God the New Year gives."

These lines appeared in a Christmas card produced by the noted chromolithography publisher Marcus Ward (figure 2.13). The Marcus Ward card designer might have created the verse or copied it from elsewhere, or perhaps it was inspired by Robert Southey's "The Holly Tree" (1832), a few lines of which appear in *Illustrated Botany*, next to Snowden's "bright berries" quotation.

Snowden, like many of her contemporaries, seems not to have distinguished between highbrow and lowbrow forms of verse, as we might judge them today. On the same page in *Illustrated Botany* that discusses violets, she quoted both Shakespeare and a poem published in agricultural magazines. Sideways in the left margin, she wrote:

"The first spring violets so sweet and white,
Recall the years that have taken flight.

The rhymed couplet is from the poem "Spring Violets," which appeared on the front page of the *Small Fruit Recorder and Cottage Gardener* for April 1, 1871, where it was noted as having come from the *Boston Cultivator*. Snowden then wrote in the right margin:

"Lay her i' the earth;
And from her fair and unpolluted flesh
May violets spring."
(Shakespeare loved violets.)

FIGURE 2.13. Christmas card published by Marcus Ward, courtesy of Bridgeman Images

These lines are from *Hamlet*, act 5, scene 1, when Hamlet mourns the dead Ophelia. Snowden might have known that "Shakespeare loved violets" from reading his works (violets appear in multiple plays), or she may have seen Jane Elizabeth Giraud's *The Flowers of Shakespeare* (1845), a collection of her flower paintings with excerpts from Shakespeare's plays. Several of Giraud's illustrations feature violets, including one that quotes from *Hamlet* (figure 2.14).[19]

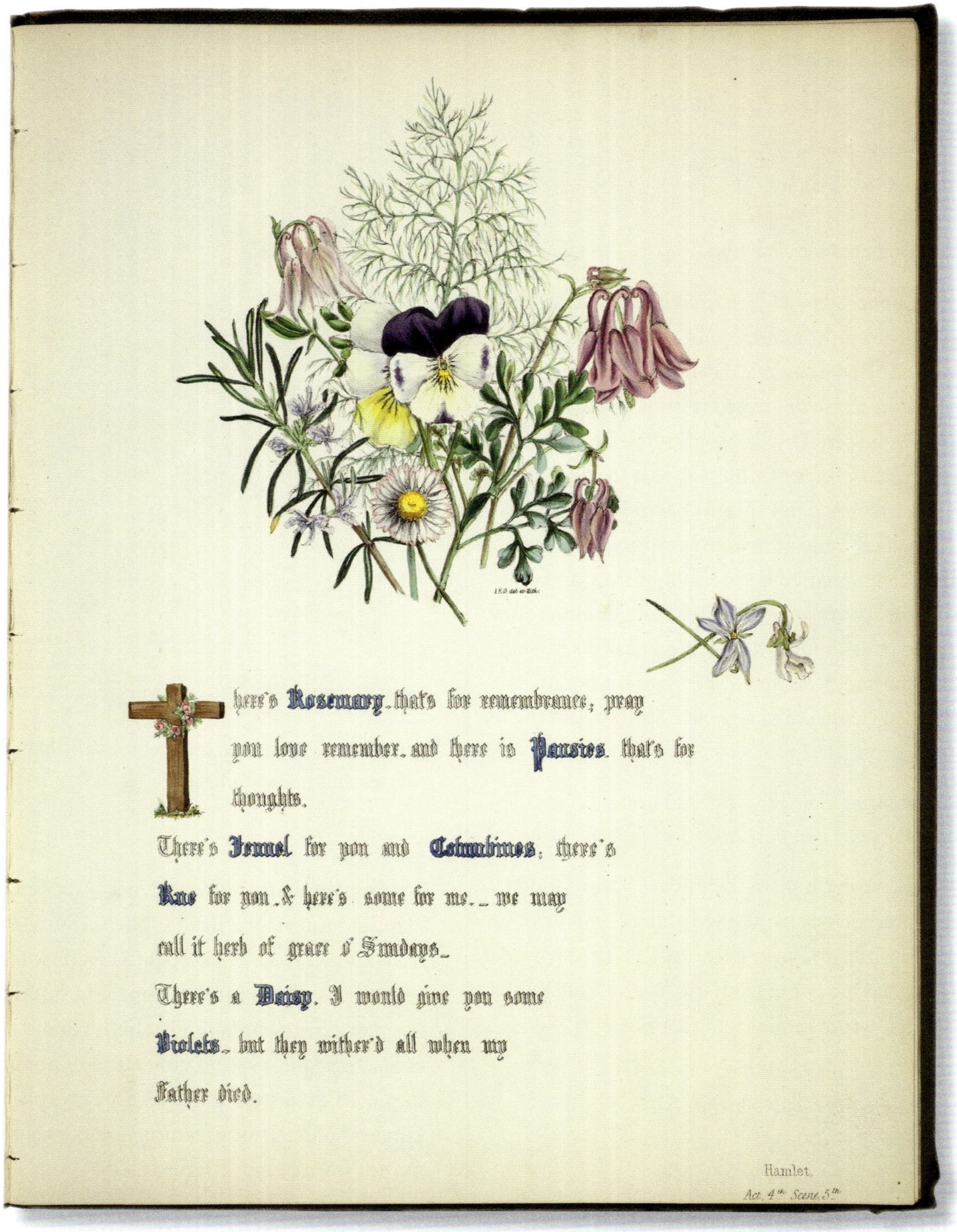

FIGURE 2.14. From Giraud, *The Flowers of Shakespeare* (1845), courtesy of the Shakespeare Birthplace Trust

Snowden quoted from prose sources less frequently. In one example, next to text that mentions geraniums, she wrote:

> *An Eastern tale says that the Geranium*
> *was formerly a mallow—but Mahomet having*
> *laid a garment upon it to dry—it was transform-*
> *ed into this beautiful plant—*
> *the two plants are botanically unlike.*

This "Eastern tale" originally appeared in Grenville Temple's *Excursions in the Mediterranean: Algiers and Tunis* (1835) and was widely reprinted in the nineteenth and early twentieth centuries, reflecting the era's fascination with Orientalism and cultures of the Middle East. Edward Mowbray Tuttle closes his entry about geranium in *The Book of Rural Life* by noting, "Unfortunately, the story assumes a trifle too much, for the mallow does not even belong to the geranium family," corroborating Snowden's assertion that "the two plants are botanically unlike."[20]

Snowden quoted from other poems without identifying them, including "The Flower" by George Herbert, *Peter Bell* by Wordsworth, "A Day of Sunshine" by Longfellow, "Among the Daisies" by Ina Coolbrith, "The Thistle Flower" by Alice Cary, and "Before Sunrise" by Celia Thaxter.[21] Cary and Thaxter published in leading periodicals such as *The Atlantic Monthly*, and their work, along with that of Coolbrith, circulated frequently in the reprinting culture typical of the time. Cary and her sister Phoebe, also a poet, often wrote about floral subjects, and they remained popular into the twentieth century (figure 2.15).

Why would Snowden have quoted all these literary excerpts in a botany instruction book? While a definitive answer is impossible, she likely also used *Illustrated Botany* as a commonplace book. This type of handwritten book, explains Michael C. Cohen, derived from the "tradition of commonplacing, or copying by hand key passages from selected and approved authors or texts." Such "copying was crucial to early American teaching, as well as to the habits of literacy inculcated among those who received even rudimentary training in reading and writing." While some commonplace books were private and individual, others circulated among groups of friends and were even coauthored, with multiple writers contributing extracts. Ellen Gruber Garvey notes that "readers use them to create a record of their reading by copying out passages from other works."[22] Snowden's literary marginalia in *Illustrated Botany*

FIGURE 2.15. Cover, *Flowers from Alice and Phoebe Cary* (1906)

recorded her reading and functioned as a handwritten scrapbook of snippets she found interesting. The scope and placement of her literary marginalia might suggest how she read and used *Illustrated Botany* more generally. Some sections have quotations on multiple pages in a row (such as pages 83–92), but once the book becomes her journal (starting on page 96), the marginalia are almost entirely diary entries, rather than remarks or quotations linked to the text. It's not clear whether Snowden's notes throughout *Illustrated Botany* were written only in the summer of 1868 or over a longer period of time, whether they appeared in this book because her journal had been stolen, or whether her literary marginalia were quoted from memory or copied from elsewhere. What is extraordinary about these examples, however, is the way Snowden made connections between botany and literature. *Sagittaria* reminded her of Longfellow's "Hiawatha"; a reference to daisies prompted an excerpt from Coolbrith's "Among the Daisies"; and next to text about thistles, Snowden quoted from Alice Cary's "The Thistle Flower." Snowden chose liberally from the wealth of floral poetry and botanical lore that proliferated in nineteenth-century popular botany, in both the United States and Europe.

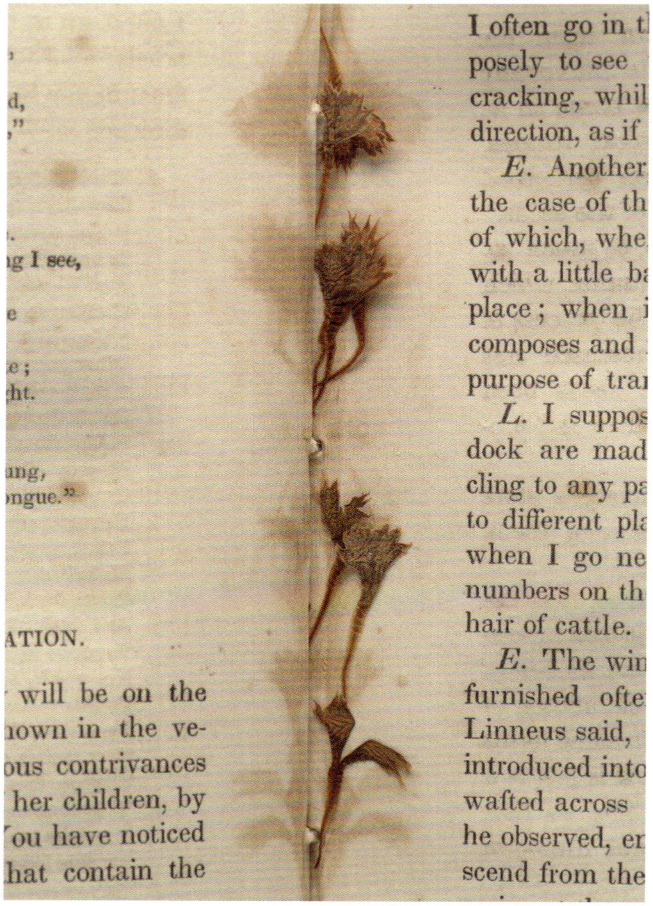

FIGURE 2.16. Detail, pages 122–123, Snowden's copy of Newman, *Illustrated Botany* (1850), GEN P. L. N46, Arnold Arboretum Archives, Harvard University

Snowden's *Illustrated Botany* as Specimen Repository

In addition to using her copy of *Illustrated Botany* as a field guide, source of medical information, journal, and commonplace book, Snowden pressed plants between its pages—likely in lieu of the herbarium that disappeared along with her journal. Several pressed specimens remain, along with shadowed indications of others that do not (figures 2.16–2.19). A specimen appears on a page with Snowden's handwritten note stating, "I was very busy collecting treasures," suggesting that the preserved specimen might have been one of those "treasures" (figure 2.16). Although Snowden noted the loss of her herbarium, these pressed flowers appear to be more mementos than scientific specimens: They are not properly mounted, do not include all parts of the plant, and are not identified in any way. Perhaps the flowers were pressed here because the text was of note; for example, given the bracket marks in figure 2.18, Snowden might have paused to mark the noteworthy text, then placed a collected "treasure" there.

Unlike the pattern of literary references appearing almost entirely on pages preceding Snowden's use of *Illustrated Botany* as her journal, pressed flowers or their traces appear throughout the book. The last specimen appears only in shadow (figure 2.19). The overall shape of the pressed specimen seems to mimic that of the snakehead (*Chelone glabra*) pictured here, but its otherwise indeterminate leaf and flower shapes mark it as likely neither snakehead nor dogbane (*Apocynum androsemifolium*), also mentioned on this page. As with so many of Snowden's additions to the printed text, there are interesting clues but still much that is left un(re)marked and thus unknown.

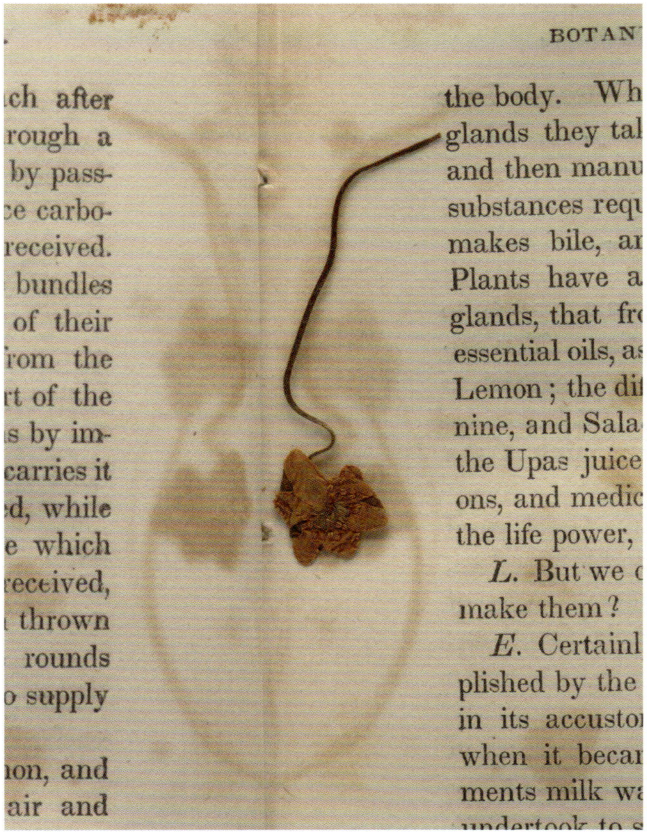

FIGURE 2.17. Detail, pages 30–31, Snowden's copy of Newman, *Illustrated Botany* (1850), GEN P. L. N46, Arnold Arboretum Archives, Harvard University

FIGURE 2.18. Detail, pages 74–75, Snowden's copy of Newman, *Illustrated Botany* (1850), GEN P. L. N46, Arnold Arboretum Archives, Harvard University

178 MEDICAL PLANTS.

SNAKEHEAD.

MEDICAL PLANTS. 179

CHELONE GLABRA—SNAKEHEAD.

Natural order, Scrophulariaceæ—Linnæan class, Dedynamia—Order, Angrospesmia.

Gen. Char. Calyx five cleft, three bracted; corol ringent, inflated; anthers woolly; capsule two celled, two valved; seeds with membranous margins. *Spec. Char.* Plant smooth; leaves opposite, oblong lanceolate, acuminate, serrate; flowers spiked. Grows two feet in height, with leaves of a shining dark green color, and sends out white flowers tinged with red, from August to September. The generic name supposes the flowers resemble a tortoise, but the common name likens them to a snake with open mouth and extended tongue. The plant has a bitter taste, and is given in the form of tea to strengthen the stomach, and is said to be useful in liver complaint, worms, and jaundice.

APOCYNUM ANDROSEMIFOLIUM—DOG S-BANE.

Natural order, Apocynaceæ—Linnæan class, Gynandria—Order, Pentandria.

Gen. Char. Calyx small; corolla bell-shaped, with small lobes; stamens inside; filament short, alternating with five glandular teeth; anthers sagittate, connivent, cohering to stigma by the middle; ovaries two; stigmas connate; follicles long, distinct. *Spec. Char.* Leaves ovate; cymes lateral and terminal; tube of corolla longer than calyx, with limb spreading; a handsome plant, attaining the height of three feet, and, in August, making a fine appearance with its dark green leaves

FIGURE 2.19. Pages 178–179, Snowden's copy of Newman, *Illustrated Botany* (1850), GEN P. L. N46, Arnold Arboretum Archives, Harvard University

The Saga of the Missing Journal Continues

In several of Snowden's journal entries, she lamented the ongoing loss of her journal. A note dated June 21 states:

> *L. C. has promised*
> *to return my journal. But*
> *"As promises so false and vain*
> *Lightly broken, though fondly spoken."*
> *Hearts we deemed that naught could sever*
> *A stolen journal may part forever*
> *(author unknown) but*
> *they declare they know.*

The quoted lines—which Snowden emended to "a *stolen journal*" (emphasis in the original)—appear in slightly different form in a stanza from "When the Autumn Comes Again" by Hattie S. Fay, printed in *Godey's Lady's Book* in 1857:

> When the autumn comes again,
> Will it witness pledges broken—
> Promises, all false and vain,
> Lightly brok'n, though fondly spoken?
> Hearts we deemed that naught could sever
> Absence chills and parts forever.[23]

A second entry, dated August 22, wondered, "*Where* oh *where is* my journal?" (emphasis in the original). Finally, a longer and undated entry began by paraphrasing a line from Robert Burns's "To a Louse" (figure 2.20).

> *Mr. Lewis C—— our [?]—"could we but see ourselves as*
> *others see us." Uncle says he*
> *watched my countenance during the*
> *ordeal, and I never betrayed myself*
> *once. Well, I am provoked that he don't*
> *return my journal—poor thing*
> *how it must have blushed*
> *to have been in such illustrious ?*
> *hands, he will never get this in*
> *his hands—Capt. N. took pity on*
> *me, and in some mysterious*
> *way—has procured me a new*
> *book. and you old Botany need*
> *do duty no longer. It is too*
> *horrid that I cannot get back*
> *my journal.* (Emphasis in the original)

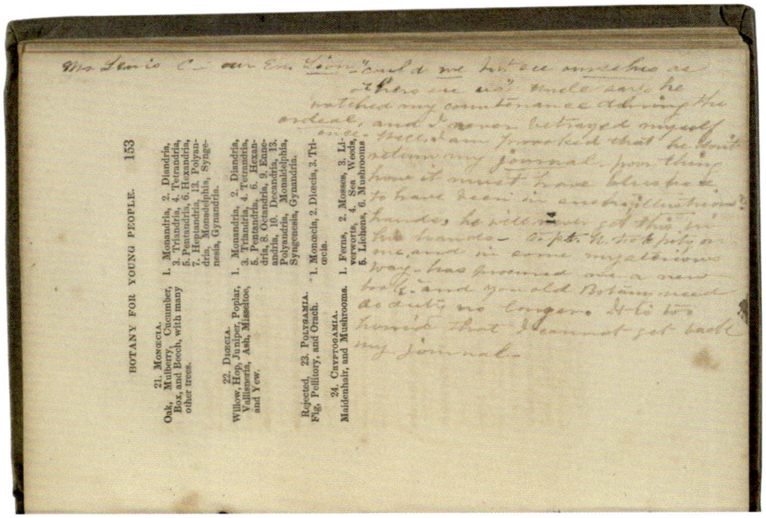

FIGURE 2.20. Page 153, Snowden's copy of Newman, *Illustrated Botany* (1850), GEN P. L. N46, Arnold Arboretum Archives, Harvard University

"Mr. Lewis C" was likely Lewis Chalmers, a mining superintendent in Alpine and staunch promoter of the area.[24] This undated journal entry is the only annotation in Snowden's *Illustrated Botany* written in ink: Perhaps when "Capt. N." procured the new journal, he also provided a pen. Snowden's depth of feeling for "you old Botany" prompts its personification: "poor thing how it must have blushed." This trusted textbook has served multiple purposes and now "need do duty no longer."

The Significance of Snowden's *Illustrated Botany*

Snowden's copy of *Illustrated Botany* is significant for several reasons. The fact that she brought a botany textbook on her excursions shows the importance of both botanizing and books in this era. The book is a very portable technology, and "marks left on their pages and spines can suggest the journeys they took."[25] This young female schoolteacher in Alpine County might have possessed only one botany book: a more popular text using a dialogic, narrative format, rather than a systematic botany with flora, which would have been more useful as a field guide. Although some botany books available at the time included species from western states, the first book-length flora of California were not published until the 1870s. William H. Brewer, Sereno Watson, and Asa Gray published *Botany, Volume 1* in 1876, based on the California Geological Survey, and Volney Rattan's *A Popular California Flora* appeared in 1879.

In the late 1860s and early 1870s, northeastern California was still sparsely populated, especially in comparison to other parts of the state. In the 1870 census, the total population of California was 560,247; those of Alpine and its five surrounding counties appear in Table 2.1.[26] Snowden's marginalia provide valuable information not only about Alpine County's botany but also about people who lived and visited there. Some of them were drawn by extractive industries such as mining and by railroad construction, and the people named

TABLE 2.1. California population by county from 1870 U.S. census, with original race categories

	AGGREGATE	WHITE	FREE COLORED	CHINESE	INDIAN
Alpine	685	676	1	8	—
Amador	9,562	7,883	72	1,627	—
Calaveras	8,895	7,405	31	1,441	18
El Dorado	10,309	8,589	132	1,582	6
Mono	430	386	—	42	2
Tuolumne	8,150	6,556	67	1,524	3

fully in Snowden's journal entries—Professor Silliman, General Rosecrans, and Governor Blasdel—were involved in one or both enterprises.

When dealing with archival materials of people who were not famous, and thus whose lives have not been documented in detail, scholars must rely on informed conjecture. Such is my narrative of Mattie Snowden's use of *Illustrated Botany*, pieced together from available sources and other comparable examples. Though rich in some kinds of detail, Snowden's marginalia in *Illustrated Botany* are silent on many topics—after all, most of us do not provide extensive background information in our private journals, recording instead what we consider most immediately noteworthy. Her dated entries all occur during summer months, from June 1 to August 22, and the only year noted is 1868. However, published sources about Silliman, Rosecrans, and Blasdel do not indicate their being in or near Alpine County in the summer of 1868, which prompts further conjecture, especially about why Mattie Snowden, a young schoolteacher in a remote small town, was in their company.

Snowden's use of *Illustrated Botany* provides a fascinating example of an amateur botanist's activities, interests, and reading. The diverse topics of her annotations—botany, literature, Indigenous cultures, contemporary exploration, and politics—convey an intrepid traveler who read widely and expressed her opinions confidently. Her botanical knowledge was both scientific and humanistic; her various uses of the book demonstrate a practical and skilled commitment to studying plants. She was a true amateur: someone who loved botany, used it for a variety of reasons and purposes, shared that interest with others, and recorded it for us to enjoy and learn from, all these years later.

FIGURE 3.1 Emily Dickinson, herbarium, ca. 1839–1846, MS Am 1118.11, seq. 49, Houghton Library, Harvard University

3

Herbaria

Preserving American Plants

TO MY HERBARIUM

Ye dry and dead remains!
Poor, wrinkled remnants of a beauteous prime!
Why, from your final doom, should I take pains
To stay the hand of time?

The world would pass you by:
For beauty, grace and fragrance all are gone.
Your age is homeliness to every eye,
And prized by me alone.

Not beautiful, but dear,
Your wrecks recall to me the happy past.
Wand-like, your stems can summon to appear
The days that could not last.

. . . .

The friend who in those years
Shared warmly in my rambles far and wide,
Back, with the same old fondness re-appears,
And trudges at my side.

These are your charms to me!
While such dear recollections ye awake,
Your ruins, blackened, crumbling though they be,
I treasure for their sake.

May I, like you, dry flowers,
When in young life I can no more engage,
A dear memento be of happy hours
To those who tend my age.

—Charles J. Sprague, *Putnam's Monthly Magazine*, December 1854

I am going to send you a little geranium leaf in this letter which you must press for me. Have you made you an herbarium yet? I hope you will if you have not, it would be such a treasure to you; 'most all the girls are making one.

—Emily Dickinson to her friend Abiah Root, May 1845

FIGURE 3.2. Game board, Sophia Thoreau, Concord, Massachusetts (ca. 1860), Concord Museum Collection, gift of Miss Sarah (Sally) R. Bartlett; Th117

FIGURE 3.3. Leaf with poem, Sophia E. Thoreau (October 13, 1868), Concord Museum Collection, gift of Mr. W. J. R. Taylor; M430b

For the amateur botanist in nineteenth-century America, pressed plant specimens were not just scientific but also aesthetic and memorial. Sophia Thoreau (1818–1876), sister of Henry David Thoreau, created the artifacts in figures 3.2 and 3.3: a checkerboard covered with ferns that she and Henry collected, and five shagbark hickory leaves inscribed with Henry's poem "Fair Haven." Sophia Thoreau was an avid botanist who kept an herbarium; the checkerboard and inscribed leaves indicate that she also used pressed plant specimens decoratively and creatively.[1] As a student at Amherst Academy, the poet Emily Dickinson (1830–1886) created an herbarium that contains 424 specimens placed over sixty-six pages in a blank book; the letter to her friend Abiah Root attests to the popularity of herbarium construction: "'most all the girls are making one." As seen in figure 3.1, Dickinson included multiple specimens per page in an artful arrangement of colors, sizes, and shapes. "To My Herbarium" by the poet and botanist Charles J. Sprague was frequently reprinted, indicating the wide appeal of both the poem and its subject, the "wrinkled remnants of a beauteous prime." These epigraphs' examples of collected plant material, though widely different, show Americans' interest in preserving botanical specimens. This preservation was both individual and national: As popular botanists physically pressed and dried plants in their material form, they also became politically and culturally dedicated to saving America's native flora.

This chapter traces the development of the herbarium in popular botany from its intended scientific purpose of classifying preserved plant species to less scientific forms of preservation such as skeletonized leaves and books sold with pressed flowers already in them. I approach the herbarium as "not just a scientific specimen" but also "a cultural expression," exploring how Americans participated in activities from formal herbarium construction to informally saving treasured blossoms between the pages of books.[2] When pressed flowers became art, they served as reminders of plants' simultaneous durability and ephemerality, an apt metaphor for environmental conservation in the latter part of the nineteenth century, when large tracts of land were set aside for national parks as Americans began to realize the limits of environmental exploitation.

Introducing the Herbarium

The herbarium is a collection of dried, pressed plant specimens that are mounted on paper and systematically arranged by species and genus, then usually stored on shelves in cabinets or portable cases. Some, like Dickinson's, use bound books instead of loose pages. The Harvard University Herbaria website explains that "some of the earliest herbaria were founded in Europe in the early 1600s, during the days of exploration, when knowledge of the earth's flora was growing at such a rapid pace that botanical gardens could no longer maintain living examples of every known species."[3] For botanists from the seventeenth century to the present, the herbarium is one of the primary tools for preserving and classifying plant species. As a widespread practice of popular botany, "collections of pressed plants show how a passion for botany brought together laypersons and scientists, women and men, in an enterprise that engaged both scientific curiosity and sentiment."[4] Individual scientists' herbaria might be as specifically focused as the plants of Warren County, Kentucky, collected by Sadie Price (figure 3.4) or as broad as the specimens collected by Carl Linnaeus in his extensive travels (figure 3.5).

America in the nineteenth century was keenly aware of its status as one of most resource-rich nations on earth: a country of vast plains and dense forests whose Manifest Destiny was to explore and settle the continent from the Atlantic Ocean at its eastern border to the Pacific Ocean at the west, from the Gulf of Mexico and Rio Grande River at the south to its northern border with Canada at the forty-ninth parallel. Government-funded expeditions such as Meriwether Lewis and William Clark's exploration of the lands along the Missouri River in 1804–1806 and John Wesley Powell's Western surveys in the late 1860s and early 1870s inspired artists, writers, and everyday Americans to extol the seemingly limitless possibilities of this seemingly limitless land.

But the very resources that lured colonizing settlers also led to widespread genocide and displacement of Indigenous peoples and destruction of natural environments. Herbaria provide a counternarrative: They showcase preservation on a small scale. Instead of looking to immense geographical wonders such as Niagara Falls or the Sierra Nevada, amateur botanists generally concentrated on their home communities, usually in nearby fields and woods. These herbarium makers devoted themselves to identifying particular plants, occasionally discovering a rare new species as worthy of preservation as Yellowstone or Yosemite. In America in the nineteenth century, the herbarium enjoyed wide

FIGURE 3.4. *Apios priceana,* Sadie Price herbarium

FIGURE 3.5. *Aquilegia canadensis*, Linnaeus herbarium, © Swedish Museum of Natural History

popularity—not only among professional botanists, whose research required that they keep detailed and precise records, but also among amateurs.

As a compilation of pages that are deliberately composed, labeled, ordered, sometimes bound, and often circulated among fellow collectors, the herbarium figures as another kind of flower book: one that was widely read, variously interpreted, preserved for posterity, and intended for educating its readers about the important native species of an increasingly agricultural and industrial nation. The comparison of herbaria to books persists to this day: For example, the aforementioned website maintained by Harvard notes that “a herbarium is analogous to a library of carefully preserved plants where the specimens themselves and the labels associated with them provide a wealth of information once they have been ‘read’ and studied by scientists.”[5] There are additional similarities: Both use paper and organize pages according to a system. For herbaria, this organization is taxonomic, by species and genus; for books, it might be topical, by chapters. Leaves of plants are pressed upon leaves of paper, and very often the paper is itself the product of plant matter such as wood, flax, cotton, or papyrus. In some instructions for making an herbarium, a book serves as a plant press, with readily available sheets of paper that will absorb the moisture of a live specimen and aid in drying and preserving it. “Paper” is both materially and etymologically derived from papyrus, just as “book” is said to have its origins in the beech tree.[6] To create their herbaria, early Americans thus relied on books for instruction *and* construction, in a complementary process of materiality and meaning.

FIGURE 3.6. *Berberis aquifolium*, Meriwether Lewis herbarium, Philadelphia Herbarium, PH00043503 at the Academy of Natural Sciences of Drexel University

The herbaria created by nineteenth-century Americans who are well known to us today include not only Dickinson’s (figure 3.1) but also those of Lewis and Clark, Henry David Thoreau, John Wesley Powell’s 1868 Colorado Exploring Expedition, John Muir, and Jane Addams. Botanists such as Meriwether Lewis and George Vasey (who accompanied Powell) pressed specimens in the field and brought their collections back

FIGURE 3.7. *Pedicularis racemosa* Douglas ex Benth., George Vasey herbarium, Brown University Herbarium

for others to study (figures 3.6 and 3.7). Nature writers Thoreau and Muir kept herbaria as part of their natural history studies (figures 3.8 and 3.9). The reformer and social work pioneer Jane Addams made an herbarium as a student at Rockford Female Seminary in the late 1870s, as Dickinson did in the early 1840s. Most of these herbarium pages remain in collections today, along with many thousands of other pressed plants collected in the United States before the twentieth century.[7] Although these herbaria fulfilled different purposes—expedition findings, natural history record, botany class assignment—they all underscore the herbarium's scientific importance and cultural popularity in nineteenth-century America.

The pictured herbaria vary in style: Dickinson mounted multiple species on one page (figure 3.1), while Price included all parts of a single plant (figure 3.4). The specimen preserved by Lewis (figure 3.6) has an interesting history of classification, as seen in the handwritten notes on the page. Thoreau often assembled several examples of the same species on a page (figure 3.8), as did Vasey (figure 3.7). In a more modern example, the northern primrose preserved by Muir (figure 3.9) has been painstakingly restored photographically by Stephen Joseph to showcase its vibrant colors. Despite their differences, each of these herbarium pages indicates a citizen scientist at work, preserving species for posterity.[8] Maura Flannery states that "each herbarium is unique" and "amounts to a private scientific space, one that includes the intuitive and aesthetic aspects of science."[9] This combination of features informed how nineteenth-century Americans created, used, and displayed herbaria and influenced a wide range of herbarium-like creations later in the century.

FIGURE 3.8. *Rosa carolina* (August 1, 1856), Herbarium of Henry David Thoreau, 1850–1862, GH872506, box 7, portfolio 2, folder 2: Specimen 15, Gray Herbarium, Harvard University

FIGURE 3.9. *Primula borealis* Duby, John Muir herbarium, photograph by Stephen Joseph

Botany Textbooks Promote Herbarium Construction

Botany instruction books exhorted students to prepare an herbarium as a crucial part of their scientific studies. In *Botany for Beginners* (1833), Almira Phelps asserts that "every child who would become a Botanist, should preserve specimens of all the plants he meets with." Similarly, Laura Johnson, in *Botanical Teacher for North America* (1834), underscores "the importance of collecting wild plants and making an herbarium with a student's own hands," because "without this exercise, no one can be a good botanist." Textbook writers such as Phelps, Johnson, Amos Eaton, Asa Gray, and Alphonso Wood all included detailed guides for collecting, preserving, and displaying plant specimens. Writers of books for children did as well, often simplifying the procedures for young learners, as seen in books by Christopher Irving, Samuel Goodrich, Jane Welsh, and Eliza Youmans. In keeping with his signature combination of education and aesthetics, John Newman lauded the importance of the herbarium in his botany books for women. In *Class-Book of Botany*, Wood summarizes the many purposes of the herbarium: "(a.) for preserving the knowledge of rare, or inaccessible, or lost species; (b.) for exchanges, enabling one to possess the flora of other countries; (c.) for refreshing one's memory of early scenes and studies; (d.) for aiding in more exact researches at leisure; (e.) for the comparison of species with species, genus with genus, &c."[10] Each of these purposes contributed to the herbarium's preservationist function: for literally preserving "rare, inaccessible, or lost species," but also for the exchanges, studies, and research that enabled nineteenth-century botanists' ongoing, developing knowledge of native plant species.

Creating an herbarium was fairly straightforward: collect live specimens, dry and flatten them, mount on paper, and label. Gray's *The Botanical Text-Book* (1850) explains that "a complete specimen consists of one or more shoots, bearing the leaves, flowers, and fruit; and, in case of herbaceous plants, a portion of the root is also desirable"; Price's herbarium (figure 3.4) provides an example of preserving all parts of the plant.[11] Wood's *First Lessons in Botany* (1851) shares these details:

> *What apparatus is necessary in collecting plants?* —A tin box of some convenient form, about 18 inches in length, shutting closely. In such a box gathered specimens may be kept several days fresh and unfading.
>
> *What apparatus for drying them?* —1st, a dozen quires of soft, unsized paper (such as printing paper or old newspapers), about 15 inches long and 11 inches wide when folded; 2nd, a small quantity of cotton batting; 3d, a small screw press with boards of the same size as the paper, or several lead weights varying from 12 to 30 pounds each. . . .
>
> *State next the process of pressing and drying.* —The specimens are to be freed from water (if gathered in the dew or in a wet day), and then spread out as nearly in their natural position as possible, between about 12 thicknesses of paper. Then let them be submitted to pressure, but the pressure should be equalized by the cotton batting, and should never be severe enough to *crush* the parts.
>
> *How often should they be changed?* —As often as once or twice each day until perfectly dry, they should be transferred to fresh and dry paper. It is also well to keep the press in a *warm* place. By carefully observing all these directions, your dried specimens will retain *almost* all the beauties of the living plant.

> *What is the next object with the collector?* —To arrange his specimens. For this purpose he procures a quantity of fine, white, rather firm paper, about 14 inches long and 10 inches wide when folded, and between the folds of each sheet places the specimens of one species, together with a card labeled with its names, &c.
>
> *Are there any directions respecting the label?* —In writing the Latin names of the species, commence them with a capital or a small letter *just as you find them* in the book you use.[12]

While Wood provides precise guidelines, including specific kinds of equipment, other botany instruction books assured students that they could make an herbarium with whatever materials were readily to hand. Instead of transporting plants in a "tin box" (or vasculum), students could use a basket; Thoreau sometimes even used his straw hat, which he called "my botany box." One didn't need a plant press and cotton batting as "apparatus for drying"; instead, Phelps notes that "all that you need in pressing plants, is some sheets of paper, (newspapers will answer, they are better than more firm and stiff paper) a board, and a stone or some other weight to press the plant," or, even more simply, "small plants may be dried between the leaves of a book."[13]

In *Leaf and Flower Pictures, and How to Make Them* (1857), the method for pressing flowers shows this conflation of artifact and activity: "Take any old book, the less stiff, and the more soft, or, to use a learned word, the more *bibulous*, the paper is, the better; for then it will *drink in* or *absorb* the moisture from the leaves. It must be a book you do not value; for the dampness that it will absorb more or less from all flowers and leaves, will make the paper yellow, and be also likely to soften the covers" (emphasis in the original). "A book you do not value" can be repurposed as a flower press, giving it new value as technology. Further instructions continue to highlight the connections: "Press your flowers in some other old book carefully. . . . In a few years you will have a charming book, especially if you make it a rule to press at least one leaf or flower in every place you visit. Your Herbarium will then not only be a beautiful collection of lovely natural objects, but a mute remembrance of all the pleasant places you have visited, and the pleasant people you have seen. These who are fond of poetry, can write on each alternate leaf, any poetical description of the flower they have pressed, that they fancy."[14] The suggestion to "write on each alternate leaf" is likely not describing what Sophia Thoreau did on shagbark hickory leaves (figure 3.3), but rather to inscribe the figurative "leaf" or page of the book. As a "charming book" and "mute remembrance," the herbarium has both aesthetic and memorial purposes. In this and many similar examples, a mise en abyme of embedded texts and objects occurs, each commenting on the materiality and referentiality of the other.

Textbook writers extolled the many benefits of making an herbarium, starting with how and where plants are collected. Phelps, in *Familiar Lectures on Botany* (1829), noted that "the love of native wild flowers, is no doubt greatly heightened by the habit of seeking them out, and observing them in their peculiar situations; a Botanist will often, at the discovery of some lowly plant, growing by the side of a brook, or almost concealed in the cleft of a rock, experience a more vivid delight than could be produced by a view of the most splendid exotic." Laura Johnson, in *Botanical Teacher for North America*, stated that "there can be no such thing in the world as a lazy botanist: and an unhealthy botanist, or a botanist of feeble health, is almost a paradoxical expression. The bodily exercise in collecting plants, and the mental exercise in studying them, seems, by experience, to secure health and cheerfulness." Students were encouraged to get outside and collect plants in their native habitats; this healthy physical and intellectual activity could lead to surprising and delightful discoveries that were well worth noting and recording. Maura Flannery asserts that "an herbarium was more than simply a kind of botanical stamp collecting, of documenting the plants a collector had 'captured.' It was a way to learn about plants and internalize that knowledge, with tactile and visual experiences reinforcing the cognitive."[15]

Even in her most basic textbook, *Botany for Beginners*, Phelps ensured that students' herbaria would provide accurate records: "By the sides of the plant should be written the *class, order, genus*, and *species*, and also the place where found, that is, whether in dry or wet ground, low or mountainous, &c. and also at what season of the year." Other books for young learners similarly emphasized the herbarium's value as a study aid: In Jane Welsh's *The Pastime of Learning* (1831), the older sister, Anne, notes that "having no fresh flowers she was obliged to exhibit specimens from her herbarium, which though not as easy to explain, [her younger sister] Caroline appeared to understand perfectly well." Irving's *A Catechism of Botany* concludes with an appendix, "On the Formation of an Herbarium, or Hortus Siccus": "The description and figure of plants, however amply and correctly they may be given in books, can be neither so instructive nor so satisfactory to the young student in Botany, as when he has the works of nature before him, and can examine and investigate for himself, without danger of being misled by the errors or misconceptions of others. It will be proper, therefore, to give a few directions, by which the plants that he has examined, may be preserved, and formed into a collection for occasional reference." Live plants encountered in nature are the optimum source for learning botany, with the herbarium—which Laura Johnson defines as "a neat, well-bound book, arranged systematically"—a close second. Margaret Coxe's instructional compendium *The Young Lady's Companion and Token of Affection* (1846) encourages herbarium construction as an act of generosity and knowledge sharing: "By preserving [your plant specimens] carefully, when dried, in an herbarium, you will be able to refer to them with satisfaction hereafter; and it is not improbable it may become an object of lively interest to some friend who may be investigating the subject of botany, without having opportunities for obtaining the same specimens."[16] Both Irving and Coxe note the herbarium's value as a reference, yet another example of the herbarium-as-book parallel.

Botanists at any age and stage of learning depended on a community of plant enthusiasts, and multiple botany textbooks underscored the importance of sharing specimens. In the "Directions for Studying Plants" at the beginning of the fifth edition of his *Manual of Botany* (1829), Eaton suggests that "duplicates, triplicates, &c. should be collected and preserved, of all scarce plants, or of those which are peculiar to a few districts. By this measure useful exchanges may be made, and the science improved." At this early period of cataloging the flora of the United States, when settler Americans were encountering many plants for the first time, botanists needed to circulate the "scarce" and "peculiar" ones, instruct others, and improve the science overall. Eaton further encourages these practices while also promoting his book: "The most important object in this extension of the Manual to the Southern and Western states, is, to furnish the student with a convenient book for examining plants received from distant states by way of exchange. For, I presume, there is not a Northern or Southern botanist, but exchanges plants with very distant correspondents." In 1829, some of those "very distant correspondents" may have lived in states such as Missouri or Louisiana, or even the territories of Arkansas, Michigan, and Florida. With this range, Eaton could reach an ever-expanding audience of amateur botanists who would use his *Manual* to identify their local flora.[17]

These plant collecting practices had negative effects, however. Overcollecting damaged fragile habitats, decimated native populations, and even led to extinction of rare and commercially exploited species. Removing an entire plant to include its roots in the herbarium specimen was a destructive practice that most botany textbook authors simply glossed over. By the turn of the twentieth century, national-level organizations were established to take action. The first circular for The Society for the Protection of Native Plants, founded in 1900, states its aim "to check the wholesale destruction to which many of our native plants are exposed,—a destruction often a matter of pure thoughtlessness in the excessive picking of flowers, and unnecessary pulling up of roots, or an extensive collecting of flowers and plants for sale."[18] In 1901, Olivia and Caroline Phelps Stokes gave $3,000 to the New York Botanical Garden "on condition that the interest of this fund should always be used for the investigation and preservation of native plants or for bringing the need for such preservation before the public." An article announcing the Stokes sisters' gift asserted that "wild flowers, once common, have become rare" because of "thoughtless and indiscriminate picking." The Stokes fund in turn inspired the creation of the Wildflower Preservation Society of America, which expanded into numerous state and local chapters. An announcement from the Washington, D.C., chapter exhorted people to "Spare the Flowers": "Many attractive wild plants have been entirely, or nearly, exterminated by thoughtless people pulling them up by the roots. . . . Please leave plenty to go to seed."[19] The unsustainable collecting practices of nearly a century earlier, as instructed by Eaton and other botanists for creating herbaria, ultimately motivated plant preservation action at the national level.

A final purpose of the herbarium was aesthetic, and a love of its beauty informs instructions in many botany textbooks. Wood's *First Lessons in Botany* (1846) noted that the preservation of plants might be regarded "almost as one of the Fine Arts; for in the practice of it there is room for the exhibition of much taste and skill, and when prepared in *a proper manner*, an HERBARIUM is a truly beautiful and interesting work" (emphasis and capitalization in the original).[20] In *Familiar Lectures on Botany*, Phelps appealed to her largely female audience's sense of taste:

> It is important to the cause of science, that it should become *fashionable*; and as one means of effecting this, the parlours of those ladies, who have advantages for intellectual improvement, should more frequently exhibit specimens of their own scientific taste. The fashionable *et ceteras* of scrap books, engravings, and albums, do not reflect upon their possessors any great degree of credit. To paste pictures, or pieces of prose or poetry, into a book; or to collect in an album the wit and good sense of others, are not proofs of one's own acquirements; and the possession of elegant and curious engravings, indicates a full purse, rather than a well

stored mind; but *herbariums* and books of *impressions of plants*, drawings, &c. show the taste and knowledge of those who execute them. (Emphases in the original)

Although Phelps implicitly categorized herbaria with other aesthetic collections (scrapbooks, albums), she emphasized "one's own acquirements," rather than simply cutting and pasting the work of others. Herbarium construction requires intellectual knowledge as well as "scientific taste." Science had indeed become "fashionable," as seen in print sources other than books.[21]

Popular Periodicals Promote Herbaria

Amateur botanists could also learn about herbarium construction from periodicals. Predictably, national scientific journals such as *The American Naturalist*, *Scientific American*, and *Bulletin of the Torrey Botanical Club* published guidelines for preserving plants. Perhaps less predictably, so did periodicals for general audiences. Many of these had a regional focus and were rural or agricultural, including *The New-York Farmer and American Gardener's Magazine*, *The Farmer's Cabinet*, *The Farmer's Monthly Visitor*, *The New England Farmer*, *The Soil of the South*, *The Country Gentleman*, *Colman's Rural World*, *Moore's Rural New Yorker*, and *The Gardener's Monthly and Horticultural Adviser*. As these titles suggest, scientists were not the only people interested in collecting and saving plant specimens; so too were those whose livelihoods depended on plants—farmers, gardeners, and people in rural communities.[22]

One of the leading magazines for women encouraged its readers to develop scientific knowledge through preserving plants. For several decades at mid-century, *Godey's Lady's Book* featured editorials, articles, and even fiction that praised herbarium construction and taught women how to make them. An article from April 1854 filled the gap "for ladies who have not the benefit of a teacher, nor friends devoted to botanical pursuits, to get an acquaintance with the method of preserving plants properly." Detailed instructions for pressing, drying, and mounting specimens are included, and perhaps this further inducement: "In these collections much taste may be displayed in the arrangement, as well as in the careful drying of the specimens; . . . the herbaria formed by ladies are, as a general rule, pre-eminent for neatness and artistic beauty in the arrangement of the specimens." *Godey's* was staunchly nonpartisan during the Civil War, and articles such as "The Collection and Preservation of Plants" in April 1863 might have provided women a worthy intellectual distraction. The author discussed careful handling of live specimens, thorough drying and pressing techniques, and precise labeling according to "botanical name, natural order, habitat, and date of collection."[23] After the war, the activity remained of interest; an article from October 1869 noted that the magazine has "received several inquiries as to the best mode of preserving and drying flowers." Quoting at length from "a very valuable little work on Elementary Botany," the article explained that "specimens which are to be dried, so that they may be kept in an herbarium, and referred to or examined at a future time, ought not to be gathered at random, but should be selected as average representatives of their species, unless they be designed to show some departure from the typical form." The focus here is clearly more than just artful arrangement or whimsical collection. One must also know how to select, identify, and then properly classify plants: "The genera should be arranged in their natural orders, the natural orders in their respective divisions and classes, and the whole placed in a suitable cabinet." When conscientiously constructed, women's herbaria would be a long-lasting resource: "When well dried, plants may be kept for hundreds of years; they are almost as indestructible as books, if properly cared for."[24]

Works of domestic sentimental fiction in *Godey's* demonstrated a protagonist's accomplishments by referencing an herbarium. As Elizabeth Keeney notes, "The literature of popular botany . . . champion[ed] botanizing [and] emphasized middle-class values," including self-improvement and the work ethic. "The Father's Revenge" by Eliza A. Dupuy, published in October 1841, features a Native American heroine, Teloula, whose "figure was delicately and perfectly proportioned" and whose voice "was soft and clear as the gushing melody of the wind harp." In keeping with these and other "signs of wealth and civilization, united with a degree of neatness and taste not often found among the aboriginal inhabitants of the country," Teloula has made an herbarium, "with flowers scattered over the leaves," as well as a portfolio of "the same, in various combinations, beautifully painted." Set in southern Tennessee just before removal, the story closes with a botanical metaphor to convey her tribe's doomed fate. When her white suitor proposes marriage, the priest who educated Teloula sees "a way by which this pretty forest flower might be transplanted to a more genial soil."[25] But as the reality of removal proved, uprooting Indigenous peoples from their native environments has fatal consequences, and Teloula dies less than a day after the wedding.

"Rizpah's Idols," a serialized novel by Marion Harland that ran from January through October 1866, traces the stormy courtship of the title character and her suitor, Gerald. In one of their early meetings, Rizpah tries to dissuade Gerald's unwelcome amorous advances by distracting him with her "gleanings from foreign shores" and "artistic gems," including her "herbarium, and sketch-book, and portfolio" and "desiccated herbs." When Gerald smiles at these items with "indulgent pity," his inferior character is clear: "In everything appertaining to scientific research, as in belles-lettres, he had retrograded since his matriculation at the University."[26] Rizpah's intellectual and artistic accomplishments, as well as her modest demeanor, mark her as Gerald's superior.

In a later story by Harland in *Godey's*, "Bred in the Bone" (1872), the protagonist Helen's scornful aunt enumerates her niece's alleged shortcomings: "She has really no time for the study of costumes and effects. 'Life is so short, and fashions so fleeting, auntie! . . . And there are so many useful and stable things I want to learn.' By 'useful things,' she meant German, and botany, and mineralogy, and chemistry. . . . Her room at home is a chaos of easels, and retorts, and herbariums, and unframed engravings, and books everywhere, especially upon the floor, and no place in particular for clothes." Throughout the story, Helen's vigorous good health and vivaciousness are associated with her disregard for fashionable dress. At the end of the story, Helen's suitor finally gives up on her, ashamed of the careless way she is attired. But Helen, who "never married," is praised by the narrator for her more meaningful achievements: "I meet her often in charitable societies and upon errands of mercy, and oftener see her name upon subscription lists and petitions for the abolition of social and political evils. . . . She has ample means, and is liberal in expenditures for the relief of the poor and the advancement of reform."[27] Harking back to Phelps's comments on the "fashionable" scientific productions of women, these stories from *Godey's Lady's Book* celebrate studying botany and making herbaria as hallmarks of the accomplished woman.

Science and Aesthetics: Herbaria and the Creations They Inspired

Even in the latter decades of the nineteenth century, when botany and other sciences had become professionalized, the herbarium remained a popular object of both intellectual utility and aesthetic display. The introduction to William Whitman Bailey's *Botanizing:*

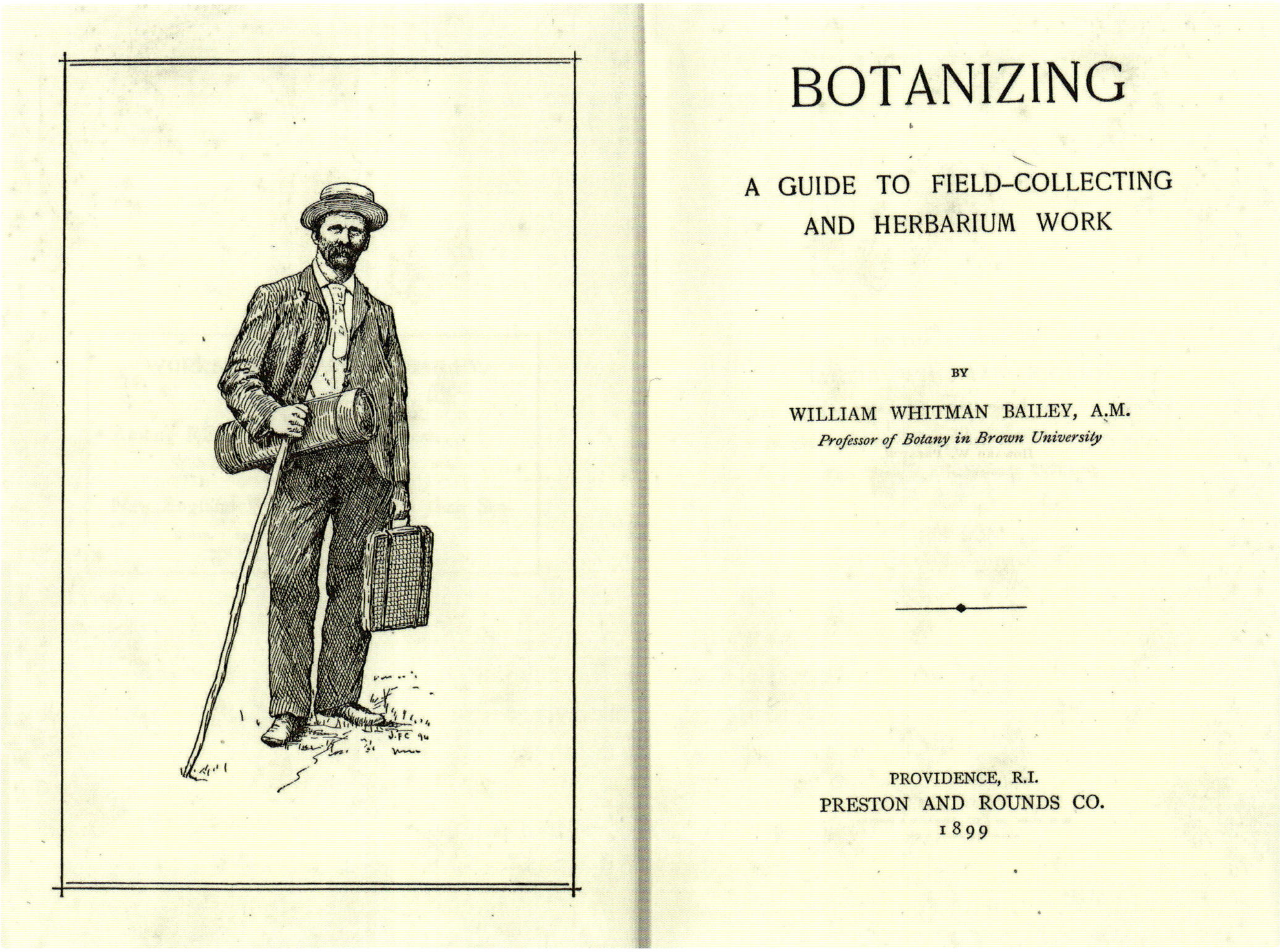

FIGURE 3.10. Title page and frontispiece illustration, Bailey, *Botanizing* (1899)

A Guide to Field-Collecting and Herbarium Work (figure 3.10) asserts that "scientific knowledge increases and develops this love of the beautiful," and "the herbarium is a cyclopædia of ultimate appeal." Similarly, Walter P. Manton's *Field Botany: A Hand-Book for the Collector* (figure 3.11) includes not only detailed instructions for "gathering and preserving plants and the formation of the herbarium," as announced on the title page, but additional chapters on leaf photography, printing plants, floral designs, and skeleton leaves. Chapter 4, "The Herbarium," begins, "There are many ways of arranging plants after they have been prepared. The question is, do you wish to make a pretty collection, or do you intend to use your specimens for study?" Manton seems not to privilege one over the other; if the former, "or if you are a traveller, and care to preserve flowers, leaves, etc. from the various places which you have visited," he recommends two forms: the scrapbook and the hollow book. He then explains what to do "for study and practical purposes," emphasizing the organization of species within cover sheets, "for the convenience of selecting any desired specimens at a glance." *Field Botany: A Hand-Book for the Collector* is small and portable, literally hand-sized (six inches tall by four inches wide) and a slim forty-one pages long. Thanks to this format, both traveler and field botanist could easily bring it along.[28]

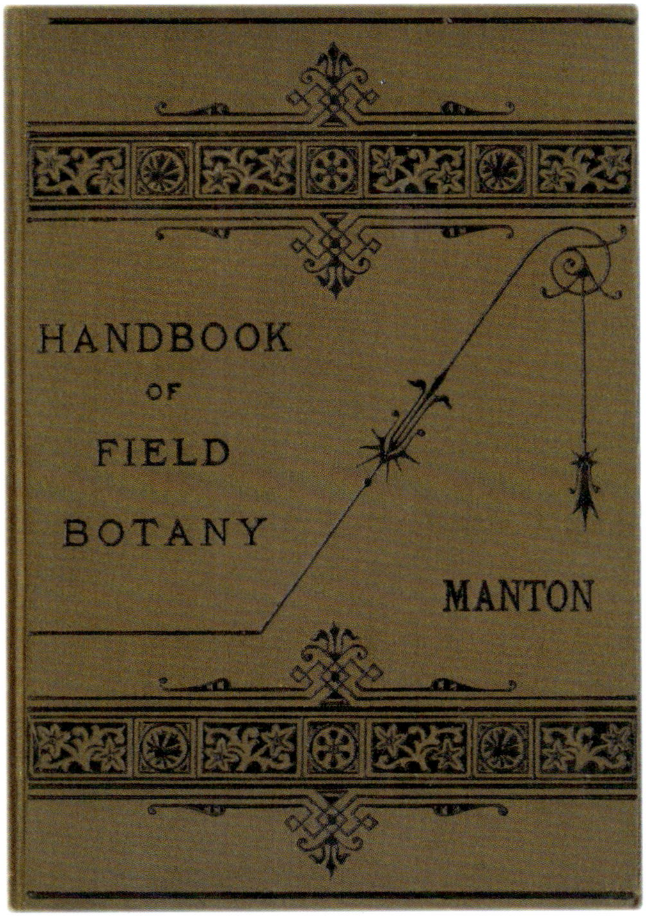

FIGURE 3.11. Cover, Manton, *Field Botany: A Hand-Book for the Collector* (1882)

The later chapters on leaf photography, "a very pretty and simple method for preserving a collection of leaves"; printing plants, "the simplest and perhaps the most satisfactory method of portraying plants"; and floral designs for fire screens, flower cards, bookmarks, and lampshades all attest to the many decorative uses of preserved plants. With skeleton leaves, "a common arrangement is to form a pretty group of the leaves and place them under a round or oval glass shade," or one can "form mottoes by making the letters of small leaves, gummed to black cardboard, which is framed under glass."[29] Skeleton-leaf works were especially popular in the latter decades of the nineteenth century.

In the 1870s, the Boston-based photographer John Soule created a series of stereoscopic images using these leaf forms to frame portraits of well-known political and cultural icons (figures 3.12 and 3.13).

FIGURE 3.12. George and Martha Washington, from the series "Skeleton Leaves," John P. Soule, creator (1874); object reference: 1976.964; The Art Institute of Chicago

FIGURE 3.13. Abraham Lincoln, from the series "Skeleton Leaves," John P. Soule, creator

Like a nineteenth-century form of virtual reality, the stereoscope transformed the two images on the card (or stereograph) into an illusion of a single three-dimensional image, enabling the viewer to perceive depth and dimensionality.[30] In Soule's skeleton leaf stereocards, the portraits of Abraham Lincoln and George and Martha Washington are small and flat in comparison to the ghostly leaves that surround them. When seen through the stereoscope, the skeleton leaf borders come to life in mounded shapes resembling an arched mausoleum (figure 3.12) or grotto (figure 3.13); both structures were featured in garden cemeteries such as Mount Auburn in Watertown and Cambridge, Massachusetts (established in 1831), and Bonaventure in Savannah, Georgia (established in 1846). Inspired by English estate gardens and the Père Lachaise Cemetery in Paris, these cemeteries were places "where nature combined with art and architecture."[31]

On a much smaller scale, the skeleton leaf memorials are similarly both human and botanical, reminding viewers that the people and leaves are dead but not forgotten. Skeleton leaf memorializing paralleled other kinds of remembrances in nineteenth-century America, such as postmortem photography.[32] The popularity of both practices reveals a cultural commitment to preserving, displaying, aestheticizing, and even commodifying the beloved in material forms that outlasted the living subject.

Edward Parrish's *The Phantom Bouquet: A Popular Treatise on the Art of Skeletonizing Leaves and Seed-Vessels and Adapting Them to Embellish the Home of Taste* (1862; figure 3.14) and the anonymously authored *Phantom Flowers: A Treatise on the Art of Producing Skeleton Leaves* (1864; figure 3.15) taught this popular practice. Throughout his brief book, Parrish emphasized the value of skeleton leaves as ornament—or, as its subtitle states, "to embellish the home of taste." Parrish did not originate the skeletonizing process; as he reported, "Hundreds of pier-tables and étagères in city and country are garnished with its airy forms and its photographic miniature, under the well-chosen

FIGURE 3.14. Cover, Parrish, *The Phantom Bouquet* (1862)

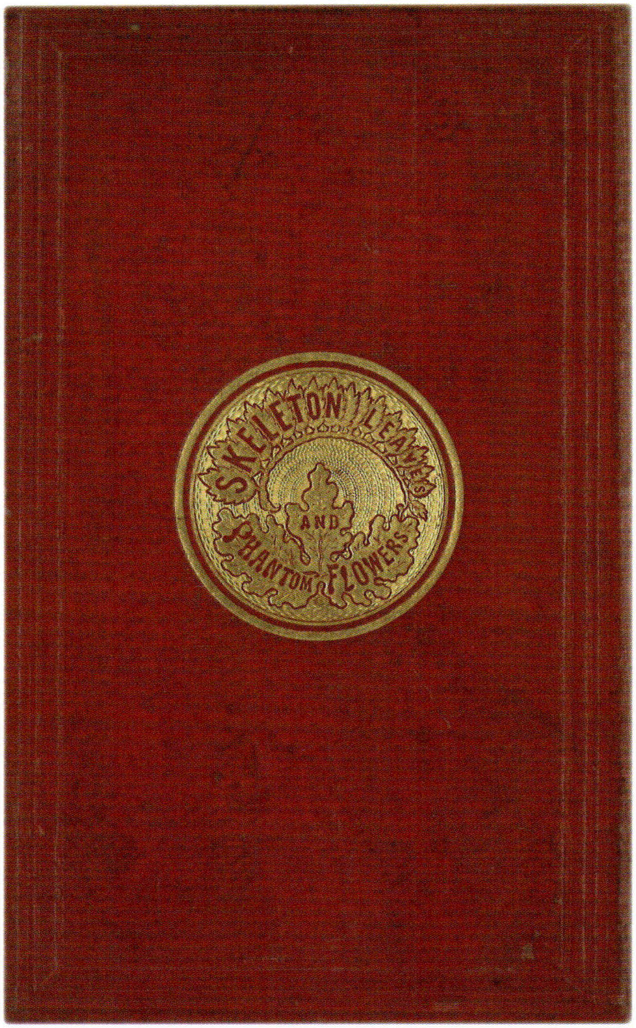

FIGURE 3.15. Cover, *Phantom Flowers: A Treatise on the Art of Producing Skeleton Leaves*, alternately titled *Skeleton Leaves and Phantom Flowers* (1864)

motto, 'Beautiful in death,' is displayed in almost every stereoscope." Parrish recommended the "Phantom Case"—"an object of unusual beauty"—for displaying skeletonized leaves. The "phantom case," like the herbarium cabinet and the earlier *Wunderkammer* or cabinet of curiosities, exhibits creations from nature. The *Wunderkammer* and herbarium were important for studying natural history, and Parrish stated that "an incidental but important result of the cultivation of this elegant pursuit is an increased attention to leaves," leading many people to "eagerly examine [them] in detail."[33]

The scientific knowledge that results from this "increased attention to leaves" further valorizes the process. An early chapter in *The Phantom Bouquet* on "Something of Science" explains the leaf's physiology, photosynthesis, and classification. Careful observations of leaves, including under a microscope, "will impart new and unexpected interest . . . and give to the process of skeletonizing, which unmasks the structure of the leaf, a practical and a scientific importance little anticipated by those who have pursued it with an exclusive view to its ornamental and aesthetic results." Parrish mentions many plant species by both common name and Latin binomial and uses precise botanical terminology such as "venation," "endogenous," and "midrib and lateral veins" (figures 3.16 and 3.17). He approaches mounting the specimens with a scientist's eye for accuracy and realism. For example, "the cushion that underlies the skeletons . . . may best be of green velvet, recalling the image of verdure in contrast with frost." The illustration of skeletonized ivy and silver poplar leaves (figure 3.18) displays this attractive contrast in black and white. Rather than the overpacked English versions of phantom cases, Parrish recommends to his American audience "a loose, airy-looking arrangement

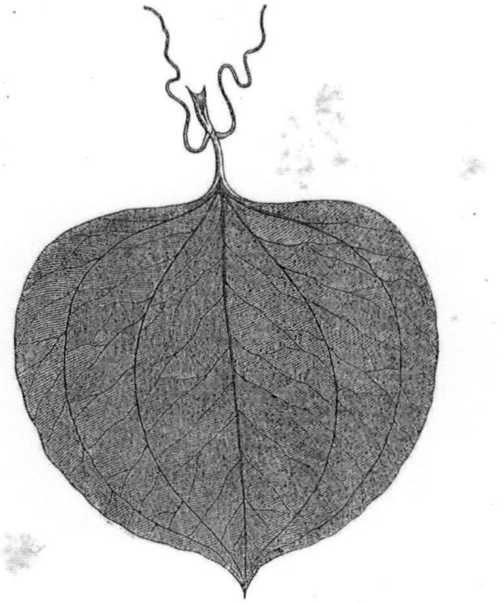

Green Briar Leaf, the under surface, showing the venation of an endogenous leaf.

Figure 3.16. Figure 1, Parrish, *The Phantom Bouquet* (1862)

Leaf of the Beech Tree, showing midrib and lateral veins, present in most exogenous leaves.

Figure 3.17. Figure 2, Parrish, *The Phantom Bouquet* (1862)

Fig. 3.

Ivy—Silver Poplar.

Figure 3.18. Figure 3, Parrish, *The Phantom Bouquet* (1862) of leaves" that better displays a plant's unique form and beauty, just as the standard herbarium page often showcases a single specimen.[34]

Phantom Flowers, published two years after Parrish's *The Phantom Bouquet*, provides much of the same information about collecting, macerating, bleaching, and preserving skeletonized plants, with similar statements about their good taste and beauty. In this longer work, the author shares more background on the science behind the skeletonizing process, that "wonderful art by which the perishable leaves and blossoms of the forest and the garden are converted into durable illustrations of the complex structure of the floral world."[35] The chapter "Illustrated List of Plants for Skeletonizing" lists optimum plant and tree species, ideal times for collecting, average times for maceration, and detailed illustrations.

FIGURE 3.19. Olive wood cover, *Flowers of the Holy Land*

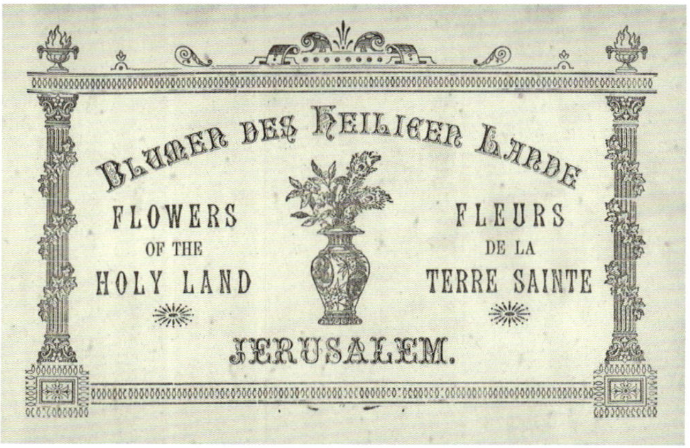

FIGURE 3.20. Title page, *Flowers of the Holy Land*

"The Wonders and Uses of a Leaf" chapter argues that whoever engages in this art will "become a botanist": "From a desire to extend her knowledge of the subject, she will consult the numerous botanical authorities which crowd the shelves of the nearest library, and thus her interest in a study so elevating and refining, will be increased. From these she will learn a multitude of facts, such as nothing short of the most elaborate chemical analysis." Some of the "wonders" of the leaf include its chemical properties, powers of absorption, and ability to draw mineral nourishment from the soil and return those nutrients once it has fallen and decomposed. Both women and men participate; just as "she will consult the numerous botanical authors," so too "he will find it worth while to become familiar with a science which, wherever his steps will lead him, . . . will still furnish him with new subjects for admiration."[36] For the authors of *Phantom Flowers* and *The Phantom Bouquet*, aesthetic creation was just one part of this activity; also worthwhile is acquiring the scientific knowledge that arises from such careful attention to plants.

By the end of the nineteenth century, preserved plant specimens appeared in still other formats. Instead of collecting and then pressing (or skeletonizing) leaves and flowers while traveling, one could buy a book with plant specimens already in them. In *The Pressed Plant*,

Andrea DiNoto and David Winter explain that "one especially charming type of herbaria is the Victorian-era grand tour souvenir. Many of these were made by individuals, but many more were produced commercially and sold to tourists in favored locales such as Italy, the Holy Land, or the Swiss Alps." Maura Flannery adds that albums sold with specimens from the Holy Land "were not only remembrances for those who traveled but also gifts for those who could not, uniting the scientific with the aesthetic and the religious in an attractive way."[37] Visitors to Jerusalem could bring home *Flowers of the Holy Land*, which was portably sized (4 inches tall by 6.5 inches wide), conveniently written in several languages, bound in needlework-embellished olive wood covers, and full of artfully composed pages of flowers (figures 3.19–3.21).

FIGURE 3.21. "Flowers of the Mount Moriah," *Flowers of the Holy Land*

Two similar books published in the United States at the end of the century are *Wild Flowers from the Holy Land* (figure 3.22) and *Wild Flowers from Palestine* (figure 3.23). Both were compiled by the Reverend Harvey B. Greene, along with *Pressed Flowers from the Holy Land* and *A Flower from the Christ Land*. Greene's books feature mounted specimens, one per page; the deluxe, larger-format *Wild Flowers from the Holy Land* sets each specimen in a double-matted frame (figure 3.24), while the smaller, more affordable *Wild Flowers from Palestine* puts them directly on the page (figure 3.25). The quality of some specimens, such as the lily of the field, is striking, considering they are more than a century old and have not been kept in climate-controlled conditions. Greene explains that once they located the plants, he and his assistants immediately pressed them in the field, "a task that demanded much delicacy and artistic appreciation, as well as wide botanical knowledge." Greene describes the Palestinian environments where they collected—the coastal region, the hill country, and the Jordan Valley—with details about topography, common trees and flowers, and climate. He closes the introduction to *Wild Flowers from Palestine* with his hope that the flowers featured will "serve to increase the study of Nature and Nature's God." Although the focus and purpose of these books were primarily religious, a bit of botanical information was also included: Each specimen was labeled with its Latin binomial as well as common name.[38]

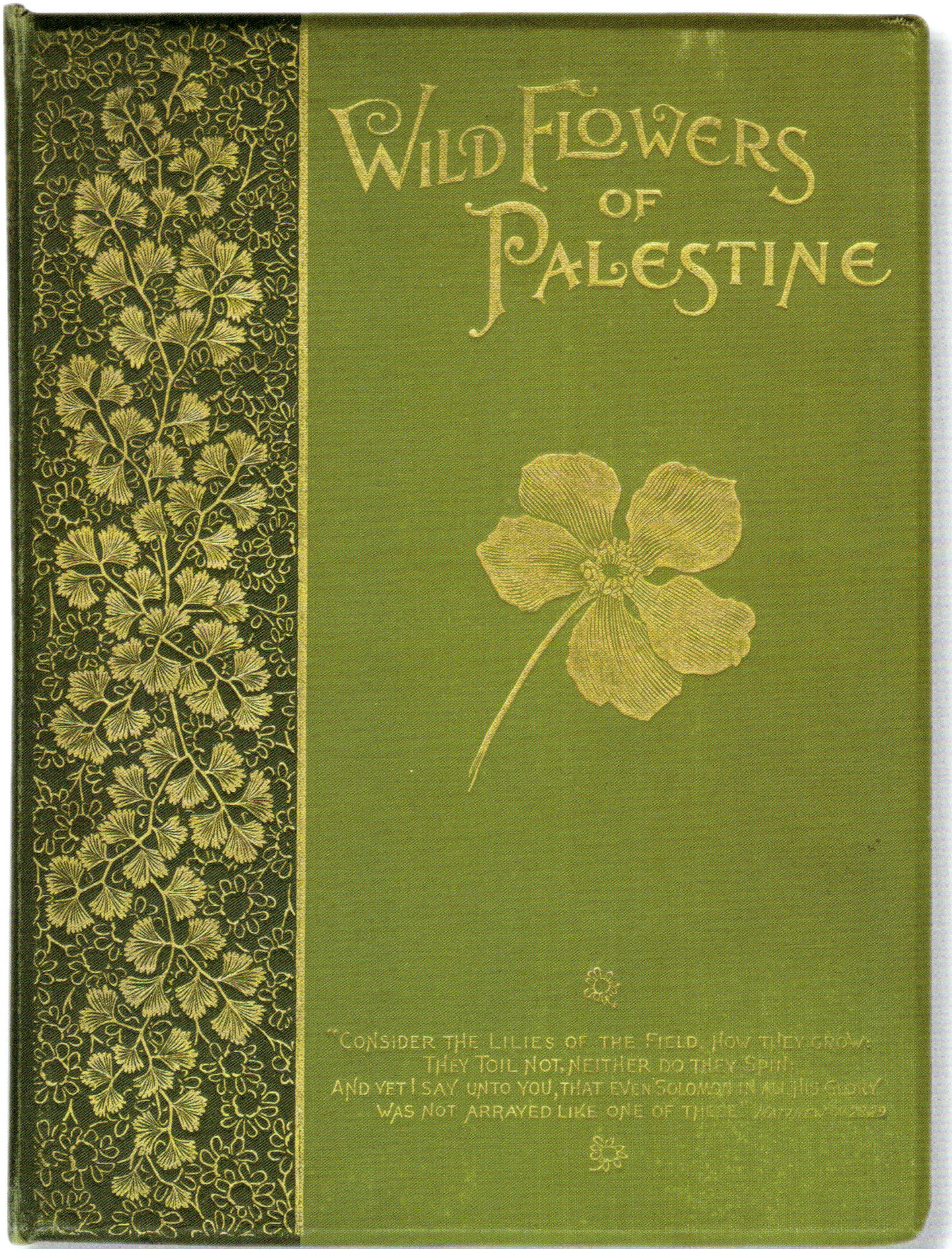

FIGURE 3.22. Cover, Greene, *Wild Flowers from the Holy Land* (1895), alternately titled *Wild Flowers of Palestine*

FIGURE 3.23. Cover, Greene, *Wild Flowers from Palestine* (n.d.)

FIGURE 3.24. Lily of the field, Greene, *Wild Flowers from the Holy Land* (1895)

Many but not all of the featured flowers have biblical references. Greene's rationale for including plants not mentioned in the Bible seems to be abundance and availability: Judean clover is "a common one in the valleys," madonna flower "grows abundantly near Nazareth," cyclamen "is very abundant in the hills about Jerusalem," and puffball "grows in every district of Palestine." There are obvious practical reasons for including plants that are easy to find, but Greene also takes a conservationist's approach, choosing these common species over rare ones. He notes that "the Hill Country of Judea, Samaria and Galilee" was "once covered with forests, and would be to-day if only a little care were used in preserving the young growth," echoing the concerns of many Americans in the late nineteenth century about the destruction of forests in the United States.[39]

FIGURE 3.25. Madonna flower, Greene, *Wild Flowers from Palestine* (n.d.)

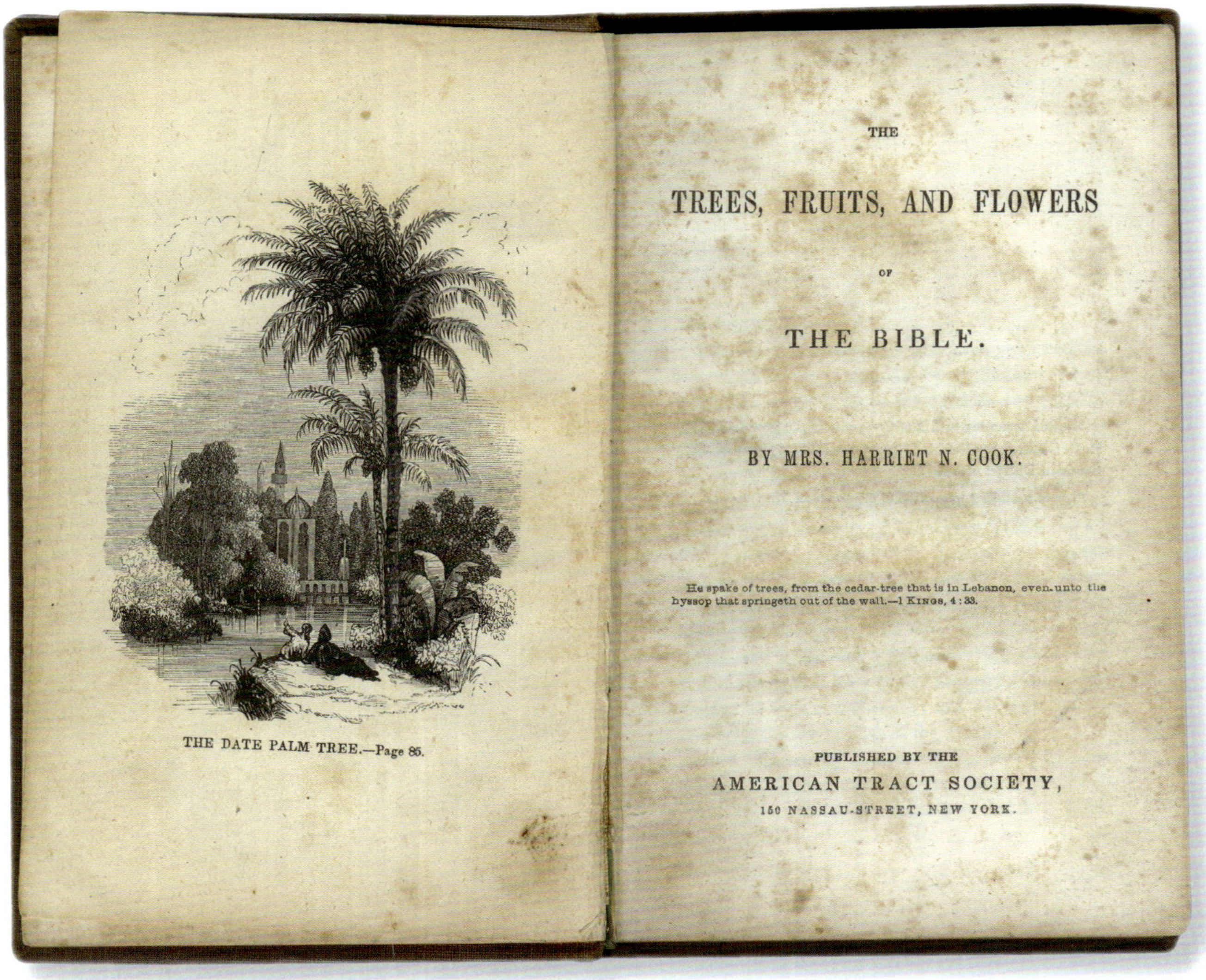

FIGURE 3.26. Title page and frontispiece, Cook, *The Trees, Fruits, and Flowers of the Bible* (1846)

Wild Flowers from the Holy Land and *Wild Flowers from Palestine* continue a long tradition of fascination with biblical natural history. Lytton Musselman points out that in 1568, the Dutch botanist Levinus Lemnius "wrote the first comprehensive treatment of all plants in the scriptures . . . that was widely circulated in several editions and languages." In the eighteenth century, systems of classification such as Linnaeus's promoted an orderly approach to organizing the thousands of plant and animal species discovered in several centuries of exploration by Europeans into other parts of the world. As travel to places such as Palestine became more practicable, natural historians eagerly researched the native flora and fauna. According to anthropologist James Bielo, Linnaeus's student Fredrik Hasselquist conducted "a four-year expedition throughout the Levant, collecting botanical samples and overseeing their transport back to Sweden," which culminated in the posthumously published *Voyages and Travels in the Levant; In the Years 1749, 50, 51, 52. Containing Observations in Natural History, Physick, Agriculture, and Commerce: Particularly on the Holy Land, and the Natural History of Scriptures* (1757). Bible natural history dictionaries by other European writers soon followed.[40]

In the United States, books such as Francis A. Ewing's *Bible Natural History; or a Description of the Animals, Plants, and Minerals, Mentioned in the Sacred Scriptures* (1835), Margaret Coxe's *Conversations on the Botany of the Scriptures* (1836), and Harriet N. Cook's *The Trees, Fruits, and Flowers of the Bible* (1846; figure 3.26) provided antebellum natural historians with copiously illustrated, scientific, and scripturally supported information about species mentioned in the Bible. Like the American Sunday School Union's series on *The Wonders of Vegetation* discussed in chapter 1, these books were published by religious organizations. Ewing's *Bible Natural History* is arranged alphabetically, from "Adamant" to "Wormwood," and while its primary purpose is "to render the acquisition of biblical knowledge easy and attractive," it also provides accurate information about the animals, plants, and minerals of the Bible (figures 3.27 and 3.28). As Ewing further explains, "To us an accurate knowledge of these [natural history] subjects is important as matter of information, and highly useful in enabling us to understand and to feel the full force, beauty, and propriety of those figures, which the sacred writers drew so abundantly from the various objects of nature."[41]

FIGURE 3.27. "Cedar of Lebanon," Ewing, *Bible Natural History* (1835)

FIGURE 3.28. "Saffron," Ewing, *Bible Natural History* (1835)

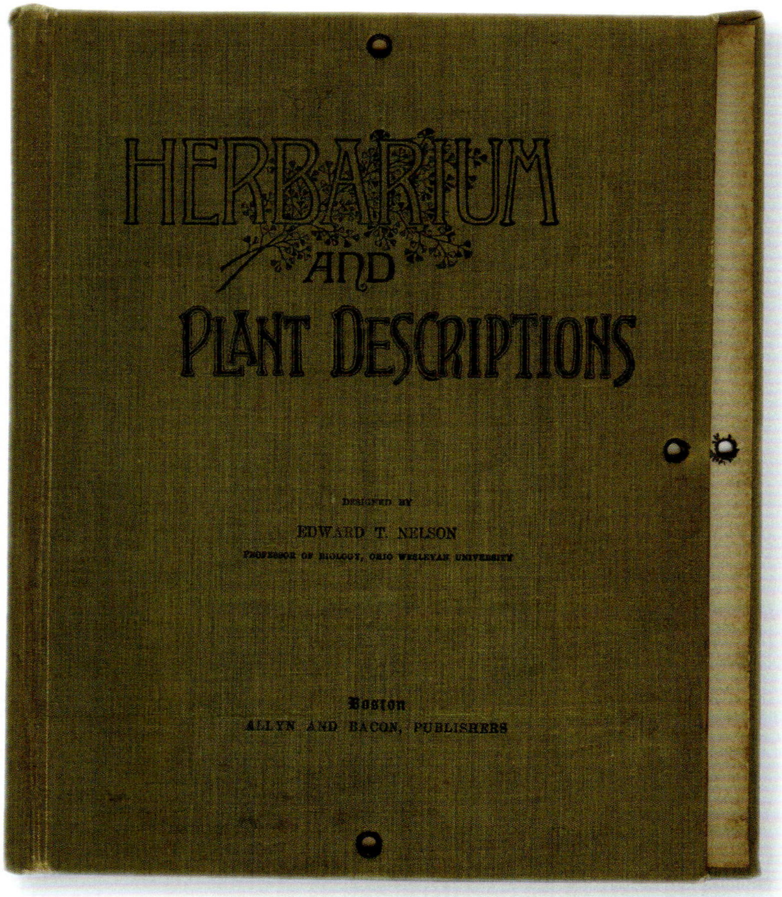

FIGURE 3.29. Outer cover, Nelson, *Herbarium and Plant Descriptions* (n.d.)

Multiple interests attracted everyday people to the preservation of plants. For some, reading about particular species in the Bible might lead them to consult a reference such as those by Ewing, Coxe, and Cook. For others, a trendy craze like skeletonizing leaves would bring them to books such as *Phantom Flowers* or *The Phantom Bouquet*, where they would learn the intricacies of both leaf anatomy and the skeletonizing process. Easily portable handbooks such as Manton's *Field Botany* and Bailey's *Botanizing* simplified the process of collecting and pressing plants, making botany accessible to those without formal scientific training. Articles in *Godey's Lady's Book* advocated for women's scientific activity, and its fiction validated that activity with protagonists who prioritized intellectual pursuits over the frippery of fashionable dress. Botany instruction books for a range of learners provided instructions for making an herbarium and the important reasons for identifying, collecting, preserving, and sharing plant specimens.

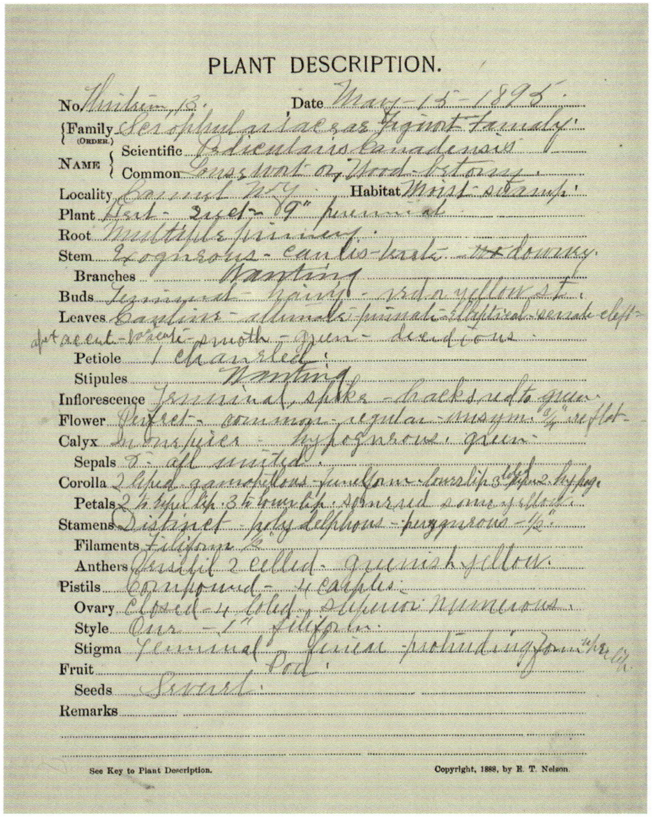

PLANT DESCRIPTION.

No. Date May 15 1895
Family (Order)
Name: Scientific / Common
Locality / Habitat
Plant
Root
Stem
Branches
Buds
Leaves
Petiole
Stipules
Inflorescence
Flower
Calyx
Sepals
Corolla
Petals
Stamens
Filaments
Anthers
Pistils
Ovary
Style
Stigma
Fruit
Seeds
Remarks

See Key to Plant Description. Copyright, 1888, by E. T. Nelson.

FIGURE 3.30. "Plant Description" cover sheet for *Pedicularis canadensis* (lousewort or wood betony), Edna Smith's copy of Nelson, *Herbarium and Plant Descriptions* (n.d.)

Two examples from the final decades of the nineteenth century show the continued evolution of the scientific, aesthetic, and preservation functions of the herbarium. *Herbarium and Plant Descriptions* by Edward T. Nelson was a do-it-yourself version: a loose outer cover, seven pages of instructions, a blank index to fill with scientific and common names, and fifty folded cover sheets for recording plant description details on the outside and preserving the specimen inside (figures 3.29–3.32). *Herbarium and Plant Descriptions* likely enjoyed wide use, judging from how many copies have survived in archives and auctions.[42] One version, compiled by Edna Smith of Carmel, New York, in 1895, is remarkably complete. She filled out all fifty covers in detail (figure 3.30), painstakingly preserved the corresponding specimens (figure 3.31), and listed her findings in the provided index (figure 3.32).

FIGURE 3.31. Pressed specimen of *Pedicularis canadensis*, Edna Smith's copy of Nelson, *Herbarium and Plant Descriptions* (n.d.)

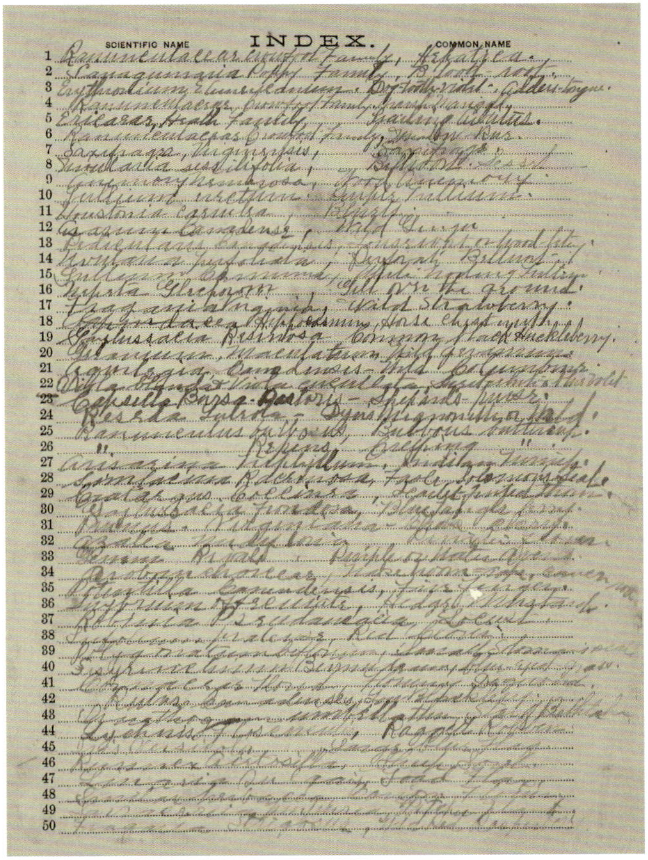
SCIENTIFIC NAME INDEX. COMMON NAME

FIGURE 3.32. Index, Edna Smith's copy of Nelson, *Herbarium and Plant Descriptions* (n.d.)

Carmel, New York, is in the scenic Hudson Valley, and many of the plants Smith collected are native to this area. Botanists like Smith were the ideal audience for *Herbarium and Plant Descriptions*. Nelson stated that he hoped "to awaken a greater interest in the study of plant life, and to introduce a little of the laboratory method of research into the first steps of Elementary Botany. The student is to be sent into the fields and woods with his eyes open . . . [and bring back] as much information as possible about their habits of growth, their appearance when in fullest blooming, and the structure and relation of stem and leaf, blossom and fruit."[43] Edna Smith certainly went into fields and woods—not to mention roadsides, swamps, and "waist places," which are some of the additional locations she recorded. As her detailed, fully filled cover sheets indicate, she also recorded "as much information as possible." Nelson's purpose at the end of the century followed those of decades of botany instructors before him. Even though "the laboratory method" is now available, the basic work of detailed observation in the field remains just as relevant as it was for Eaton and Phelps in the 1820s.

A second late-nineteenth-century example extends the reach of the herbarium even further, into what Rochelle Johnson has termed "the aesthetic pursuit of nature's physicality."[44] *Flowers of the Field and Forest* (figure 3.33) is a beautifully bound, folio-sized volume with illustrations by Isaac Sprague and text by the Reverend A. B. Hervey, the same team that created *Beautiful Wild Flowers of America* (see figure I.2). The book features fourteen flowers native to the United States, with chapters of about ten pages for each species. Like an herbarium specimen, Sprague's drawings were created "after nature," a phrase signaling that the artist worked with an actual plant, rather than copying it from another source. Also like an herbarium, Sprague shows only one species per page and includes the root, leaves, bud, and full flower of the plant (figures 3.34 and 3.35). Hervey's text generally compiles information from other sources, most frequently from Asa Gray, George Goodale (professor of botany at Harvard), and Thomas Meehan (editor of horticulture journals). He also quotes poetry: Each chapter begins with a title poem, and excerpts from others appear throughout the chapter. Most of the poets he cites are American, including William Cullen Bryant, Alice and Phoebe Cary, Ralph Waldo Emerson, Oliver Wendell Holmes, Henry Wadsworth Longfellow, James Russell Lowell, and Henry Timrod. In addition to botanical details, poetry, and flower lore, Hervey discusses the scientists who are memorialized in the flower's Latin binomial. For example, partridge berry, *Mitchella repens*, was named by Linnaeus for "Dr. John Mitchell of Virginia . . . one of our best known botanists." The carnivorous pitcher plant, *Sarracenia purpurea*, honors "Dr. Sarrazin, of Quebec, who . . . first sent specimens of this plant, with some account of its habits, to European botanists."[45] The breadth of Hervey's coverage of each plant contrasts the close detail of Sprague's drawings, but in ways analogous to the herbarium, Hervey's text situates the plant in its cultural taxonomy.

In one copy of *Flowers of the Field and Forest*, Sprague's visual and Hervey's textual representations of native American wildflowers are accompanied by another herbarium feature: actual pressed specimens. Unlike those included in *Wild Flowers from the Holy Land*, these have been added by an unknown reader or owner of the book (figures 3.36–3.39). In each case, the preserved plant does not correspond to the text of the page: The specimen in figure 3.36 looks like a

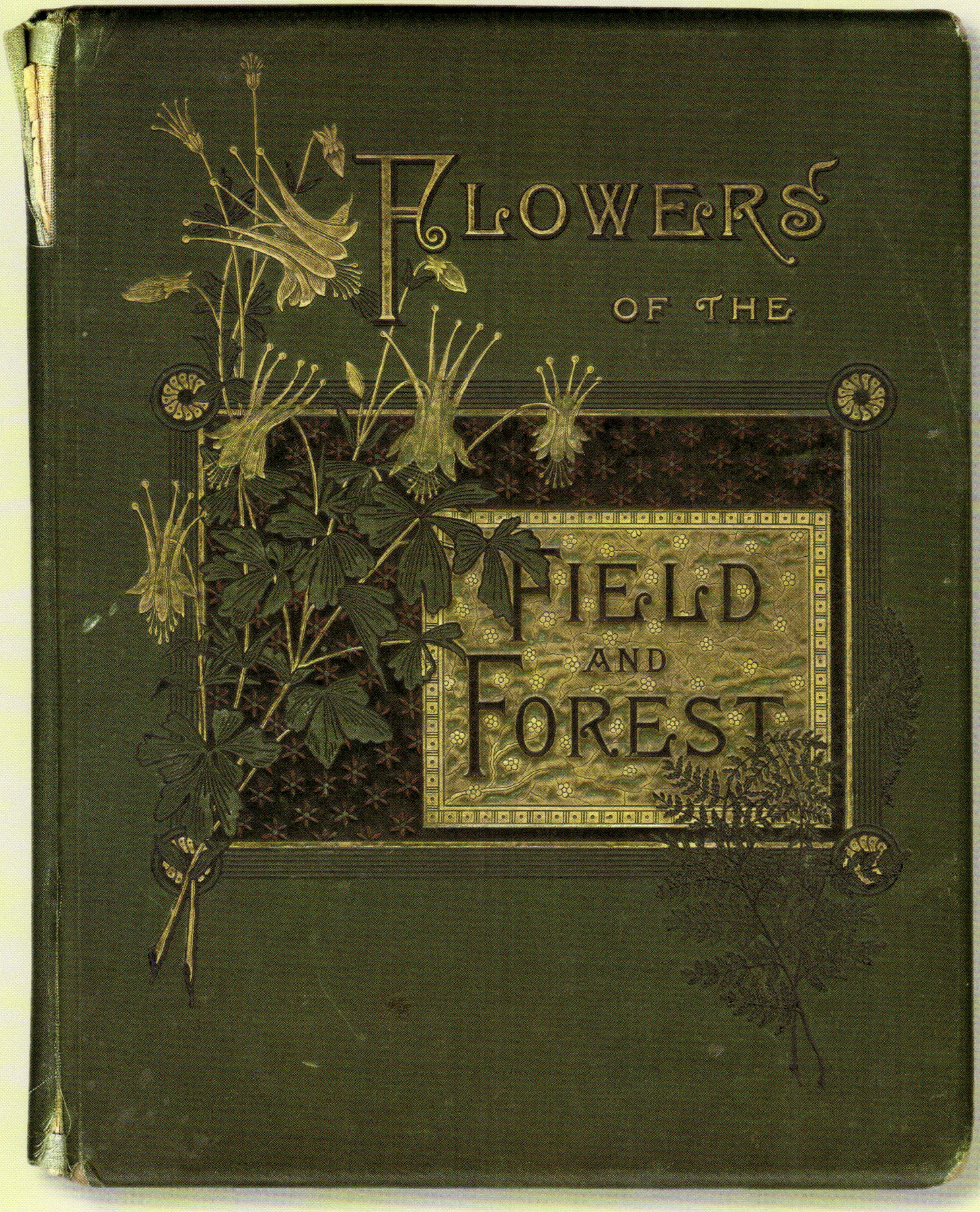

FIGURE 3.33. Cover, Sprague and Hervey, *Flowers of the Field and Forest* (1882)

FIGURE 3.34. Bloodroot, Sprague and Hervey, *Flowers of the Field and Forest* (1882)

FIGURE 3.35. Arethusa, Sprague and Hervey, *Flowers of the Field and Forest* (1882)

pansy or violet, definitely not the arethusa that appears at the right. The aster-like flower in figure 3.37 is not bloodroot; the fern in figure 3.38 is nothing like the cardinal flower; and while the squashed material in figure 3.39 could be many things, it is not a galax-leaved shortia. When pressed specimens like these do not correspond to visual or textual information on the page, they signify in other ways. The pressed flower in figure 3.37 stands out because it is surrounded by blank space; this single specimen, with its three additional shadowed forms, announces itself like the single verbal specimen "BLOOD-ROOT" does elsewhere on the page. Although the flower in figure 3.36 is not arethusa, it does mimic the shape: The mirrored image of the specimen and its shadow becomes two stems, flowers, and sets of roots—similar to the four stems, leaves, and roots of Sprague's arethusa drawing. The fern frond in figure 3.38 looks particularly striking against a page full of text: Its multiple pinnate leaves along a curved stalk both replicate and contrast with the many straight lines of sentences. As much as this copy of *Flowers of the Field and Forest* might seem to diverge from the standard

FIGURE 3.36. *(top left)* Pressed specimen near "Arethusa" page; FIGURE 3.37. *(top right)* Pressed specimen at "Blood-root" page; FIGURE 3.38. *(left)* Pressed specimen, "Cardinal-flower" section. Sprague and Hervey, *Flowers of the Field and Forest* (1882)

FIGURE 3.39. Pressed specimen, page 65, Sprague and Hervey, *Flowers of the Field and Forest* (1882)

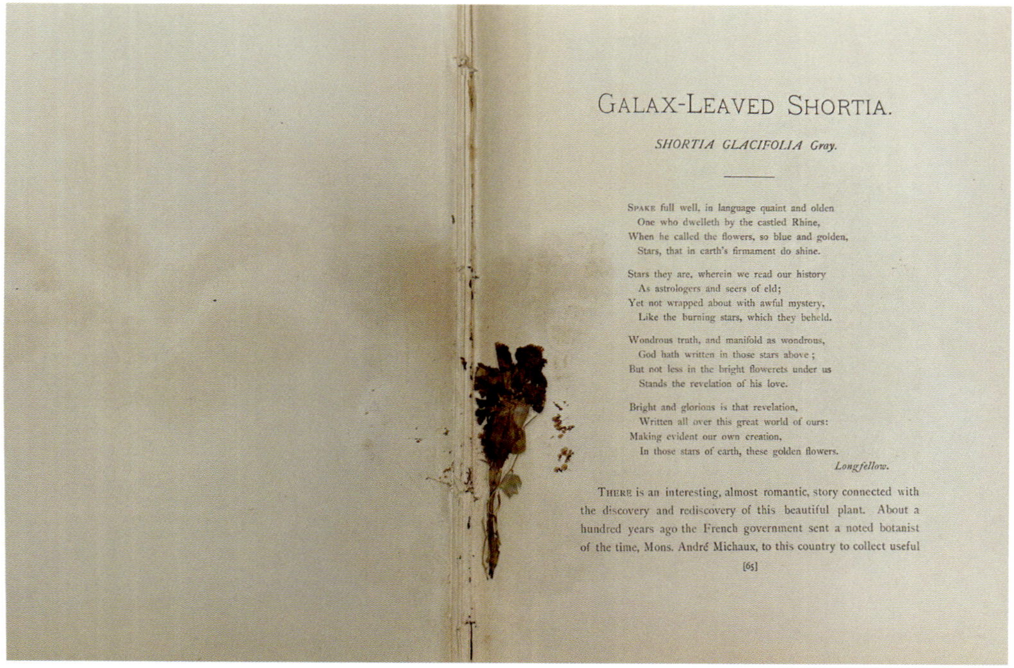

herbarium, it retains several features of its more scientific predecessor.[46] It too was a useful reference, with a rich variety of botanical information and beautiful, lifelike images.

The Preservation of Plants

As noted at the beginning of this chapter, preservation of flowers in the United States at this time was both individual and national, scientific and aesthetic. Everyday Americans created herbaria, shared plant specimens, decorated with dried flowers and skeletonized leaves, and treasured beautiful books with full-color, full-page plates of the flowers they might encounter in fields and forests—including by pressing their own specimens in them. They also became concerned about preserving native flora, including both trees and wildflowers.

In 1847, George Perkins Marsh, then a senator from Vermont, gave a lecture to the Agricultural Society of Rutland County that warned of the "evil resulting from the injudicious destruction of the woods." An article in *Godey's Lady's Book* from 1841 (likely by its editor, Sarah Josepha Hale) on "The Preservation of Wild Flowers" urged its women readers to consider it "a subject worthy the attention of our sex," because "the diminishing and fading wild flowers of our land call to them for protection." The article then encourages women to transplant native wild flowers to their home gardens, where they will be safe from the ravages of "the remorseless plough" of large-scale agriculture. Planting a native flower garden enables them to "hand down a floral treasure to the admirers of nature, who shall come after them."

The author continues, "I do not mean a collection of dried plants, although that is also pleasing and useful; but an Herbarium of living specimens"—thus explicitly linking the protection of native wildflowers to the preservationist purpose of the herbarium. In 1882, the year Sprague and Hervey published *Flowers of the Field and Forest*, the first American Forestry Congress was held in Cincinnati in conjunction with the first national Arbor Day tree planting, with thousands attending, including twenty thousand schoolchildren.[47]

The herbarium remains a crucial tool in the conservation of plants. The specimen in figure 3.4 is *Apios priceana*, named for the botanist who discovered it, Sadie Price. Price wrote books, articles, and notes for botanical journals, and, Carol Crowe-Carraco points out, her herbarium contained "almost 3000 specimens, approximately one-third of which she illustrated in watercolor, pen and ink, and pencil sketches." *Apios priceana*—also known as Price's potato-bean, Price's groundnut, and traveler's delight—is now federally threatened, due primarily to habitat destruction. Although the U.S. Fish and Wildlife Service is actively working to restore *Apios priceana* with some success, the service explains it "will be considered for delisting when 25 geographically distinct, self-sustaining populations are adequately protected and have been maintained for 10 years."[48] Historical herbaria provide data that can be used not only to track climate change, but also to "document the decline of rare and endangered species and the arrival and spread of non-native invasive species."[49]

Herbaria have immense value, and preserving these collections is crucial, especially when they are threatened by closures and termination of funding. The February 2024 case of the Duke University Herbarium is but one example.[50] Maura Flannery asserts that the stories of herbaria "and the stories of those who care about them are well worth telling."[51] Each herbarium page has a story, as shown by the example of Sadie Price's *Apios priceana*, and there are *millions* of herbarium specimens catalogued in collections throughout the world. The next chapter tells the story of another herbarium-like creation, a "diary kept in flowers" by one of nineteenth-century America's best-connected literary figures.

FIGURE 4.1. Undated entry, Fields, "Fragrant Memories," Fields Addenda, box 6, folder 1, The Huntington Library, San Marino, California

4

FROM THE ARCHIVES

"Diary Kept in Flowers"

Annie Adams Fields, "Fragrant Memories"

One of the first violet leaves,
Una sent it from Concord.

Dear Mrs. Hawthorne came to make her spring visit on the 8th of May. She brought this "Heart's Ease" which Julian has painted because she says it is a portrait of Mr Fields in his velvet coat.

FIGURE 4.2. Cover, Fields, "Fragrant Memories," Fields Addenda, box 6, folder 1, The Huntington Library, San Marino, California

In this second "From the Archives" chapter, the focus is again on a single object: a diary by Annie Adams Fields (1834–1915).[1] Unlike Mattie Snowden, however, Annie Fields is well known: Her husband was James T. Fields (1817–1881), senior partner of Ticknor and Fields, one of the leading literary publishers in mid-nineteenth-century America. Annie Fields was a prolific diarist who filled more than sixty unpublished volumes during her lifetime. She titled one of them "Fragrant Memories," a diary "kept in flowers," where brief written entries join pressed flowers on nearly every page (figure 4.2). What makes Snowden's *Illustrated Botany* distinctive are her manuscript additions and its polyvocality as journal, commonplace book, specimen repository, and more—all delivered to the pages of a botany instruction book. What makes Annie Fields's "Fragrant Memories" distinctive is her use of pressed specimens to preserve memories of travels, social interactions, and important events. While its material format might resemble an herbarium, its purpose departs from the strictly botanical. Instead, Fields used it as a parallel diary to mark both quotidian and significant aspects of her life for almost a decade. For this prolific diarist, the flowers she preserved were cherished gifts, objects of beauty, and intentional reminders of events that sometimes defied expression in words.

Annie Adams married James Fields in 1854, when she was twenty and he was thirty-seven and a widower. They lived in Boston and regularly entertained a who's who of American authors: Lydia Maria Child, Ralph Waldo Emerson, Nathaniel Hawthorne, William Dean Howells, Henry James, Lucy Larcom, Henry Wadsworth Longfellow, Harriet Prescott Spofford, Harriet Beecher Stowe, Celia Thaxter, John Greenleaf Whittier, and many others. Fields was an avid gardener, poet, novelist, editor, biographer, philanthropist, and, after James's death, the partner of fellow writer Sarah Orne Jewett.[2]

She wrote detailed journals for decades, beginning in 1859 on a trip to Europe with James; some of her last diaries record trips with Jewett to the Caribbean in 1896 and to France in 1898. In between, she wrote most often about the people that she hosted, visited, and read. She started writing a regular diary in the early 1860s and titled several such volumes "Journal of Literary Events and glimpses of Interesting people" (figure 4.3).[3] For these diaries, she mostly used faintly ruled, blue paper–covered copybooks measuring 8 inches tall by 6.75 inches wide and generally wrote in ink, with entries averaging several pages each. She began "Fragrant Memories" in 1862. An inscription, likely written later, states:

> *Diary Kept in flowers but begun a little too late in the summer of life.*
>
> *Too much has passed unrecorded except in memory.*
> *That is the warmest place however—*

Fields wrote in "Fragrant Memories" most actively from 1862 to 1865, then sporadically from 1866 to 1870. Like many of her other diaries, this one is a blue paper–covered copybook, but smaller, measuring six inches tall by four inches wide. It contains about 150 faintly ruled pages. When new, the notebook was probably only a few sixteenths of an inch thick; now, it's a bulging several inches (figure 4.4).

In contrast to her other diaries, most of the entries in "Fragrant Memories" are very brief and often written informally in pencil, and they do not show the same awareness of a possible audience.[4] While Fields usually filled her other diaries from cover to cover, "Fragrant Memories" has about fifty blank pages, mostly toward the end. It is a parallel diary: a less detailed, intermittent record that sometimes covers the same events as in the more detailed diaries, but not always. It contains

FIGURE 4.3. Title inscription, Fields, Diary 4, Collection of the Massachusetts Historical Society

FIGURE 4.4. Side view, Fields, "Fragrant Memories," Fields Addenda, box 6, folder 1, The Huntington Library, San Marino, California

eighty-two pressed plants, which account for its bulk. Fields identified only thirty-five, representing twenty-one different species. Those mentioned most often are violets, roses, and sweetbriar (*Rosa rubiginosa*), which was Fields's favorite flower. The earlier years of the diary coincided with the Civil War, and Fields occasionally wrote about the conflict. Most often, she wrote about people she knew, both close friends and the authors James published; these categories certainly overlapped. She pressed and recorded the flowers people gave her—which happened often—and usually noted the giver, the date, and sometimes the type of flower. During the years she kept this diary James was still alive, and she wrote about the happiness of their marriage. There's not much reflection in "Fragrant Memories," perhaps because Fields was writing more substantively in her other diaries. But even in their brevity, many entries convey her thoughtful and emotional responses to events.

Flowers, Friends, and Famous People

The first full entry (figure 4.5) describes a visit:[5]

> *May 28th 1862*
> *Called at the Thaxters in Watertown*
> *this afternoon—We brought home a*
> *carriage full of wildflowers and*
> *the sweet fragrance of a new friendship*
>
> *Blueberry blossom given*
> *me by Mr. Thaxter and his*
> *wife—Nature's pet—*

FIGURE 4.5. May 28, 1862, entry, Fields, "Fragrant Memories," Fields Addenda, box 6, folder 1, The Huntington Library, San Marino, California

Inserted on the page is a small bit of blueberry leaf, tucked into a slit cut into the paper, in the manner used by many botanists for their herbaria. "The Thaxters" were Levi and his wife, Celia, a poet who published her first poem in 1861 in the *Atlantic Monthly*, which James Fields edited from 1861 to 1871. Annie Fields was Celia Thaxter's "closest confidante and her most solicitous (and most professionally helpful) friend." Like Fields,

Thaxter loved growing flowers, and later in her life, she published the memoir *An Island Garden* (1894), with illustrations by the painter Childe Hassam (figures 4.6 and 4.7).[6] Fields's use of zeugma in this entry—"We brought home a carriage full of wildflowers and the sweet fragrance of a new friendship"—is a rhetorical move that recurs in "Fragrant Memories," where Fields connects material flowers to her musings.

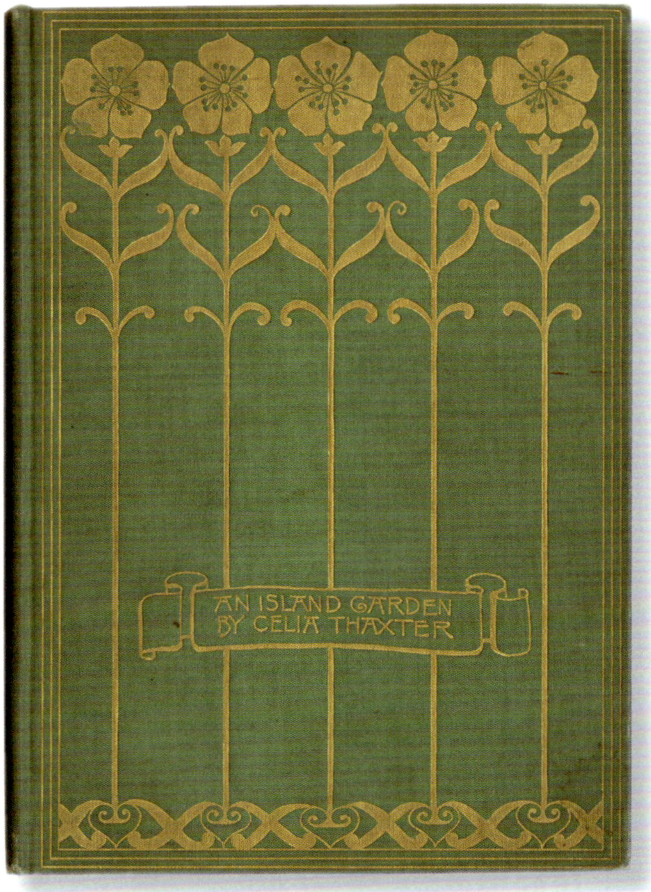

FIGURE 4.6. Cover, Thaxter, *An Island Garden* (1894), Boston Public Library

FIGURE 4.7. Frontispiece, Thaxter, *An Island Garden (1894)*; "In the Garden" by Childe Hassam, Smithsonian American Art Museum, gift of John Gellatly

Two example entries show features typical of "Fragrant Memories": records of visits, evidence of friendships, and comments about flowers (figures 4.8 and 4.9).

> *Laura found this first violet*
> *on Staten Island and sent to me*
> *Sunday March 19th 1865.*
> *I thank this dear friend for my*
> *first taste of spring after*
> *the crocuses.—The latter I*
> *saw March 20th in the*
> *Public Garden when walking*
> *with Sophia Hawthorne.*

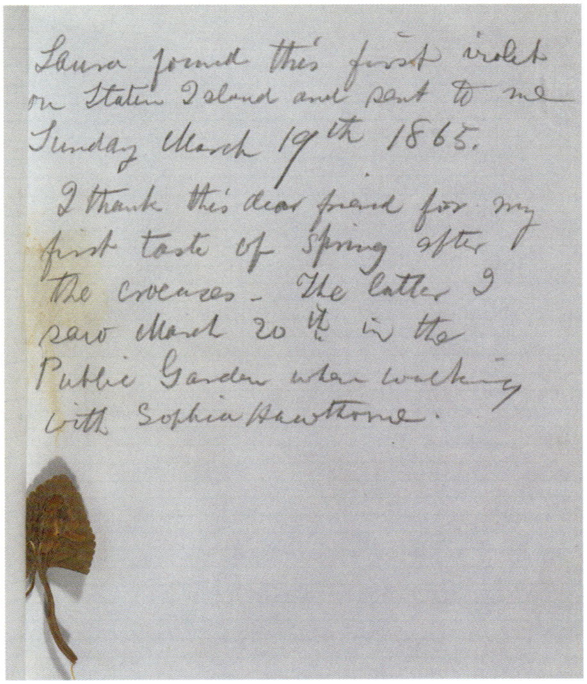
Laura found this first violet
on Staten Island and sent to me
Sunday March 19th 1865.
I thank this dear friend for my
first taste of Spring after
the crocuses. The latter I
saw March 20th in the
Public Garden when walking
with Sophia Hawthorne.

FIGURE 4.8. Ca. March 1865 entry, Fields, "Fragrant Memories," Fields Addenda, box 6, folder 1, The Huntington Library, San Marino, California

> *May 7—In Staten Island where*
> *the Westons were in perfection.*
> *Jamie went to see poor Edwin Booth*
> *and Laura parted from us near*
> *the Academy which we enjoyed*
> *together*

> *Quince blossom from*
> *sweet Bessie Weston*
> *It is not unlike her.*

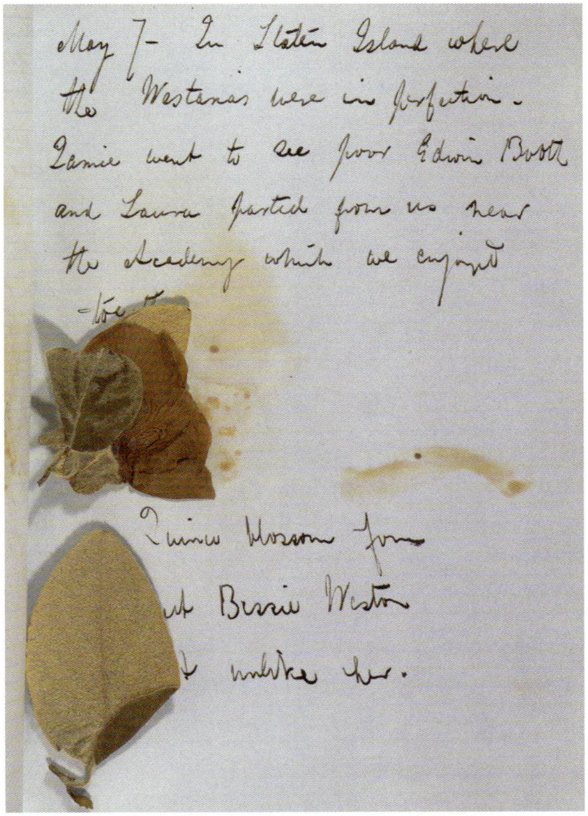
May 7- In Staten Island where
the Westons were in perfection.
Jamie went to see poor Edwin Booth
and Laura parted from us near
the Academy which we enjoyed
toge
Quince blossom from
t Bessie Weston
t unlike her.

FIGURE 4.9. May 7 (1865) entry, Fields, "Fragrant Memories," Fields Addenda, box 6, folder 1, The Huntington Library, San Marino, California

Annie and James Fields had the leisure to spend time visiting friends, including the Westons and Laura Winthrop Johnson, Annie's dear friend and correspondent for twenty-seven years.[7] Sophia Hawthorne, the wife of author Nathaniel Hawthorne, was another close friend. "Poor Edwin Booth," one of the most famous actors of the time and a friend to both James and Annie, was clearly in need of sympathy in early May 1865: His brother John Wilkes Booth had assassinated Abraham Lincoln just a few weeks earlier, on April 15. People often gave Fields flowers, either directly (as did Bessie Weston, figure 4.9) or in letters (as did Laura Johnson, figure 4.8). Although she clearly treasured these flowers, she did not preserve or label them as a botanist would in an herbarium. Many of the specimens have lost their color or shape, but because Fields frequently noted dates and locations, a partial phenological record remains. She commented on the seasonality of flowers such as crocuses (figure 4.8), and when her friend Mr. Bartol gave her a flower on May 8, 1865, she noted, "This blossom marks the spring to be 12 days earlier than the last."

While many of the plants in "Fragrant Memories" were local to Fields's New England home environment, others came from much farther away. She included some from her travels abroad, such as a "White Jasmine given me by Tennyson while we were at Farringford in July 1859" (figure 4.10). Alfred, Lord Tennyson was one of the best-known English poets of his day, serving as poet laureate from 1850 until his death in 1892. Annie and James met him during their trip to Europe in 1859, and Fields wrote a memoir of him for *Harper's Magazine* in 1893.[8] Beneath the inserted pressed jasmine, Fields wrote, "Fading relics of imperishable memories," a reminder of a flower's ephemerality and the durability of remembrance. Fields likely pressed the flower from

FIGURE 4.10. "White Jasmine from Tennyson," Fields, "Fragrant Memories," Fields Addenda, box 6, folder 1, The Huntington Library, San Marino, California

Tennyson during her trip in 1859 and kept it elsewhere, then inserted it into her diary "kept in flowers," which she began several years later.

Other flowers from abroad were sent by her sister Elizabeth ("Lissie") Adams, an artist who studied in France and Italy. Some were "from the Pamfili Doria" (figure 4.11), the Villa Doria Pamphili in Rome, which the Fieldses had also visited in 1860. Another came from Sorrento a few months later.[9] Curiously, Fields kept a "Leaf from Athens from plant brought by the Italian ship 'Il Galantuomo'" (figure 4.12). The ship was actually the *Re Galantuomo*, an Italian line-of-battle ship that left from New York in March 1864. It's unclear how Fields might have gotten the leaf that she pressed, nor what kind of plant the ship had brought. But given its location in the diary—near entries dated February 1864 and March 25, 1864—Fields was likely caught up in the media storm about the *Re Galantuomo* throughout March, when it was feared to be wrecked on its return to Italy.[10]

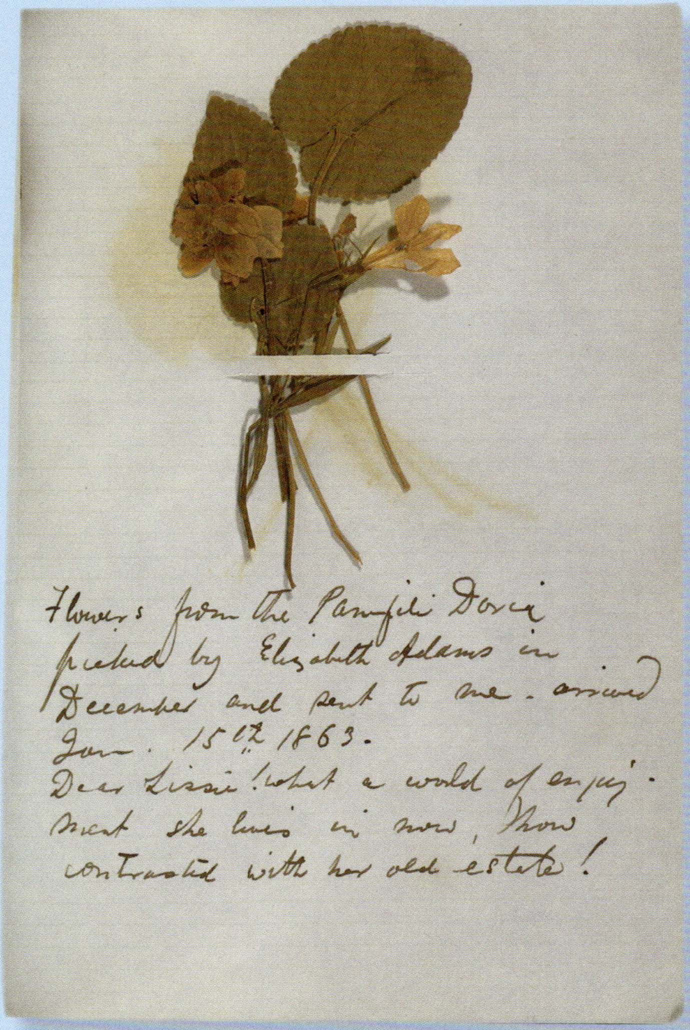

Flowers from the Pamfili Doria
picked by Elizabeth Adams in
December and sent to me - arrived
Jan. 15th 1863.
Dear Lizzie! what a world of enjoyment
she lives in now, how
contrasted with her old estate!

FIGURE 4.11. "Flowers from the Pamfili Doria," Fields, "Fragrant Memories," Fields Addenda, box 6, folder 1, The Huntington Library, San Marino, California

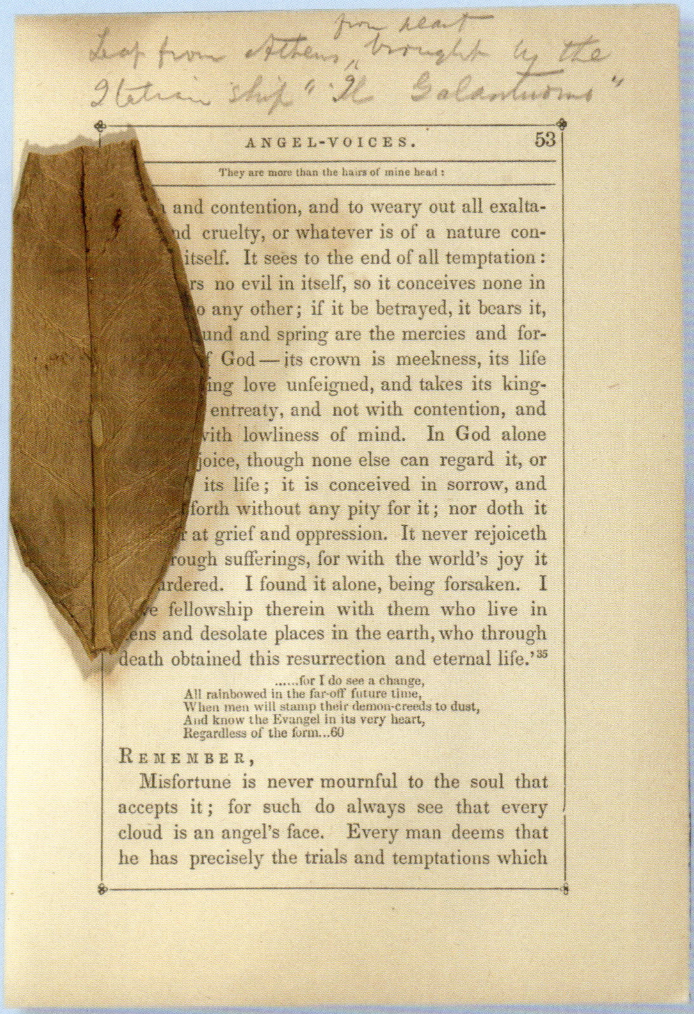

Leaf from Athens, [illegible] brought by the
Italian ship "Il Galantuomo"

ANGEL-VOICES. 53

They are more than the hairs of mine head:

n and contention, and to weary out all exalta-
nd cruelty, or whatever is of a nature con-
itself. It sees to the end of all temptation:
rs no evil in itself, so it conceives none in
o any other; if it be betrayed, it bears it,
und and spring are the mercies and for-
f God—its crown is meekness, its life
ing love unfeigned, and takes its king-
entreaty, and not with contention, and
vith lowliness of mind. In God alone
joice, though none else can regard it, or
its life; it is conceived in sorrow, and
forth without any pity for it; nor doth it
at grief and oppression. It never rejoiceth
rough sufferings, for with the world's joy it
rdered. I found it alone, being forsaken. I
e fellowship therein with them who live in
ens and desolate places in the earth, who through
death obtained this resurrection and eternal life.'[35]

......for I do see a change,
All rainbowed in the far-off future time,
When men will stamp their demon-creeds to dust,
And know the Evangel in its very heart,
Regardless of the form...60

REMEMBER,

Misfortune is never mournful to the soul that accepts it; for such do always see that every cloud is an angel's face. Every man deems that he has precisely the trials and temptations which

FIGURE 4.12. "Leaf from Athens," Fields, "Fragrant Memories," Fields Addenda, box 6, folder 1, The Huntington Library, San Marino, California

Far more prevalent are the flowers that Fields preserved from interactions with American writers, many of whom were published by Ticknor and Fields, including the poets Oliver Wendell Holmes, John Greenleaf Whittier, and Henry Wadsworth Longfellow (figures 4.13–4.15). These pressed specimens are the botanical parallel to the "Glimpses of interesting People" that she recorded in other diaries: Just as botanizers' herbaria became a record of the plants they encountered and collected, the flowers in "Fragrant Memories" were Fields's record of the interesting people she encountered. She and James frequently socialized with many of these writers, who were also personal friends.

From the garden of Professor Holmes's
old house in Cambridge

We sat there under the trees with them
in the afternoon when he gathered for me
a bouquet—Sep. 1863

Sent with a box full of companions
from Mr Whittier and his sister May 6
1864—The spring she was ill
They came from Folly Mill he writes which I have mentioned in the lines
"And still the May Day flowers make sweet
The woods of Folly Mill

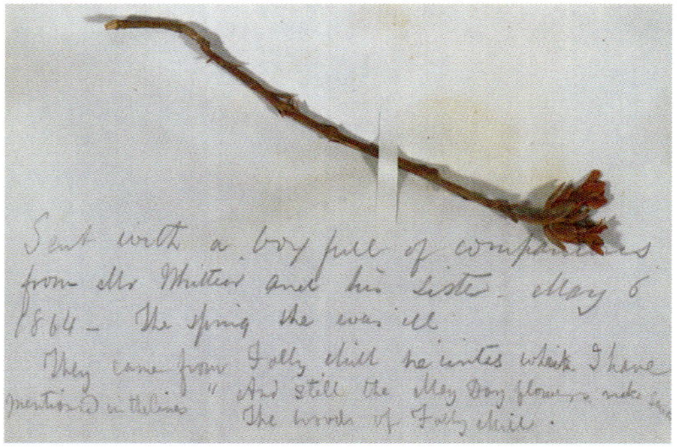

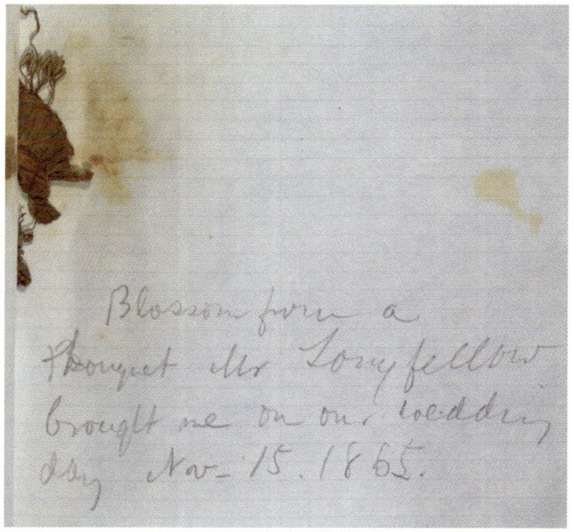

FIGURE 4.13. (*left*) September 1863 entry;
FIGURE 4.14. (*above top*) May 6, 1864 entry;
FIGURE 4.15. (*above*) November 15, 1865 entry. Fields, "Fragrant Memories," Fields Addenda, box 6, folder 1, The Huntington Library, San Marino, California

Blossom from a
bouquet Mr Longfellow
brought me on our wedding
day Nov. 15, 1865.

Fields identified another pressed flower, "from the dinner given to Longfellow on completion of the Dante Nov. 6th 1867," marking the occasion of the poet's finishing the first American translation of Dante's *Divine Comedy*. Fields did not attend the dinner, but she, along with Harriet Beecher Stowe and Lady Amberley, wife of the British ambassador, sent a wreath of flowers. Longfellow reciprocated by sending her flowers "from the table."[11] Fields's love of flowers was widely known, and on the diary pages in figures 4.13–4.15, the pressed specimens appear first, signaling their importance; the comments are written below in whatever space is left.

Fields befriended Stowe and Harriet Prescott, both of whom published with Ticknor and Fields, and other women writers whom James Fields featured in the *Atlantic Monthly*, including Thaxter, Lucy Larcom, and Rebecca Harding Davis. When Fields attended a Palm Sunday service with Stowe, she preserved a flower from the church.[12] Larcom sent her "May flowers" on one occasion and violets and anemones on another. Fields's friendships with women were an important part of her life. Biographer Rita Gollin notes that "for most of the women she befriended in the [1860s], Annie served as a confidante and a dispenser of social privilege. . . . Annie's rewards included their admiration, gratitude, and affections, vicarious entry into their lives, and knowing that she had served the deserving."[13] Prescott spent the night with the Fieldses in Boston after they all had attended Nathaniel Hawthorne's funeral in 1864 (figure 4.16). By then, Ticknor and Fields had published her short story collection *The Amber Gods* (1863) and her novel *Azarian: An Episode* (1864).

from Harriet Prescott.
She past the night here after
our return from Concord.
She said Julian seemed to her
like the expiated spirit,
the suffering one having [just?]
past before.

FIGURE 4.16. Undated entry, Fields, "Fragrant Memories," Fields Addenda, box 6, folder 1, The Huntington Library, San Marino, California

The preserved flower is likely lily of the valley, which might have grown in the Hawthornes' flower garden, and "Julian" is Hawthorne's son, who was a month shy of his eighteenth birthday when his father died.

FIGURE 4.17. July 10, 1862, entry, Fields, "Fragrant Memories," Fields Addenda, box 6, folder 1, The Huntington Library, San Marino, California

Both Annie and James Fields had close friendships with Nathaniel and Sophia Hawthorne, who appear several times in "Fragrant Memories." The couples visited one another regularly, either at the Hawthornes' in Concord, Massachusetts, or with the Fieldses in Boston (figure 4.17). Nathaniel's first three novels—*The Scarlet Letter* (1850), *House of the Seven Gables* (1851), and *The Blithedale Romance* (1852)—were all published by Ticknor and Fields.

> *Gathered while walking in the path worn*
> *by Mr. Hawthorne's foot-steps upon the hill back*
> *of his cottage "The Wayside" in Concord*
> *July 10th 1862*
>
> *As we walked we talked, and as we talked we*
> *mused, the beloved wife with J, upon the*
> *genius of the place and upon the holy aspirations*
> *permitted even to the least, nay, held out to*
> *us through all beauty as well as by the hand of*
> *awe-inspiring genius itself.*
> *White lilies were in bloom in the garden*
> *but better, those eternal white lilies of love which*
> *never fade out of the bosom of this family.*

"The Wayside" is a stately house along one of Concord's main streets and was the Alcott family's home from 1845 to 1852; Hawthorne bought the house in 1852 and spent the last four years of his life there. Fields noticed the white lilies in the Wayside garden and extended them into a metaphor of the Hawthornes' familial love.

The diary page that serves as this chapter's epigraph (figure 4.1), undated but likely from 1863, mentions both Julian and his sister Una, the Hawthornes' oldest child. The arrangement of the page—with specimen at the top, painted image in the middle, and note at

FIGURE 4.18. Detail, undated entry, Fields, "Fragrant Memories," Fields Addenda, box 6, folder 1, The Huntington Library, San Marino, Californiao, California

FIGURE 4.19. Detail, undated entry, Fields, "Fragrant Memories," Fields Addenda, box 6, folder 1, The Huntington Library, San Marino, California

the bottom—implicitly privileges the specimen over its representation, and both visual artifacts over the few written lines crammed at the bottom. The violet leaves and "heart's ease" are carefully mounted, using slits in the paper (see red arrows in figures 4.18 and 4.19). The "earliest of Julian's surviving artworks, a floral watercolor," also depicted a pansy; he painted it in 1859 in Rome, when the Hawthorne family was living abroad.[14]

When Nathaniel Hawthorne died, Fields wrote only a brief entry:

> *May 23. 1864*
>
> *How can we tell the sadness*
> *of this day. Hawthorne's body*
> *was buried.*

A note is tucked into that page:

> *Annie—I see why you could*
> *not find a lily for the*
> *wreath that must fade.*
> *I have told your husband*
> *that it is because you are*
> *my lily of the wreath—*
> *and you are a lily that*
> *will never fade!*

The inserted note is from Sophia Hawthorne, who called Fields "dearest Lily." In nineteenth-century America, lilies were a common funeral flower. When Sophia's note identifies Annie as "my lily of the wreath" and "a lily that will never fade," she not only references a typical mourning custom but also makes Fields the embodiment of the lily, which in nineteenth-century floriography signified purity, beauty, modesty, and sweetness—all traits that others associated with Fields.[15] This mention of lilies recalls Fields's remark two years earlier about how "white lilies were in bloom in the garden" at the Hawthornes' house (figure 4.17).

Fields's sparse commentary about Hawthorne's death provides an example of how "Fragrant Memories" operates as a parallel diary to her others. In the copybook that she labeled "Diary no. 8 Feb. 1864," there is an even briefer mention of Hawthorne's death, along with entries for other dates on the same page. The final line on the page reads simply "May 19 Hawthorne is dead." In the immediate aftermath of Hawthorne's passing, Fields seemed stunned into near silence. Years later, in *James T. Fields: Biographical Notes and Personal Sketches* (1881), she reflected that Hawthorne's "passing was like losing a portion of our own household, so closely interwoven had become the interest and affection of the two families."[16]

The other entries on the page deal with entirely different topics: At the top, in a note that continues from the previous page, Fields cites Oliver Wendell Holmes's comments about two "blackguards": the "whipper snappers of NY" Thomas Bailey Aldrich, "a stick of sugar-candy—and William Winter, who is not even a gum-drop." The middle entry, dated Sunday, May 14, begins, "Terrible days of war and change. Dr. Holmes came in in the afternoon to correct his Shakespeare poem. said he was ashamed to be troubled about so slight a thing when the battles were raging." Fields is likely referring to the Battle of Spotsylvania Court House, which occurred in Virginia from May 8 to 21, 1864, and was one of the bloodiest battles of the Civil War, with about eighteen thousand Union and eleven thousand Confederate casualties. In a single page, three distinctly different entries briefly address some literary gossip, "terrible days of war and change," and Hawthorne's death. Unlike in "Fragrant Memories," where all the entries are short, such brevity stands out in this diary. Fields often reflected on an event or conversation for several pages, but these terse comments suggest that war and death weighed heavily on the usually prolix diarist.

The Civil War

The Civil War had been underway for over a year when Fields first wrote in "Fragrant Memories" in May 1862. Her entries about the war are among the diary's most touching, although she rarely identifies these events by name. The first reference to the war is actually a newspaper clipping affixed to the top of a page, a memorial for Lieutenant James Jackson Lowell (figures 4.20 and 4.21). J. J. Lowell belonged to the Massachusetts Twentieth Regiment, nicknamed the Harvard Regiment for the number of its graduates who served together. He was wounded at the Battle of Glendale on June 30, 1862, and died there on July 4. In one entry, Fields noted the death of Willie Putnam—"he was killed at Ball's Bluff"—when she visited his mother eighteen months later, calling Mrs. Putnam "a noble martyr" whose faith and "self-renunciation" Fields deeply admired. On October 20, 1864, she wrote,

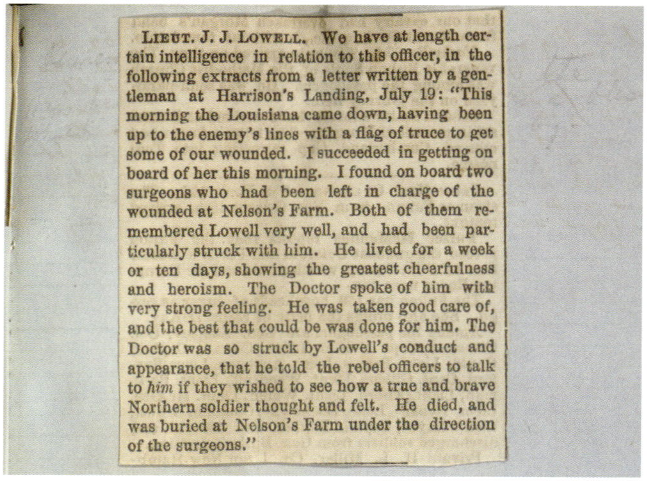

LIEUT. J. J. LOWELL. We have at length certain intelligence in relation to this officer, in the following extracts from a letter written by a gentleman at Harrison's Landing, July 19: "This morning the Louisiana came down, having been up to the enemy's lines with a flag of truce to get some of our wounded. I succeeded in getting on board of her this morning. I found on board two surgeons who had been left in charge of the wounded at Nelson's Farm. Both of them remembered Lowell very well, and had been particularly struck with him. He lived for a week or ten days, showing the greatest cheerfulness and heroism. The Doctor spoke of him with very strong feeling. He was taken good care of, and the best that could be was done for him. The Doctor was so struck by Lowell's conduct and appearance, that he told the rebel officers to talk to *him* if they wished to see how a true and brave Northern soldier thought and felt. He died, and was buried at Nelson's Farm under the direction of the surgeons."

FIGURE 4.20. Newspaper clipping, "Liet. J. J. Lowell," detail, undated entry, Fields, "Fragrant Memories," Fields Addenda, box 6, folder 1, The Huntington Library, San Marino, California

FIGURE 4.21. Photograph of J. J. Lowell, Collection of the Massachusetts Historical Society

"Colonel Lowell died, in the Victory at Cedar Creek." What Fields does not write anywhere is that Lieutenant J. J. Lowell and Colonel (Charles Russell) Lowell were brothers, and Lieutenant William Lowell Putnam was their cousin. All were nephews of the poet and diplomat James Russell Lowell, who became the first editor of the *Atlantic Monthly* in 1857 and was succeeded by James Fields in 1861. The deaths of these three young men remind us that even in the rarified environment of literary Boston that the Fieldses and Lowells inhabited, the war took its toll. There were conscription laws in both the Union and the Confederacy; one way to avoid serving—if you could afford it—was to purchase a substitute, someone who would fight in your place for a fee. But many more were volunteers, like the Lowells.

Fields was a staunch supporter of the Union and its troops. On October 9, 1862, she wrote, "All went to Readville this afternoon to see the camp" (Camp Meigs in nearby Readville, Massachusetts, where the Union army trained) and "We found our young soldiers enjoying their out of door life and the novelty of the camp, but behind all this we saw a spirit which said 'I am at work for my country and a great cause[.]' Little things cannot shake the mettle of those young men—It is inspiring to look at them even though the tears do start." Fields's response is both emotional ("the tears do start") and patriotic ("a great cause"). On January 1, 1863, she wrote: "Today the oppressed are made free—all nature rejoices and the star in the East has at last arisen to guide our suffering land! Hallelujah! Such music! Such joy! Let us embalm the memory of each hour." She did not identify it by name, but this was the day Lincoln issued the Emancipation Proclamation.

As "a resolute abolitionist" Fields marked similar occasions, such as "June 24, 1864: Slavery abolished in Maryland! We praise thee, O God!" In the following entry she enthusiastically championed the African American soldiers of the famed Massachusetts Fifty-Fourth regiment:

> *April 30—Fast Day. Wonderful sermon from*
> *Dr. Putnam at Mr. Hale's.*
> *Drove to Readville and saw the colored*
> *regiment now 850 men strong reviewed*
> *by Lt. Col. Hallowell. Took tea with*
> *Miss Baker. Lilac leaves half out.*

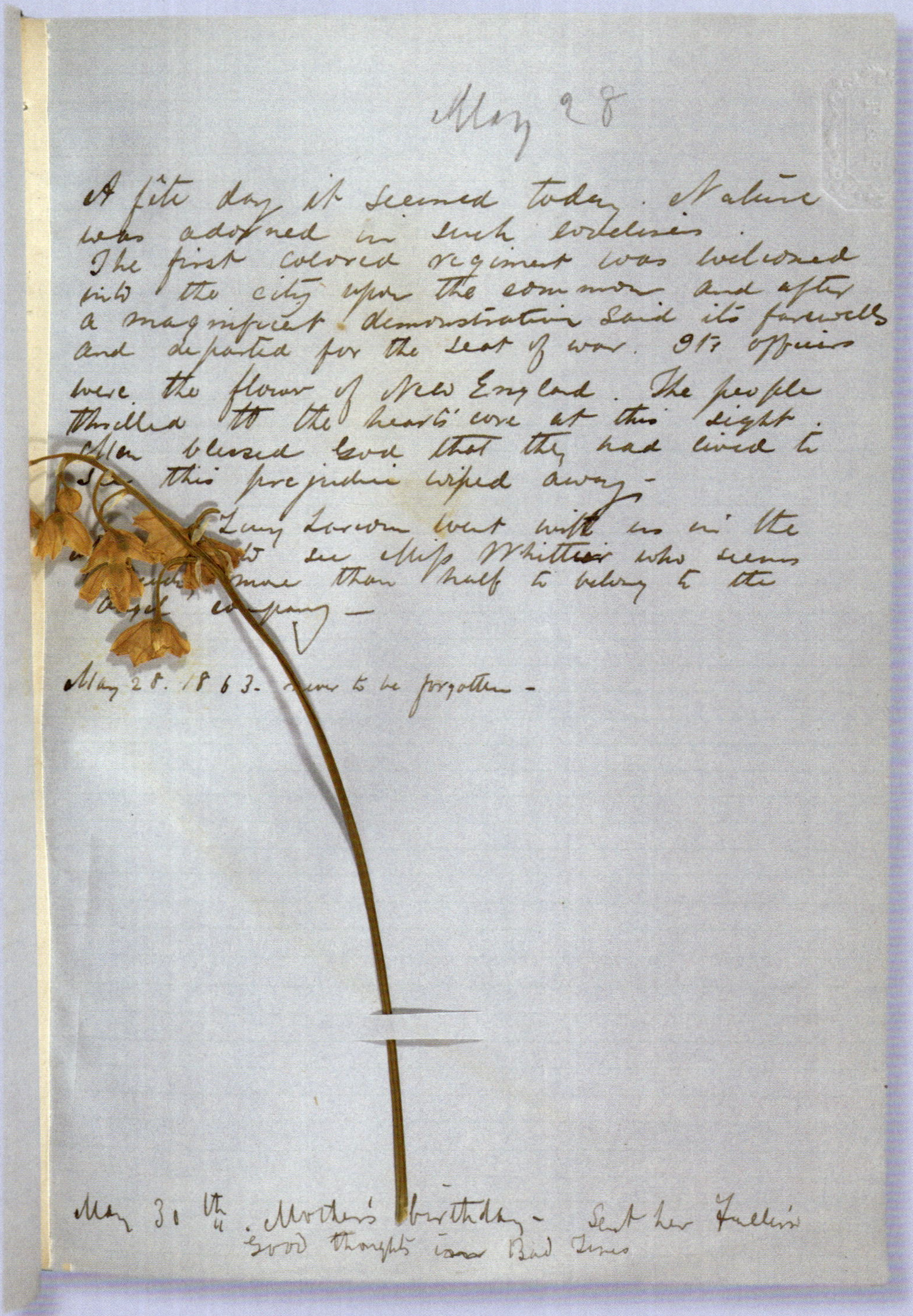

FIGURE 4.22. May 28, 1863, entry, Fields, "Fragrant Memories," Fields Addenda, box 6, folder 1, The Huntington Library, San Marino, California

April 30, 1863, was decreed a "day of national humiliation, fasting and prayer" by President Lincoln, who urged "all the People to abstain, on that day, from their ordinary secular pursuits, and to unite, at their several places of public worship and their respective homes, in keeping the day holy to the Lord."[17] After attending Edward Everett Hale's South Congregational Church to hear the Reverend George Putnam's sermon, Fields again went to the training camp at Readville, this time to see the first all–African American regiment—perhaps her way to continue "keeping the day holy."

A month later, she mentioned this regiment again in a longer entry (figure 4.22).

May 28

A fête day it seemed today. Nature
was adorned in such loveliness.
The first colored regiment was welcomed
into the city upon the common and after
a magnificent demonstration said its farewells
and departed for the seat of war. Its officers
were the flower of New England. The people
thrilled to the heart's core at this sight.
Men blessed God that they had lived to
see this prejudice wiped away.
Lucy Larcom went with us in the
afternoon to see Miss Whittier who seems
already more than half to belong to the
angel company—

May 28. 1863. never to be forgotten—

May 30th Mother's birthday. Sent her Fuller's
Good Thoughts in Bad Times.

When the Fifty-Fourth was welcomed at Boston Common on May 28, 1863, Fields joined a crowd of thousands. The *Boston Evening Transcript* called it "an event of momentous importance" and noted that "no single regiment has attracted larger crowds into the streets than the 54th."[18] Fields also praised the regiment's officers, calling them "the flower of New England," and she marked the day with a pressed flower, although she does not mention it. The lily of the valley sprig might have commemorated the regiment, related to the visit she and Lucy Larcom paid to Whittier's sister Elizabeth, or marked her mother's birthday—all of which are mentioned on that page. These two entries from the spring of 1863 combine events of the war and more mundane occurrences, as well as a botanical detail: "lilac leaves half out."

Writing about multiple topics on a single page—also seen when she noted Hawthorne's death in "Diary no. 8"—shows Fields's multiple interests and activities, and hints at how she might have used "Fragrant Memories." In figure 4.22, "May 28" is written in pencil at the top of the page; the remainder is in ink. The date appears again in the middle of the page, with the year and the phrase "never to be forgotten." The slit for the pressed flower is toward the bottom of the page; the staining in the top middle portion suggests that the flower originally lay higher on the page. Squeezed at the bottom of the page are two lines about her mother's birthday. She may have penciled in "May 28" sometime after, perhaps as a quick reference, as the inked date is buried lower on the page with its weighty phrase of remembrance, where it seems to conclude the entry about the fête day. But why write about her mother's birthday on that same page, given that it was two days later? There were many blank pages remaining in the diary, so Fields was not tight on space. The book she gave as a birthday present, *Good Thoughts in Bad Times*, published by Ticknor and Fields in 1863, was a collection of spiritual meditations by Thomas Fuller (1608–1661), "a chaplain in the army during the great civil war in England." The book's preface drew parallels between the English and American civil wars: It "is reprinted now in this country because there is much in it of a nature relevant to our own disturbed state."[19] Perhaps Fields was linking the two civil wars, or perhaps her mother's birthday was another kind of "fête day." Such connections may have been conscious or unconscious, or simply not connections at all. That Fields chose to place all this information on the same page does suggest they were somehow related.

She marked another important event of the war with even less commentary. On April 3, 1865, she entered only the date and underscored it twice, leaving the rest of the page blank (figure 4.23). A full-page entry dated April 3, 1865, in her parallel diary for "[18]63 and 65" began: "Forever memorable! Our armies entered Richmond." The capital of the Confederacy had finally fallen to Union troops, signaling the beginning of the end of the war. Robert E. Lee surrendered to Ulysses Grant six days later at Appomattox. Fields's war-related entries in "Fragrant Memories," though brief, still

FIGURE 4.23. April 3, 1865, entry, Fields, "Fragrant Memories," Fields Addenda, box 6, folder 1, The Huntington Library, San Marino, California

resonate powerfully: the grief of Willie Putnam's mother, pride in the all–African American regiment, joy at the Emancipation Proclamation and abolition of slavery in Maryland, and importance of a double-underscored date in large letters. Fields's experience "allows us to see how the Civil War was lived as part of everyday life," as Judith Ann Giesburg contextualizes the Civil War–era diary of Emilie Davis, a free Black woman in Philadelphia. "We generally study the war as a series of big events," Giesburg continues, "of victories and setbacks, turning points and proclamations, but it was lived as a periodic and oftentimes unwelcome visitor."[20] Fields's abolitionism and patriotism informed her experience of the war, as did her privileged status as a white woman of means in New England. She recorded events of significance to her, such as the deaths of the Lowells, and she responded to events significant to the nation, sometimes commemorating them with flowers.

Angel Voices: The Diary Thickens

Another curiosity of this diary are the tipped-in pages from *Angel Voices, or, Words of Counsel for Overcoming the World*, a pocket-sized book published by Ticknor and Fields (figures 4.24 and 4.25).[21] The *Angel Voices* pages appear sporadically in the blue copybook and are the same size (six inches tall by four inches wide); some remain, while others have been cut out (figure 4.26; see also figure 4.22). Fields almost never wrote on the *Angel Voices* pages; one exception is the page labeling the leaf from the *Re Galantuomo* (figure 4.12). More often than not, the *Angel Voices* pages that have been removed occur where there are pressed flowers, as seen in figures 4.26 and 4.22, suggesting that Fields may have used them as drying sheets—as herbarium instructions often

FIGURE 4.24. Cover, *Angel Voices* (1855)

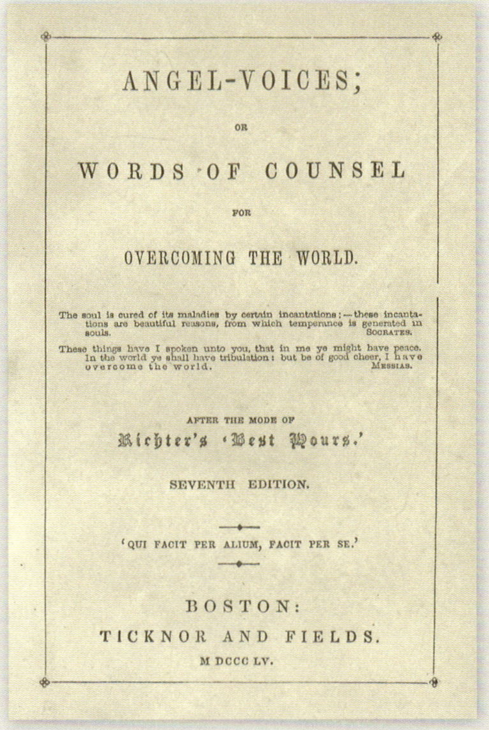

ANGEL-VOICES;

OR

WORDS OF COUNSEL

FOR

OVERCOMING THE WORLD.

The soul is cured of its maladies by certain incantations;—these incantations are beautiful reasons, from which temperance is generated in souls. SOCRATES.

These things have I spoken unto you, that in me ye might have peace. In the world ye shall have tribulation: but be of good cheer, I have overcome the world. MESSIAS.

AFTER THE MODE OF

Richter's 'Best Hours.'

SEVENTH EDITION.

'QUI FACIT PER ALIUM, FACIT PER SE.'

BOSTON:

TICKNOR AND FIELDS.

M DCCC LV.

FIGURE 4.25. Title page, *Angel Voices* (1855)

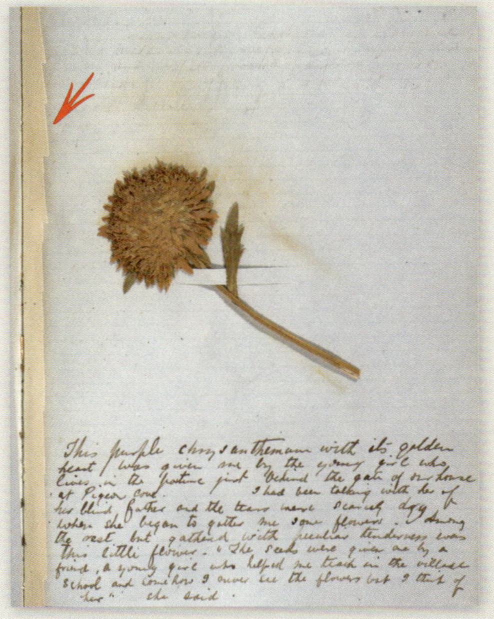

This purple chrysanthemum with its golden heart was given me by the young girl who lives in the pasture just behind the gate of our house at Pigeon Cove. I had been talking with her of her blind father and the tears were scarcely dry when she began to gather me some flowers. Among the rest but gathered with peculiar tenderness was this little flower. "The seeds were given me by a friend, a young girl who helped me teach in the village school and somehow I never see the flowers but I think of her" she said.

FIGURE 4.26. Example of pages from *Angel Voices* cutout, Fields, "Fragrant Memories," Fields Addenda, box 6, folder 1, The Huntington Library, San Marino, California recommended, one could use any old book for this purpose—and then removed them when the specimen was dry. A few *Angel Voices* pages remain intact, with the stain of a pressed flower on them and the actual specimen on the facing page of the diary (figure 4.27).

> *Sent by dear Mr Bartol from his*
> *garden May 17 1864*

The pages that have been set into the diary are thicker and stiffer than the pages in my copy of *Angel Voices*; they might have been the equivalent of page proofs, or perhaps from a deluxe edition, either of which Fields could have easily obtained from the publisher. My copy has a single pressed flower and is inscribed "For Anne from Lizzie March 31st /56"—not for Annie Fields from her sister Lissie, but tantalizingly close (figures 4.28 and 4.29).

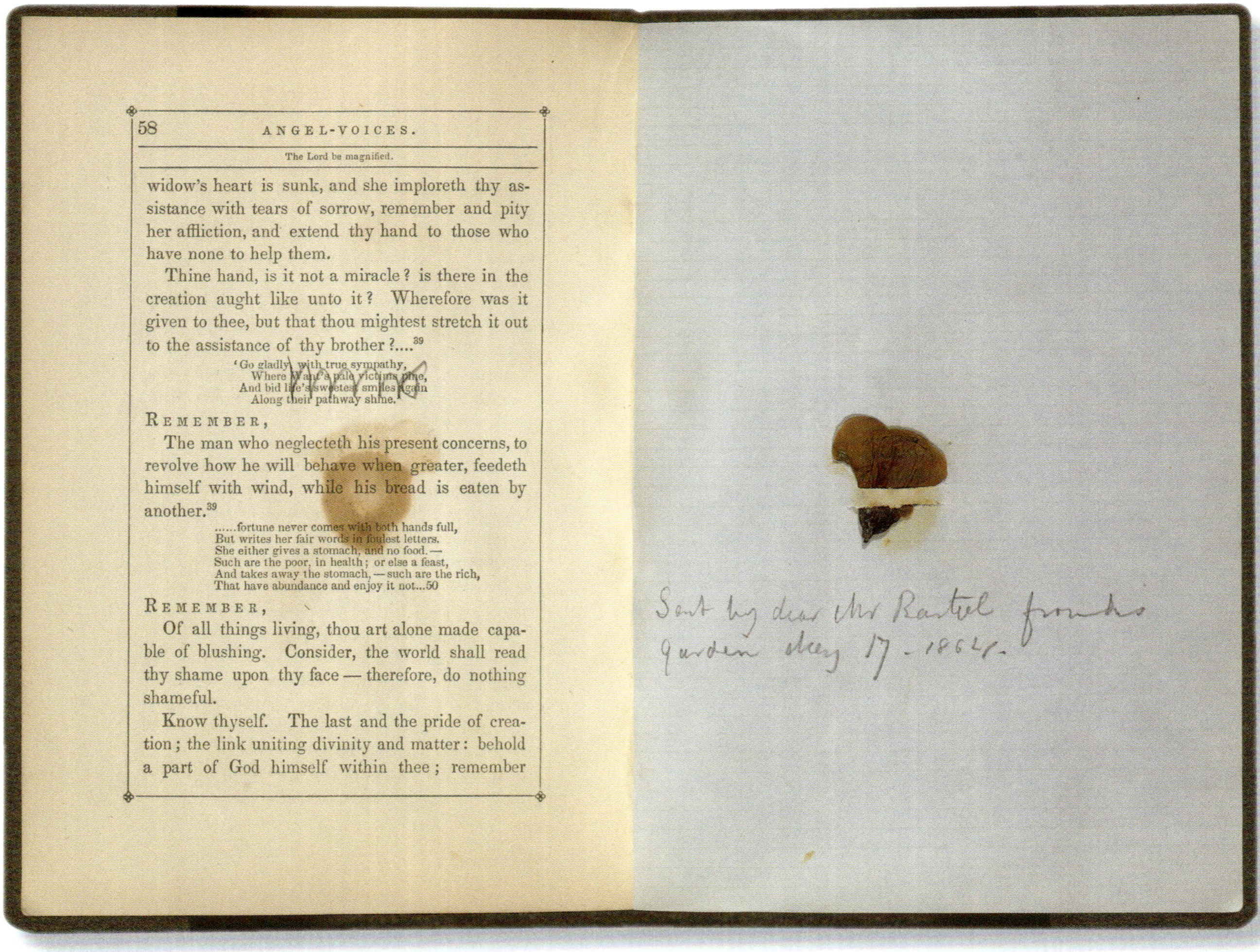

58 ANGEL-VOICES.

The Lord be magnified.

widow's heart is sunk, and she imploreth thy assistance with tears of sorrow, remember and pity her affliction, and extend thy hand to those who have none to help them.

Thine hand, is it not a miracle? is there in the creation aught like unto it? Wherefore was it given to thee, but that thou mightest stretch it out to the assistance of thy brother?....[39]

'Go gladly with true sympathy,
Where Want's pale victims pine,
And bid life's sweetest smiles again
Along their pathway shine.'

REMEMBER,

The man who neglecteth his present concerns, to revolve how he will behave when greater, feedeth himself with wind, while his bread is eaten by another.[39]

......fortune never comes with both hands full,
But writes her fair words in foulest letters.
She either gives a stomach, and no food.—
Such are the poor, in health; or else a feast,
And takes away the stomach,—such are the rich,
That have abundance and enjoy it not...50

REMEMBER,

Of all things living, thou art alone made capable of blushing. Consider, the world shall read thy shame upon thy face—therefore, do nothing shameful.

Know thyself. The last and the pride of creation; the link uniting divinity and matter: behold a part of God himself within thee; remember

Sent by dear Mr Bartol from his
garden May 17 - 1864.

FIGURE 4.27. Pressed flower, May 17, 1864, Fields, "Fragrant Memories," Fields Addenda, box 6, folder 1, The Huntington Library, San Marino, California

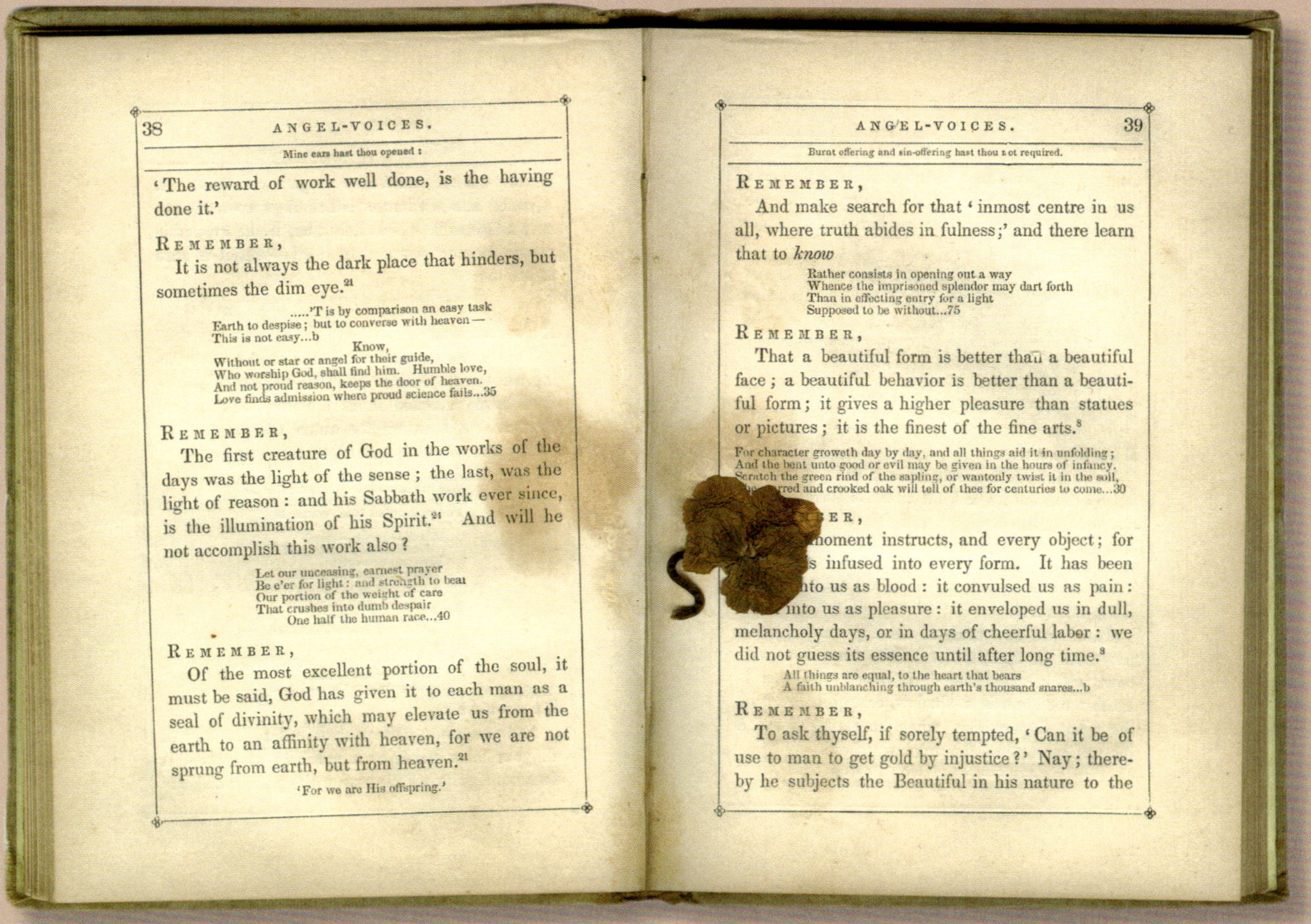

38 ANGEL-VOICES.

Mine ears hast thou opened :

'The reward of work well done, is the having done it.'

REMEMBER,

It is not always the dark place that hinders, but sometimes the dim eye.[21]

.....'T is by comparison an easy task
Earth to despise; but to converse with heaven —
This is not easy...b
Know,
Without or star or angel for their guide,
Who worship God, shall find him. Humble love,
And not proud reason, keeps the door of heaven.
Love finds admission where proud science fails...35

REMEMBER,

The first creature of God in the works of the days was the light of the sense; the last, was the light of reason: and his Sabbath work ever since, is the illumination of his Spirit.[21] And will he not accomplish this work also?

Let our unceasing, earnest prayer
Be e'er for light: and strength to bear
Our portion of the weight of care
That crushes into dumb despair
One half the human race...40

REMEMBER,

Of the most excellent portion of the soul, it must be said, God has given it to each man as a seal of divinity, which may elevate us from the earth to an affinity with heaven, for we are not sprung from earth, but from heaven.[21]

'For we are His offspring.'

ANGEL-VOICES. 39

Burnt offering and sin-offering hast thou not required.

REMEMBER,

And make search for that 'inmost centre in us all, where truth abides in fulness;' and there learn that to *know*

Rather consists in opening out a way
Whence the imprisoned splendor may dart forth
Than in effecting entry for a light
Supposed to be without...75

REMEMBER,

That a beautiful form is better than a beautiful face; a beautiful behavior is better than a beautiful form; it gives a higher pleasure than statues or pictures; it is the finest of the fine arts.[8]

For character groweth day by day, and all things aid it in unfolding;
And the bent unto good or evil may be given in the hours of infancy.
Scratch the green rind of the sapling, or wantonly twist it in the soil,
...rred and crooked oak will tell of thee for centuries to come...30

...ER,

...moment instructs, and every object; for ...s infused into every form. It has been ...nto us as blood: it convulsed us as pain: ...into us as pleasure: it enveloped us in dull, melancholy days, or in days of cheerful labor: we did not guess its essence until after long time.[8]

All things are equal, to the heart that bears
A faith unblanching through earth's thousand snares...b

REMEMBER,

To ask thyself, if sorely tempted, 'Can it be of use to man to get gold by injustice?' Nay; thereby he subjects the Beautiful in his nature to the

FIGURE 4.28. Pressed flower, *Angel Voices* (1855)

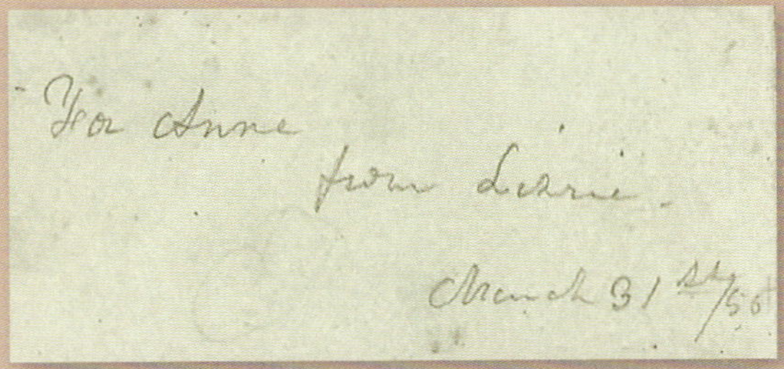

FIGURE 4.29. Inscription, *Angel Voices* (1855)

But how did the *Angel Voices* pages get into "Fragrant Memories" in the first place? The published book is 117 pages long, and the last of its pages in "Fragrant Memories" is numbered 92; even with the tipped-in pages that were cut out of the diary, it's unlikely that the entirety of *Angel Voices* was set into the copybook. This interleaving of printed pages from *Angel Voices* into the blank copybook is akin to a process called grangerizing, where additional pages are added to a book. While the process generally adds one print source to another, the example of "Fragrant Memories" can be similarly interpreted. The authors of *Interacting with Print* explain the process as "the practice of 'thickening' by pasting or 'tipping in' prints, autograph letters, printed pages, maps, original art, or other materials into printed books. To describe these processes simply as 'adding' misses the point." They argue that books thickened this way are "dialogic, reflexive, and interactive" and "function in ways akin to an individual reader's marginalia: inserted materials variously reiterate, elucidate, emend [and] recast . . . the existing text."[22] If one purpose of "Fragrant Memories" was to preserve flowers, then the grangerized pages from *Angel Voices* reiterated its analogy to an herbarium by providing drying sheets, while also extending it to the dialogic and interactive form of a parallel diary.

Perhaps Fields herself constructed the smaller copybook with pages from *Angel Voices*, knowing these would be handy for absorbing moisture from flowers she intended to preserve. Or maybe she designed the concept and got someone at Ticknor and Fields to make it for her. But why choose *Angel Voices* out of all the books the publisher offered in this standard size? The "Words of Counsel" it provides might have been a comfort, perhaps like Fuller's *Good Thoughts in Bad Times*, the book she gave her mother. Only once does she respond in words to *Angel Voices* in the diary's extant pages (figure 4.30). The *Angel Voices* page ends, "What is the source of sadness, but feebleness of" and the marginalia reads, "the soul? Rouse thyself to the combat, and she quitteth the field before thou striketh." Fields has continued the printed text in her own hand, writing what appears at the top of the next page of *Angel Voices*. The passage might have been meaningful to her, with its exhortation to "rouse thyself to the combat"—inspiring words for an abolitionist in the 1860s. Her direct interaction with this printed page suggests the choice of *Angel Voices* for interleaving in her diary could have been deliberate.

From a material culture perspective, the "thickening" of Fields's diary occurs in several ways. In addition to the interleaved pages from *Angel Voices*, many pressed specimens thicken the copybook quite considerably (figure 4.4), as do a few newspaper clippings, the note

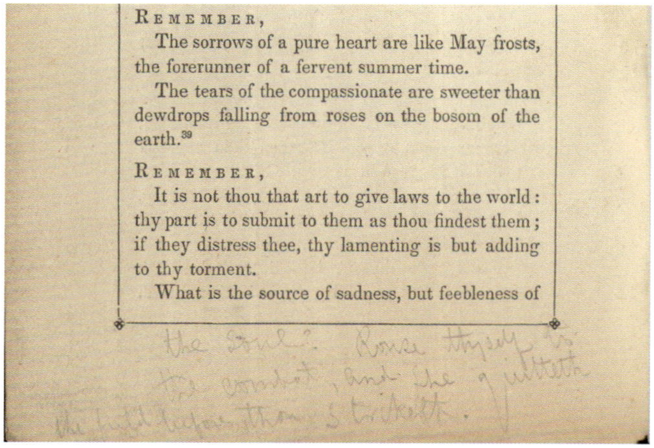
REMEMBER,
The sorrows of a pure heart are like May frosts, the forerunner of a fervent summer time.
The tears of the compassionate are sweeter than dewdrops falling from roses on the bosom of the earth.[39]

REMEMBER,
It is not thou that art to give laws to the world: thy part is to submit to them as thou findest them; if they distress thee, thy lamenting is but adding to thy torment.
What is the source of sadness, but feebleness of

FIGURE 4.30. Marginalia, *Angel Voices*, page 60, Fields, "Fragrant Memories," Fields Addenda, box 6, folder 1, The Huntington Library, San Marino, California from Sophia Hawthorne calling Fields "a lily that will never fade," and additional loose items that may or may not have been placed there by Fields herself, including concert programs, locks of hair, and a few manuscript pages. The "dialogic, reflexive, and interactive" aspects of thickening provide other interpretive possibilities. "Fragrant Memories" appears to be a combination of entries written both immediately and retrospectively; the dialogue between these perspectives also thickens its significance. Fields's interactions with other media in the diary, both print and nonprint materials—and those texts' own voices—contribute to the diary's polyvocality, a trait common to the flower book genre. So, too, does the interplay of momentous and mundane events, small flowers and a big war, and the "interleaving" of printed pages and pressed specimen leaves that fill the diary.

This small diary has been largely overlooked by scholars. The Huntington Library labels it a "scrapbook containing some diary entries and many pressed flowers and leaves with notes," even though Fields herself called it a diary. Fields's biographers Judith A. Roman and Rita K. Gollin mention it only briefly and do not discuss its parallels to Fields's other diaries, nor the significance of its many pressed flowers. "Fragrant Memories" is a fascinating piece of nineteenth-century American popular botany. As a historical document, it responds to events of the Civil War and provides glimpses into the lives of some of the nineteenth century's major literary figures. As a material artifact, it shows that the process of collecting, saving, preserving, and cherishing flowers extended beyond the herbarium and into a textual environment as intimate and intricate as the flowers themselves. Although the start date of "Fragrant Memories" marks it as one of Fields's earliest known diaries, she framed it—perhaps much later—with a sense of time already passed, not passing. One of the first pages is dated "May, 1862," in ink, with no other commentary, but the penciled inscription below it seems retrospective: "begun a little too late in the summer of life." This dialogue of pencil and ink is one kind of material text to decode; the many flowers she pressed are another: gift, souvenir, commemoration, and memorial. In the next chapter, the language of flowers opens up an even larger array of decoding possibilities, in books that celebrate the bounty and variety of American flora.

FIGURE 5.1. Osgood, *The Flower Alphabet, in Gold and Colors* (1845)

5

Flower Language Dictionaries

Cataloging American Abundance

I hope the endeavor to select, and incorporate with our love of nature and flowers, some of the finest specimens of American poetry, breathing of the affections in their purity, tenderness, triumph, or desolation, will be acceptable to our community. [. . .] And it is not without pride as well as pleasure that I have found so many fair specimens of this kind flourishing in a land that has been stigmatized as producing nothing but corn and cotton, the tobacco and potatoe. If we shelter and cherish our flowers they will soon beautify our Republic.

—Sarah Josepha Hale, *Flora's Interpreter, or, the American Book of Flowers and Sentiments* (1832)

Flowers are types of our brightest hopes, they are emblems of joy, they have been even called "the alphabet of angels." And the harmony of their colors, the variety of their forms, the profusion with which they are scattered over every solitary place, make us consider them as fragrant gems of the earth, beautiful ministers of winged and spiritual thoughts.

—Lucy Hooper, *The Lady's Book of Flowers and Poetry* (1842)

The richly varied and magnificent flora of our American continent offers our sons and daughters a floral vocabulary capable of almost unlimited application.

—*Flowers, Their Language, Poetry, and Sentiment* (1870)

FIGURE 5.2. Dinnies, *The Floral Year* (1847), Bouquet 9

BOUQUET FOR SEPTEMBER, COMPOSED OF THE SUNFLOWER—ZINNIA—LARKSPUR—COLTSFOOT—SWEET BRIAR—AND BACHELOR'S BUTTON.

SIGNIFICATIONS:—*Lofty and Pure Thoughts—Absent Friends—Conscious Attachment—Maternal Love—Poetry—I with the morning's love have oft made sport.*

The language of flowers, also known as floriography, is a system of symbolism that pairs a plant with an "emblem": an emotion, concept, or personality trait. In nineteenth-century America, flower language books proliferated in a variety of styles, topics, and purposes. In Catharine Waterman's *Flora's Lexicon* (figure 5.3), the live oak stands for liberty, uniting a native tree with one of America's highest ideals.[1] Lucy Hooper, in *The Lady's Book of Flowers and Poetry,* celebrates the many fine qualities of flowers: their ability to convey meaning as "emblems of joy," their material "variety" and "profusion," even their spiritual qualities.[2] The poet Frances Sargent Osgood created an alphabet book for children in rhymed couplets based on the language of flowers (figure 5.1), literally from

A—brought the loved acacia flower
that round him played a sunlit shower

to

Z—breathed low a last farewell
while at her feet a zinnia fell.[3]

The "eloquent silence" of the forget-me-not shows that flowers signify not through sound or word but through physical features such as the size and color of this "little blue" flower. Anna Peyre Dinnies, in *The Floral Year* (figure 5.2), composed twelve monthly bouquets, provided lush illustrations of each, and detailed the "significations" of flowers from a wide variety of American flora.

Sarah Josepha Hale twice uses the term "specimens" for the American poems she selected for *Flora's Interpreter*, implicitly associating them with botanical specimens. American poetry is "flourishing"—more plant terminology, with a floral etymology—even in a nation "stigmatized as producing nothing but corn and cotton, the tobacco and potatoe." Hale counters the critics who denigrate the solely utilitarian production of these New World native crops with the loftier aim to "shelter and cherish our flowers." The anonymous author of *Flowers, Their Language, Poetry, and Sentiment* (1870) boasts that "the richly varied and magnificent flora of our American continent" are "capable of almost unlimited application," nearly four decades after Hale articulated her hopes for beautifying the republic.

Waterman's entry for live oak (figure 5.3) encapsulates the content of many nineteenth-century U.S. flower language dictionaries, starting with the species' Latin binomial (*Quercus virens*) and Linnaean classification ("Class 21, Monœcia. Order: Polyandria"), and continuing with its native habitat "from 37° [north latitude] to Florida, and westward to the Sabine river," size, and use as "the finest material we have for ship-building." An assertion of American pride follows—"no way inferior to the European species"—as well as a cautionary note about the tree's possible fate: "In consequence of its narrow limits and the more profitable culture of Cotton in the districts where it abounds, its total extinction is considered certain at no distant day." The entry's combination of botanical specifics, physical description, habitat, uses, and relevant ecological issues is not only typical of other American flower language books; it also reflects the place-based emphasis that runs throughout this "lost subgenre of American nature writing."[4]

This chapter analyzes how flower language books from the 1820s to the 1880s catalogued the significance, beauty, and variety of American flowers. Whether in their briefest form as a glossary of species and emblems, or their longer dictionary and anthology formats, they promoted everyday Americans' appreciation of native and cultivated plant species in a genre that is also unabashedly literary and sentimental. Flower language books foreground affective experiences of beauty and art in their poetry, illustrations, and bindings; they emphasize intellectual learning by including botany instruction and scientific information; and, above all, they showcase the abundance of America. It might seem paradoxical to claim abundance as an environmental ideal: The nineteenth century was also a time of growing awareness that American natural resources were finite. Even when they acknowledged possibilities of "total extinction," as Waterman's did, flower language books simultaneously enacted a different form of environmental awareness, one based on aesthetic appreciation of profusion. Herbaria preserved flowers in their

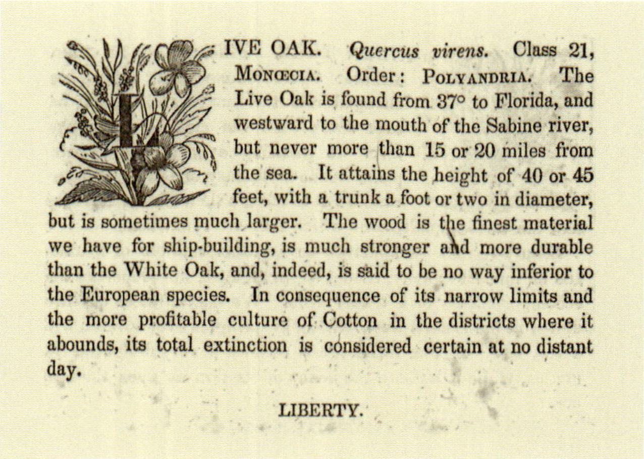

LIVE OAK. *Quercus virens.* Class 21, Monœcia. Order: Polyandria. The Live Oak is found from 37° to Florida, and westward to the mouth of the Sabine river, but never more than 15 or 20 miles from the sea. It attains the height of 40 or 45 feet, with a trunk a foot or two in diameter, but is sometimes much larger. The wood is the finest material we have for ship-building, is much stronger and more durable than the White Oak, and, indeed, is said to be no way inferior to the European species. In consequence of its narrow limits and the more profitable culture of Cotton in the districts where it abounds, its total extinction is considered certain at no distant day.

LIBERTY.

FIGURE 5.3. Waterman, *Flora's Lexicon* (1840)

material form; floriography gave them an immortality of meaning and symbolism. The systematic botany of instruction books disciplined plants into taxonomic hierarchies of classes, orders, genera, and species; flower language dictionaries arranged them in bountiful bouquets and sang their praises in poems. Because their copiousness was representational and informational, and the resources they extracted were figurative and literary, the abundance they celebrated countered the prevailing forces of commodification and utilitarianism that were leading inexorably to the depletion of American flora.

Introducing Flower Language Books

Nineteenth-century authors of flower language books traced the phenomenon's origin to cultures of the Near East, where it was allegedly a form of communication between harem women and their would-be suitors on the outside, or a form of amusement among harem women themselves. In nineteenth-century Europe, the earliest flower books were published in France: B. Delachénaye's *Abécédaire de Flore ou Langage des Fleurs* (1811) and Charlotte de Latour's *Le Langage de Fleurs* (1819). These were soon followed in England by Henry Phillips's *Floral Emblems* (1825) and Frederic Shoberl's adaptation of Latour, *The Language of Flowers; With Illustrative Poetry* (1834).[5]

Many nineteenth-century flower language books are organized as dictionaries, where one could look up a particular plant and find its corresponding emblem, or vice versa. The connection between a species and its trait was largely arbitrary: Originally, Persian poets used words that rhymed with particular flowers as their emblems. Later books in France and England

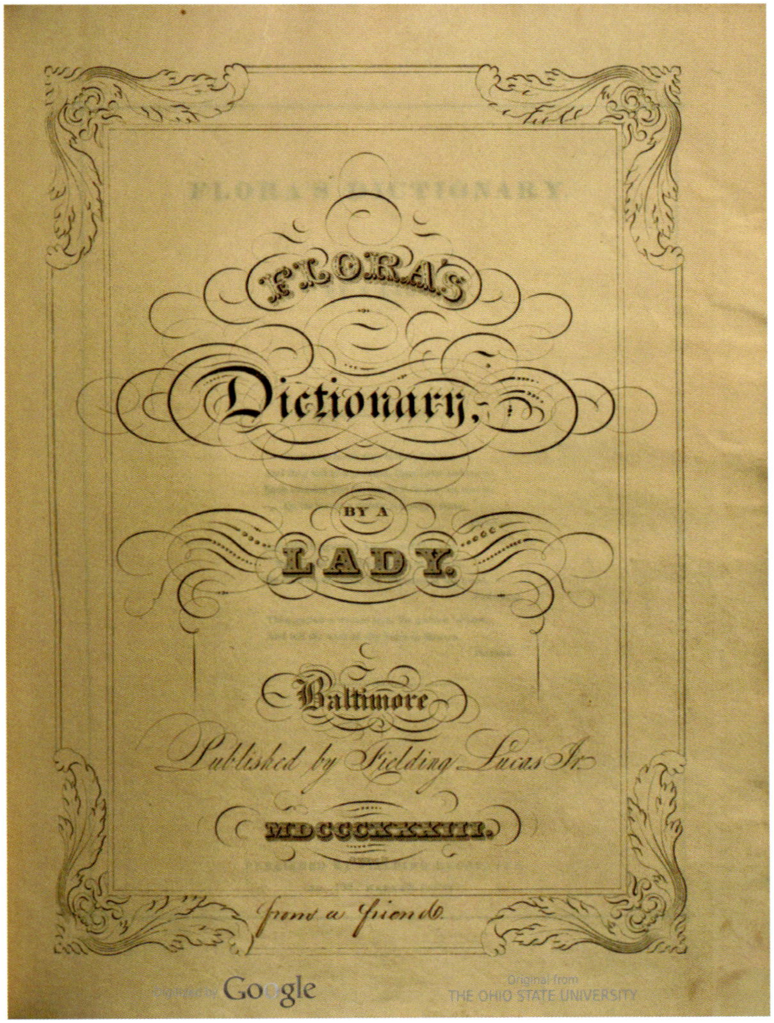

FIGURE 5.4. Title page, Wirt, *Flora's Dictionary* (1832), courtesy of HathiTrust

sometimes based the correspondences on etymology (for example, pansies, like *pensées*, stand for thoughts) or botanical characteristics such as a plant's form, color, scent, habitat, or time of blooming. In the American *Flora's Dictionary* (first published in 1829; figures 5.4 and 5.5), Elizabeth Wirt explains her selection process

FIGURE 5.5. Title page, Wirt, *Flora's Dictionary* (ca. 1837)

for emblems: "In general, they have been suggested either by some allusion to the specific flower in British poets, or by its botanical, or its popular name, or by some property peculiar to it, such as its hue, form, odour, place and manner of growth, sensibility, medicinal virtue, or some other circumstance connected with its history or character."[6] There is no single, stable language of flowers: Consult five different books, and you are likely to find at least three distinct meanings for the same species (see table 5.1). This inconsistency drives floriography's staying power: Each new book on the market could assert the relevance of its emblems, perhaps by appealing to past sources, or by conferring a trait suggested by the flower's physical features, or by whatever other method the author could devise.

American flower language dictionaries differed not only in their emblems but also in their organization and content, as seen in a comparison of five of the most popular books from the 1830s to 1850s (table 5.1).[7] For the same species, aloe, there are four different emblems—religious superstition, grief, imagination, and acute sorrow. Hale and Waterman include aloe's classification according to Linnaeus's artificial system (figures 5.6 and 5.9). Waterman and Osgood borrow from French explorer and naturalist François Le Vaillant, as does Henrietta Dumont, without attribution (figures 5.8–5.10). Their entries convey aloe's physiology, habitat, and growth characteristics—and perpetuate Le Vaillant's stereotypes of "barbarous Africa" and misinformation about its nonexistent tigers. All five

TABLE 5.1. Comparison of Hale, Waterman, Hooper, Osgood, and Dumont

AUTHOR, TITLE (Date of First Edition)	ORGANIZATION OF BOOK	EMBLEM FOR ALOE	SOURCES CITED IN ENTRY
Hale, *Flora's Interpreter* (1832)	alphabetical by species, one per page	religious superstition	*Flora's Dictionary*, James A. Hillhouse
Waterman, *Flora's Lexicon* (1839)	alphabetical by species, one per page	grief	Shakespeare
Hooper, *The Lady's Book of Flowers and Poetry* (1841)	by species but not alphabetical, continuous	acute sorrow	Mark Akenside, Shakespeare, *A Winter's Tale*
Osgood, *The Poetry of Flowers and Flowers of Poetry* (1841)	alphabetical by emblem, continuous	imagination	F. S. O. [Osgood herself]
Dumont, *The Floral Offering* (1851)	by species but not alphabetical, continuous	grief	R. M. Milnes, Anon., Mrs. Hale

include excerpts from both English and American literature, only one of which is repeated, a few lines from Shakespeare's *A Winter's Tale*. As collections of botanical science, natural history, literature, folklore, and more, flower language dictionaries are true anthologies: from the Greek ανθολογία (*anthos* + *logia*), "flower collection." One way these books distinguish themselves

16 FLORA'S INTERPRETER.

ALOE. *Class* 6. *Order* 1. Native of the Cape of Good Hope, Egypt, etc. The flower of the Aloe has no calyx. A bitter and medicinal juice is extracted from the leaves.
Aloe.

RELIGIOUS SUPERSTITION.

In climes beneath the solar ray,
Where beams intolerable day,
And arid plains in silence spread,
The pale green *Aloe* lifts its head—
The mystic branch at Moslem's door
Betokens travel long and sore
In Mecca's weary pilgrimage.
Flora's Dictionary.

SENTIMENT.

All tenderness you seemed,
Gentle and social as a playful child;
But now in lonely *superstition* wrapped,
As on an icy mountain-top thou sittest
Lonely and unapproachable, or tossed
Upon the surge of passion, like the wreck
Of some proud Tyrian in the stormy sea.
Hillhouse.

FIGURE 5.6. "Aloe," page 16, Hale, *Flora's Interpreter, and Fortuna Flora* (1850)

THE ALOE.

THE aloe is made the emblem of acute sorrow, on account of its painful bitterness. The bitter of the aloe affects the body, that of affliction reaches the soul.

SORROW that locks up the struggling heart.
AKENSIDE.

IF you do sorrow at my grief in love
By giving love your sorrow and my grief
Were both extermined.
SHAKSPEARE.

BESIDES, you know,
Prosperity's the very bond of love;
Whose fresh complexion, and whose heart together,
Affliction alters.
WINTER'S TALE.

(155)

FIGURE 5.7. "Aloe," page 155, Hooper, *The Lady's Book of Flowers and Poetry* (1858)

IMAGINATION.

ALOE.

THE aloe is said to thrive best in the desert, and is only attached to the soil by a very slender fibre. Its taste is very sharp and bitter. This plant derives its support almost entirely from the air, and assumes very singular and fantastic shapes. Le Vaillant found many species very numerous in the deserts of Namaquoise; some of them six feet long, which were thick and armed with long spines. From the centre of these a light twig shoots forth to the height of a tall tree all garnished with flowers. Others exalt themselves like the cactus, bristling with thorns. Others again are marbled, and seem like serpents creeping upon the earth.

Brydone saw the ancient city of Syracuse entirely covered by great aloes in flower; their elegant branches giving to the promontory which bounded the coast, the appearance of an enchanted forest.

These magnificent and monstrous members of the vegetable kingdom are also found in barbarous Africa. There they grow upon the rocks in arid and sandy soil, in the midst of that burning atmosphere in which scarce aught but tigers and lions can breathe alive.

For rich Imagination's jewelled wand,
With living forms can fill the lonely hall;
With glorious bloom, enwreath the desert-sand,
And crown again, with sculpture's grace, the ruined temple's wall!
F. S. O.

FIGURE 5.8. "Aloe," pages 117–118, Osgood, *The Poetry of Flowers and Flowers of Poetry* (1841)

from one another is with their literary extracts, chosen to align with the "sentiment" or trait. Osgood quotes her own poem "Imagination," while Hooper's excerpts align with "sorrow" and its "affliction" (figures 5.8 and 5.7). They also support one another's work: Hale quotes from Wirt's *Flora's Dictionary,* and Dumont cites a verse by Hale.

LOE. Class 6, HEXANDRIA. Order: MONOGYNIA. The aloe is said to thrive best in the desert, and is only attached to the soil by a very slender fibre. Its taste is very sharp and bitter. So sorrow drives us away from the world, detaches our hearts from the earth, and fills them with bitterness. This plant derives its support almost entirely from the air, and assumes very singular and fantastic shapes. Le Vaillant found many species very numerous in the deserts of Namaquoise; some of them six feet long, which were thick and armed with long spines. From the centre of these a light twig shoots forth to the height of a tall tree, all garnished with flowers. Others exalt themselves like the cactus, bristling with thorns. Others, again, are marbled, and seem like serpents creeping upon the earth. Brydone saw the ancient city of Syracuse entirely covered by great aloes in flower; their elegant branches giving to the promontory which bounded the coast, the appearance of an enchanted forest. These plants also prosper well in our gardens. The collection in the museum of Paris is said to be the most complete in the world.

These magnificent and monstrous members of the vegetable kingdom are also found in barbarous Africa. There they grow upon the rocks in arid and sandy soil, in the midst of that burning atmosphere in which scarce aught but tigers and lions can breathe and live. Let us bless Providence, then, for raising in our climate verdant bowers over our heads, and for spreading under our feet the soft carpet of grass, ornamented with saffron, violets, and daisies.

GRIEF.

Besides, you know,
Prosperity's the very bond of love;
Whose fresh complexion, and whose heart together,
Affliction alters.

SHAKSPEARE.

FIGURE 5.9. "Aloe," page 17, Waterman, *Flora's Lexicon* (1840)

ALOE.... *Grief.*

THE Aloe is attached to the soil by very feeble roots; it delights to grow in the wilderness, and its taste is extremely bitter. Thus grief separates us from earthly things, and fills the heart with bitterness. These magnificent and monstrous plants are found in barbarous Africa: they grow upon rocks, in dry sand under a burning atmosphere. Some have leaves six feet long, and armed with long spires. From the centre of these leaves shoots up a slender stem covered with flowers.

Sister Sorrow! sit beside me,
Or, if I must wander, guide me:
Let me take thy hand in mine,
Cold alike are mine and thine.
Think not, Sorrow, that I hate thee,—
Think not I am frightened at thee,—
Thou art come for some good end;
I will treat thee as a friend.

R. M. Milnes.

And this is all I have left now,
Silence and solitude and tears;
The memory of a broken vow,
My blighted hopes, my wasted years!

Anon.

It may be that I shall forget my grief;
It may be time has good in store for me;
It may be that my heart will find relief
From sources now unknown. Futurity
May bear within its folds some hidden spring
From which will issue blessed streams; and yet
Whate'er of joy the coming year may bring,
The past—the past—I never can forget.

Mrs. Hale.

FIGURE 5.10. "Aloe," pages 28–29, Dumont, *The Floral Offering* (1853)

American flower language dictionaries came in sizes from miniatures and pocket-sized books to the octavos featured in table 5.1, and even larger formats intended for display (figures 5.11 and 5.12). They were beautifully bound and many included full-page color illustrations. They were also part of a larger trend of gift books that burgeoned in the nineteenth-century United States. As Kathryn K. Shinn explains, "More and more publishers emerged to meet the growing demand for literature in the nineteenth century. During a time when literary publication grew faster than the population, the circulation of newspapers and magazines increased significantly, but they could not meet the demands for pleasure reading fully. In the 1820s, the gift book emerged as one of the most popular kinds of books available." Gift books met this demand for literature and pleasure reading in their compilations of poetry, short fiction, and essays. Also known as literary annuals, many of them featured floral titles, reinforcing the gift value of flowers: *Aloe: A Gift for All Seasons*, *Amaranth: A Literary and Religious Offering*, *Crocus: A Fresh Flower for the Holidays*, *Dahlia or Memorial of Affection*, and *Forget Me Not*—not to mention *Primrose*, *Rosebud*, *Violet*, and *Woodbine*. Alexandra Urakova argues that

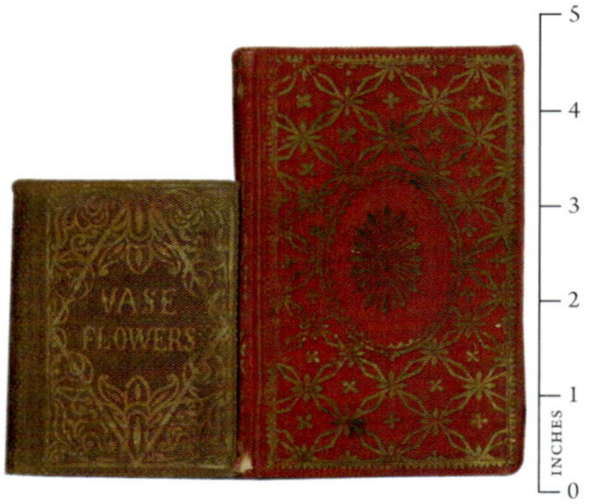

FIGURE 5.11. Miniature and pocket-size language books

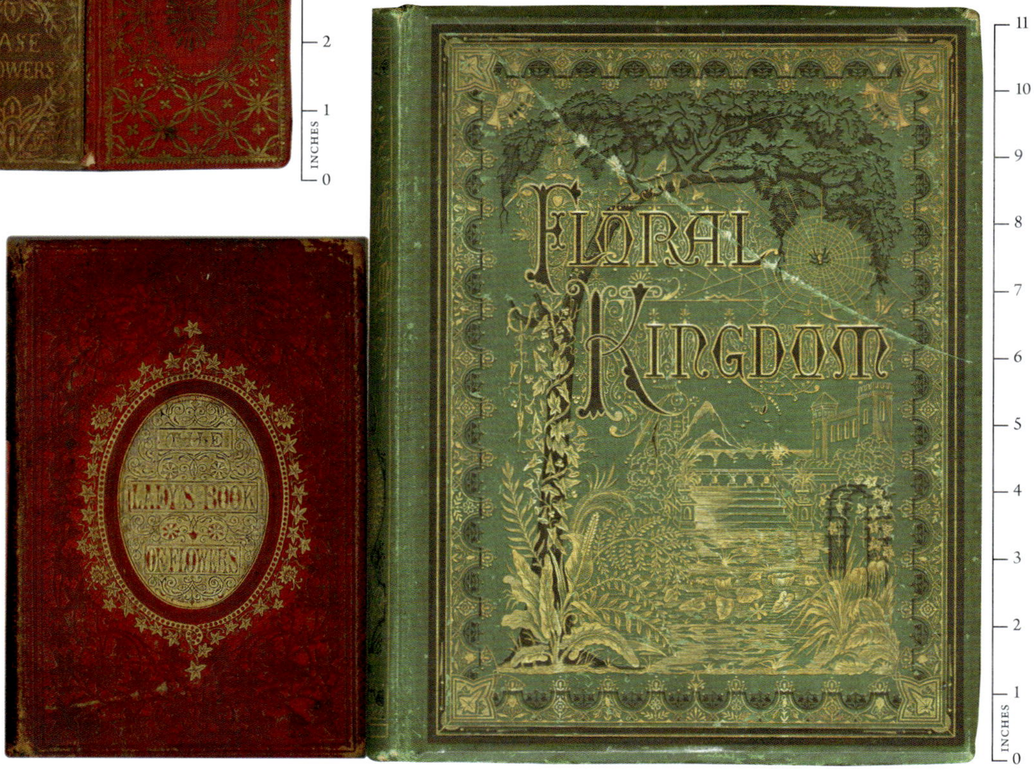

FIGURE 5.12. Octavo and larger-format flower language books

gift books occupied an "intermediary or transitional position between different textual forms: periodical, book, anthology, friendship album, and almanac"; as such, they featured both new compositions and reprinted material, as did flower language books. The publisher Samuel G. Goodrich's description of gift books as "messengers of love, tokens of friendship, signs and symbols of affection, and luxury and refinement" directly parallels the way flower language dictionary authors such as Hale and Hooper characterized flowers, as seen in this chapter's epigraphs.[8]

The titles of flower language books reveal their various purposes and contents. Some are explicitly dictionaries: *Flora's Dictionary*, *Flora's Lexicon*, *Lexicon of Ladies' Names with their Floral Emblems*, and *Flora's Pocket Dictionary*. Others denote the ways that flowers can be assembled, contained, or displayed: *The Garland of Flora*, *The Flower Vase*, *The Bouquet*, *Flora's Album*, *The Floral Wreath*, *Vase of Flowers*, and *The Rural Wreath*. Many of them highlight the poetry they contain: *The Lady's Book of Flowers and Poetry*; *The Poetry of Flowers and Flowers of Poetry*; *The Language of Flowers, Poetically Expressed*; *Songs of the Flowers*; and *Flowers, Their Language, Poetry, and Sentiment*. One could make a present of *The Floral Offering* or *The Floral Gift, from Nature and the Heart*, or keep track of the passing days with *Flora's Dial* or *The Floral Year*. Hale's *Flora's Interpreter* foregrounds floriography as a language to be decoded.[9]

While there is little evidence that people actually used these books to construct and send meaning-laden floral missives, the flower language book was a popular genre, with nearly one hundred different works published in the United States throughout the nineteenth century.[10] As seen in other genres of flower books, inscriptions can suggest how individuals interacted with their flower language dictionaries. As gift books, they were presented for birthdays and holidays, to cherished friends, by teachers as prizes to students, and for more esoteric reasons, as in this tiny inscription in J. S. Adams's pocket-sized *The Language of Flowers* (figure 5.13):

To S. A. Cummings
For freezing to death.

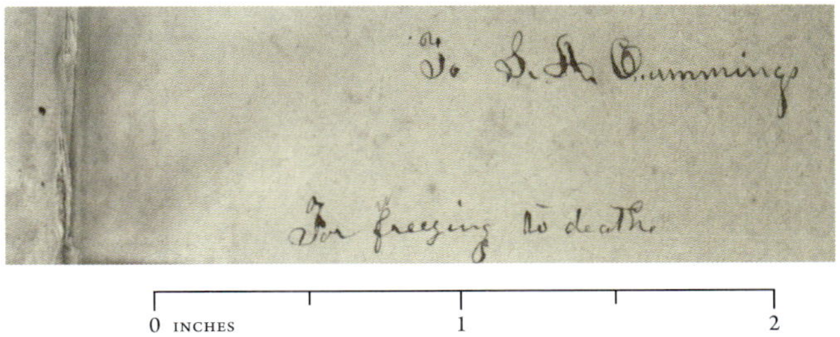

FIGURE 5.13. Inscription, J. Adams, *The Language of Flowers, Poetically Expressed* (1852)

Another curious inscription appears in *The Bouquet: Containing the Poetry and Language of Flowers* (1845): "Presented to Miss Mary H. Morrill By her friend Wm W. Heart as a Philopene present."[11] John Russell Bartlett's *Dictionary of Americanisms* (1848) provides this definition for "fillipeen": "There is a custom common in the Northern States at dinner or evening parties when almonds or other nuts are eaten, to reserve such as are double or contain two kernels, which are called fillipeens. If found by a lady, she gives one of the kernels to a gentleman, when both eat their respective kernels. When the parties again meet, each strives to be the first to exclaim, *Fillipeen*! for by so doing he or she is entitled to a present from the other" (italics in the original).[12]

FIGURE 5.14. Cover, *The Philopena* (n.d.)

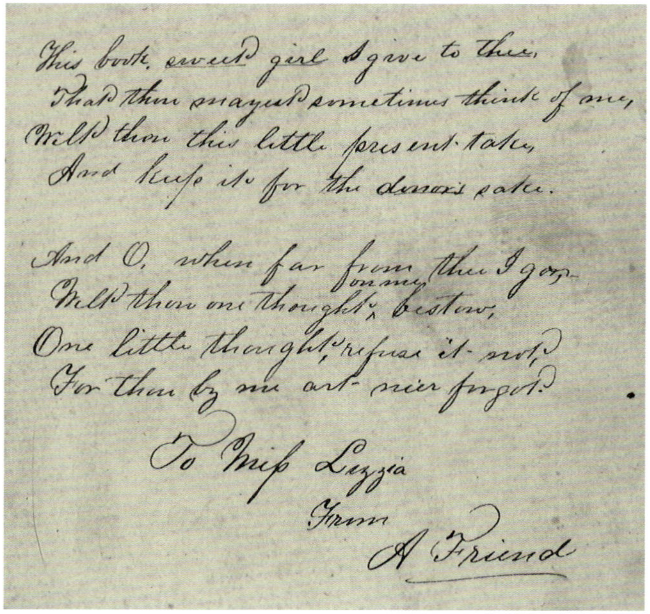
This book, sweet girl I give to thee,
That thou mayest sometimes think of me,
Wilt thou this little present take,
And keep it for the donor's sake.

And O, when far from thee I go,
Wilt thou one thought on me bestow,
One little thought, refuse it not,
For thou by me art neer forgot.

To Miss Lizzia
From
A Friend

FIGURE 5.15. Inscription, Hale, *Flora's Interpreter, and Fortuna Flora* (1850)

The "Philopene present" custom became widespread enough that one publisher capitalized on the phenomenon with a miniature gift book titled *The Philopena* (figure 5.14). A later edition of Hale's *Flora's Interpreter, and Fortuna Flora* is inscribed on the front flyleaf in a neat cursive hand with two quatrains of poetry (figure 5.15).

This book, sweet girl I give to thee,
That thou mayest sometimes think of me,
Wilt thou this little present take,
And keep it for the donor's sake.

And O, when far from thee I go,
Wilt thou one thought on me bestow,
One little thought, refuse it not,
For thou by me art ne'er forgot.
To Miss Lizzia
From
A Friend

By inscribing the book in verse, the giver contributes another "specimen" of poetry to Hale's collection. Whether cryptic, curious, or lyrical, inscriptions indicate how flower language books circulated in the gift-giving culture of the mid-nineteenth century.

Scholars generally assume that the language of flowers was the pastime and creation of middle-class white women, part of an extensive culture of sentiment in the nineteenth-century United States. While this is largely the case, material evidence shows a wider range of floriography's appeal. Flower language dictionaries were available not just in expensive gold-stamped cloth bindings but also in more affordable paper-cover editions, including Waterman's *Flora's Lexicon* (1840) and Hale's *Flora's Interpreter* (1842), following trends in publishing practices and expanding literacy in both middle- and working-class readers. Sarah Carter Edgarton begins the preface to *The Flower Vase* (1844) by asserting, "This little volume is put forth, not for its originality of design or arrangement, but because it embraces the language and poetic sentiments of flowers in a smaller and less expensive form than any similar work hitherto before the public." *The Flower Vase* was published in Lowell, Massachusetts, as was *The Lowell Offering*, a periodical written by and for the women who worked in its textile mills. Sylvia Jenkins Cook notes that in the early nineteenth century, women factory workers were "acquiring the tools of literacy in an expanding public school system as well as through libraries, lyceums, and improvement societies. Indeed Lowell operatives were manifesting such literary enthusiasm that, by 1838, 'reading in the mill' became official grounds for dismissal."[13]

The December 1840 edition of *The Lowell Offering* included "The Hill Side and the Fountain Rill," an essay in which author Adelaide described how she "gathered the choicest flowers from nature's garden, and sat down to weave into a chaplet the fragrance of the pink, the blush of the rose, the half-hid innocence of the mountain daisy, the meekness of the modest violet, the emblematic purity of the water lily, the richly-varied hue and simple sweetness of the forest wild flower. The stately oak reared its head, [her] only symbol of dignity, and spread out its long-sturdy arms in [her] defence." Adelaide echoes here the commonplaces of sentimental flower rhetoric—"the fragrance of the pink, the blush of the rose"—as well as common emblems for the daisy (innocence) and violet (modesty). She uses the specific terminology of flower language by calling the oak a "symbol of dignity" and noting the "emblematic purity of the water lily."[14] Adelaide's awareness of plants' symbolic capabilities suggests the availability of the language of flowers beyond the middle class. The title page of this issue of *The Lowell Offering* also featured a floral epigraph and two other flower-related contributions: "Spirit Flower Transplanted" and "Flowers in the Mills, &c.," confirming flowers' popularity and significance for these working women (figure 5.16).

In a different form of textual circulation, friendship albums owned by Amy Matilda Cassey and sisters Martina and Mary Anne Dickerson, free African

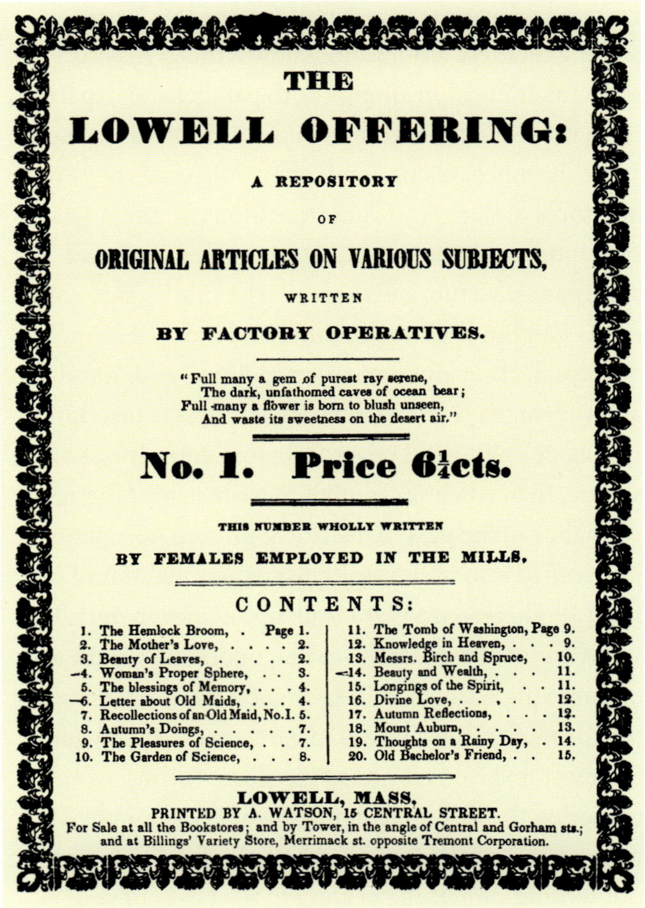
THE

LOWELL OFFERING:

A REPOSITORY

OF

ORIGINAL ARTICLES ON VARIOUS SUBJECTS,

WRITTEN

BY FACTORY OPERATIVES.

"Full many a gem of purest ray serene,
The dark, unfathomed caves of ocean bear;
Full many a flower is born to blush unseen,
And waste its sweetness on the desert air."

No. 1. Price 6¼cts.

THIS NUMBER WHOLLY WRITTEN

BY FEMALES EMPLOYED IN THE MILLS.

CONTENTS:

1. The Hemlock Broom, .	Page 1.	11. The Tomb of Washington,	Page 9.
2. The Mother's Love, ...	2.	12. Knowledge in Heaven, . . .	9.
3. Beauty of Leaves, ...	2.	13. Messrs. Birch and Spruce, .	10.
–4. Woman's Proper Sphere, . .	3.	←14. Beauty and Wealth, . . .	11.
5. The blessings of Memory, . . .	4.	15. Longings of the Spirit, . .	11.
—6. Letter about Old Maids, . . .	4.	16. Divine Love, ...	12.
7. Recollections of an Old Maid, No. I.	5.	17. Autumn Reflections, . . .	12.
8. Autumn's Doings, ...	7.	18. Mount Auburn, ...	13.
9. The Pleasures of Science, . .	7.	19. Thoughts on a Rainy Day, .	14.
10. The Garden of Science, . . .	8.	20. Old Bachelor's Friend, . .	15.

LOWELL, MASS.
PRINTED BY A. WATSON, 15 CENTRAL STREET.
For Sale at all the Bookstores; and by Tower, in the angle of Central and Gorham sts.; and at Billings' Variety Store, Merrimack st. opposite Tremont Corporation.

FIGURE 5.16. Title page, *The Lowell Offering* (December 1840)

American women in antebellum Philadelphia, contain rich visual and textual examples of their inscribers' participation in sentimental flower culture. Jasmine Nichole Cobb, in *Picture Freedom: Remaking Black Visuality in the Early Nineteenth Century*, states that "not only did flowers have a significant symbolic value among Black women but floral motifs are also a common component of the friendship album." Cobb argues that through their use of sentimental flower language and imagery, free Black women participated in cultural expectations of gentility and respectability otherwise targeted primarily toward white women. Scholars writing on these albums note how African American women's friendship albums both parallel and depart from those of white women in this era; as Katherine Isabel Bondy explains, while they "mirrored the content and form of European American albums," they "[took] on a more explicitly political tint: poems on friendship, female virtue, and flowers exist side by side with entries on the abolitionist cause and moral reform." The connection among flowers, politics, and friendship albums can be seen in Frederick Douglass's inscription in Cassey's album: "I never feel more entirely out of my sphere, than when presuming to write in an Album. . . . my habits of life passed history & present occupation have called into exercise all the sterner qualities of my head and heart—so that I walk upon uneven uncultivated and stony ground—gazing upon huge rocks with far more pleasure, than I experience, while promenading the most richly cultivated garden and gazing upon the most luxurious flowers." He closes the inscription by noting that his more familiar work "to grapple with huge wrong, with gigantic tyranny" is his "apology for not writing something becoming the pages of your precious Album."[15]

Some inscriptions in these albums have flower books as their possible source material. The sketch in figure 5.17 portrays a flower-de-luce, or iris, and the unsigned inscription quotes from act 4, scene 4 of Shakespeare's *The Winter's Tale*. These same lines appeared in a section "On the Lily" in *Flora and Thalia; or, Gems of Flowers and Poetry*, a flower book published in Philadelphia in 1836 that might have been available to the 1839 inscriber.

> *"The crown imperial, lilies of all kinds,*
> *The Flower-de-luce being one. Of these I lack*
> *To make you garlands of, and my sweet friend*
> *To strew thee o'er and o'er."*
> *New York*
> *June 14th*
> *1839*[16]

Another inscription signed "L. A. B." (Lydia A. Bustill) in Cassey's album excerpts the first stanza and final five lines of "Flowers," an unattributed poem in Osgood's *The Poetry of Flowers and Flowers of Poetry* (1841) that appears just before the section on botany:

> *Oh! they looked upward in every place*
> *Through this beautiful world of ours,*
> *And dear as a smile on an old friend's face*
> *Is the smile of the bright, bright flowers!*
> *They tell us of wanderings by woods and streams;*
> *They tell us of lanes and trees;*
> *But the children of showers and sunny beams*
> *Have lovlier [sic] tales than these—*
> *The bright, bright flowers!*
> *They were left as the proof of loveliness*
> *Of Adam and Eve's first home:*
> *They are here as a type of the joys that bless*
> *The just in the world to come—*
> *The bright, bright flowers!*[17]

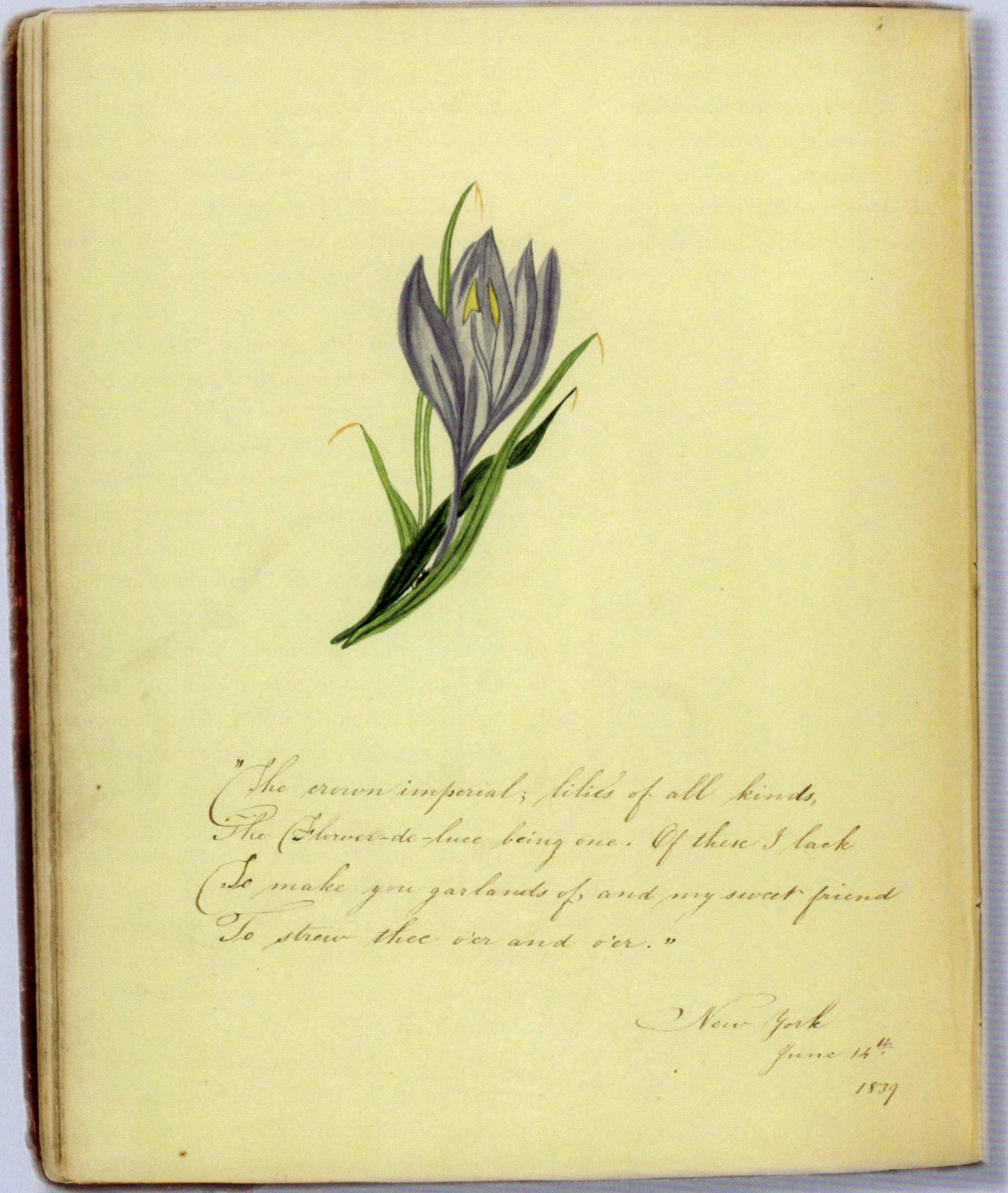

FIGURE 5.17. Sketch and inscription, Mary Anne Dickerson album, Library Company of Philadelphia

Lydia Bustill, "an anti-slavery activist and cousin of educator, artist, and anti-slavery activist Sarah Mapps Douglass," also provided a pencil sketch in Cassey's album (figure 5.18).[18] The detailed rosebuds and flowers, sweet peas with tendrils, hellebores, harebells, and other species might have been copied from a flower painting guide, as Sarah Douglass did in her watercolor of a fuchsia in Mary Anne Dickerson's album, according to Cobb. Britt Rusert, in *Fugitive Science: Empiricism and Freedom in Early African American Culture*, notes that Sarah Douglass's "contributions to the friendship albums [of the Dickersons and Cassey] joined the aesthetics of flower culture to the science of natural history," as flower language dictionaries more generally did. As I will discuss later in this chapter, illustrations of bouquets and other floral arrangements like Bustill's exuberant compilation of flowers showcased the variety of American flora.[19]

FIGURE 5.18. Sketch by L. A. B., Philadelphia, October 1841, Amy Matilda Cassey album, Library Company of Philadelphia

White working-class and middle-class African American women's knowledge of the language of flowers is, like the phenomenon itself, largely dependent on literacy and print. While botany students engaged in hands-on learning and fieldwork, and herbaria preserved live specimens collected from their native habitats, the nineteenth-century American language of flowers was almost entirely a print phenomenon. As far as we know, there were no flower language clubs where like-minded people gathered to discuss the meanings of flowers in their yards or on their walks, nor did floriography enthusiasts trade examples of new emblems they discovered, as was the case with organizations such as the Torrey Botanical Club and the pervasive exchange of botanical specimens. Instead, individuals encountered the language of flowers primarily in books. The popularization of flower language coincided with and was facilitated by a combination of the expanding book trade, printing technologies that enabled mass production, and wider distribution networks for both books and periodicals.

Research about the language of flowers in the United States has generally taken one of two approaches: surveying the historical development of flower language books or using flower language to interpret flower symbolism in literature. Beverly Seaton's *The Language of Flowers: A History* (1995) is still the most complete of the former type: Foundational and thorough, it covers flower language in nineteenth-century France, Britain, and the United States. Scholars of nineteenth-century American women's writing have used the language of flowers to analyze floral symbolism in the works of Mary Austin, Willa Cather, Lydia Maria Child, Emily Dickinson, Fanny Fern, Margaret Fuller, Susan Glaspell, Harriet Prescott Spofford, Elizabeth Stoddard, Harriet Beecher Stowe, and Edith Wharton, among others. While this approach creates productive new interpretations, it can sometimes apply the language of flowers out of context, as when scholars overlook whether the author in question owned or consulted an appropriately contemporary flower language source. The notorious inconsistency of flower language meanings also complicates, if not invalidates, this approach.[20]

Instead, I analyze U.S. flower language books as environmental and scientific texts, repositories of place-based knowledge, literary anthologies, and material artifacts rich in visual imagery and manuscript evidence. This interdisciplinary method does not denigrate their sentimentality, nor does it stretch the decoding possibilities of flower language past a point of reasonable evidence. Flower language books were profuse by definition: Just as lexical dictionaries aim to be comprehensive in their coverage of individual words, floriographic dictionaries included many different flower species, emblems, and literary excerpts. Emblems are only one part of a flower's definition, however: Botanical classification, cultivation and environmental information, poetic extracts, and visual art all contribute to the catalog of abundance that nineteenth-century American flower language dictionaries created.

Promoting Scientific Literacy and the Specifics of Place

The authors of U.S. flower language books were not scientists, but they did encourage women to develop scientific literacy. Dorothea Dix, author of *The Garland of Flora* (1829), spent several summers in Newport, Rhode Island, as governess and tutor to the children of Unitarian minister William Ellery Channing, leading them on "marches to examine plants and insects." She celebrated the results of her botanizing by enclosing flower petals in letters to friends. After publishing *The Garland of Flora*, she studied algae and collected seaweeds, sending them to Yale Professor Benjamin Silliman Sr. and contributing to the growing archive of knowledge about native plant species.[21] Hale, author of *Flora's Interpreter*, wrote articles championing the study of botany and the preservation of native wildflowers during her decades as editor of *Godey's Lady's Book*, source of the fiction discussed in chapter 3.

Over half of the nineteenth-century U.S. flower language dictionaries that I have analyzed contain at least some scientific botany. At one end of the range are books that include brief taxonomic information for each species, such as Edgarton's *The Flower Vase* and the anonymously authored *Flowers, Their Language, Poetry, and Sentiment*, which include Latin binomials (genus and species) for each entry (figures 5.19 and 5.20). Others, such as Osgood's *The Poetry of Flowers and Flowers of Poetry* and Waterman's *Flora's Lexicon*, have about ten pages of basic botany instruction at the end of the book. One of the century's best-selling flower language dictionaries, Hale's *Flora's Interpreter* (1832), begins with sections on "Botanical Explanations," "Inflorescence, or Manner of Flowering," "Classes and

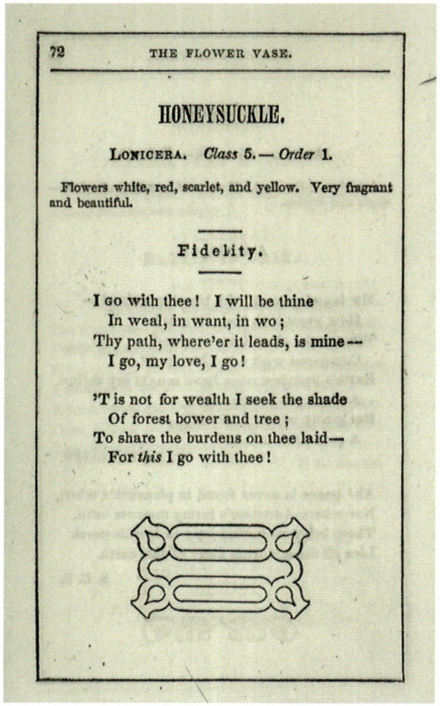

72 THE FLOWER VASE.

HONEYSUCKLE.

LONICERA. *Class* 5.—*Order* 1.

Flowers white, red, scarlet, and yellow. Very fragrant and beautiful.

Fidelity.

I go with thee! I will be thine
In weal, in want, in wo;
Thy path, where'er it leads, is mine—
I go, my love, I go!

'T is not for wealth I seek the shade
Of forest bower and tree;
To share the burdens on thee laid—
For *this* I go with thee!

FIGURE 5.19.
"Honeysuckle," page 72,
Edgarton, *The Flower Vase*
(1844)

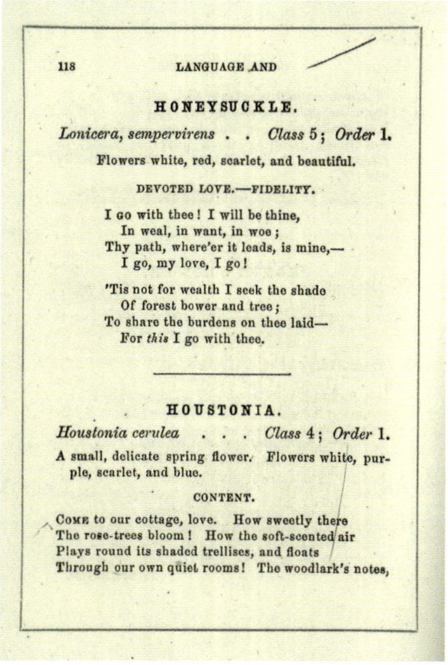

118 LANGUAGE AND

HONEYSUCKLE.

Lonicera, sempervirens . . *Class* 5; *Order* 1.

Flowers white, red, scarlet, and beautiful.

DEVOTED LOVE.—FIDELITY.

I go with thee! I will be thine,
In weal, in want, in woe;
Thy path, where'er it leads, is mine,—
I go, my love, I go!

'Tis not for wealth I seek the shade
Of forest bower and tree;
To share the burdens on thee laid—
For *this* I go with thee.

HOUSTONIA.

Houstonia cerulea . . . *Class* 4; *Order* 1.

A small, delicate spring flower. Flowers white, purple, scarlet, and blue.

CONTENT.

COME to our cottage, love. How sweetly there
The rose-trees bloom! How the soft-scented air
Plays round its shaded trellises, and floats
Through our own quiet rooms! The woodlark's notes,

FIGURE 5.20.
"Honeysuckle," page 118,
Flowers, Their Language,
Poetry, and Sentiment (1870)

Orders," and "Poisonous Plants"; starting with the sixth edition (1838), she also included a section on how "To Preserve Flowers and Plants." In the preface to *Flora's Interpreter*, Hale hopes that her book will "give an increased interest to botanical researches among young people, at least," and further states that she has "given the *generic* and usually the specific name, also the Class, Order, and native country of each flower. These particulars will be of some use if the study of botany is pursued, or at any rate, they must associate in the mind of the reader some notion of the science" (italics in the original). Wirt's *Flora's Dictionary* is unique for its inclusion of scientific botany both before and after the main section on the language of flowers, and she devotes almost as many pages to the study of flowers as to their symbolism. In the preface, Wirt describes the joys of science. "This blooming earth: what an interest has the science of botany thrown over it!" she writes. "Yet how few are there, among us, who are disposed to taste of the banquet which this science affords!" The vast scale of the "blooming earth" and the many choices afforded by the "banquet" of science signal the abundance of knowledge that awaits her readers.[22]

In their prefaces and introductions, compilers of flower language dictionaries encouraged competence in identifying and classifying species, thus bringing scientific insight to what would otherwise be mere entertainment. Waterman explained that "the outline of Botany placed at the end of the volume will be found to contain a sufficiently clear exposition of the Linnean system to explain fully the scientific terms and the classification used in the body of the work." In "A New Introduction" to the expanded *Flora's Interpreter, and Fortuna Flora* (1849), Hale wrote that she "endeavored to unite real knowledge with this fanciful language; the arrangement of each flower with its botanic as well as common name, and also its class and order, will be found of much utility by familiarizing or fixing these terms in the mind of the reader." Real *and* fanciful, botanic *and* common, utility *and* the mind: Hale has described the multiple aims of flower language books. The "Botanical Introduction" at the end of Lucy Hooper's *Lady's Book of Flowers and Poetry* also taught classification according to the Linnaean system, which categorizes plants based on the number of stamens or their configuration. The intricacies of this system make it occasionally "difficult to ascertain, as in the pea for instance," in which case "it is necessary to take a pin and separate the filament which is alone, in order to perceive that it is not united to the other nine." Although this detailed level of dissection may seem daunting, the fact that it can be done with a readily available pin makes it more accessible.[23]

A knowledge of plant anatomy opened up other kinds of scientific information. The flower horologe, popularized by Linnaeus, enabled one to tell time according to the hour when particular flowers opened. After several pages of quotations about this "dial of flowers," Dix stated that "if flowers furnish us with a beautiful means of counting the fast flying hours, they afford a much more certain means of determining the state of the atmosphere. Many species are admirable barometers. Most of the bulbous rooted flowers contract, or wholly close their petals on the approach of rain. The African marygold [*sic*] indicates rain, if the corolla is closed after seven or eight in the morning." The passage then describes the barometrical behavior of bindweed, scarlet pimpernel, barley, geranium, chickweed, oxalis, and convolvulus. Knowing the specific parts of a flower—bulbous roots, corolla, the awn of barley, the capsule of geranium—was crucial to understanding plants as barometers.[24]

Just as many flower language dictionaries contained sections on botany, many botany textbooks included a section on the language of flowers. Alphonso Wood's *Class-Book of Botany* and *Leaves and Flowers; or, Object Lessons in Botany* both include a "Language of Flowers" section at the end. Wood wrote that he included it "at the request of many readers, . . . regarding it as by no means an unsuitable *appendix* to the Flora, although it might not be entitled to a place in the body of a scientific work" (italics in the original). Each entry includes the order, genus, and species of the flower (figure 5.21). Almira Phelps's *Familiar Lectures on Botany* and *Botany for Beginners* also have flower language appendices (figures 5.22 and 5.23). In the 1836 edition of *Familiar Lectures*, she cites as her sources two French books as well as the American *Flora's Dictionary* and *Garland of Flora*, both published just a few years earlier in 1829. For readers of *Botany for Beginners*, Phelps recommends "the article, 'Symbolical Language of Flowers'" in her *Familiar Lectures on Botany* to those wishing to pursue the subject; by the 1867 edition, she has omitted the two French sources, recommending instead the "more full explanations" available in *Flora's Dictionary* and Hale's *Flora's Interpreter*.[25] John Newman's *Illustrated Botany* (1850) prominently mentions the inclusion of "a floral dictionary" on its title page. Mattie Snowden, discussed in detail in chapter 2, directly interacted with the language of flowers in her copy of *Illustrated Botany*, with several comments in the floral dictionary section. For example, beside the flower poet's narcissus and its emblem egotism, she writes, "C. C. wants to find *one* at least for Mr. S."—perhaps her nemesis, Benjamin Silliman Jr.?

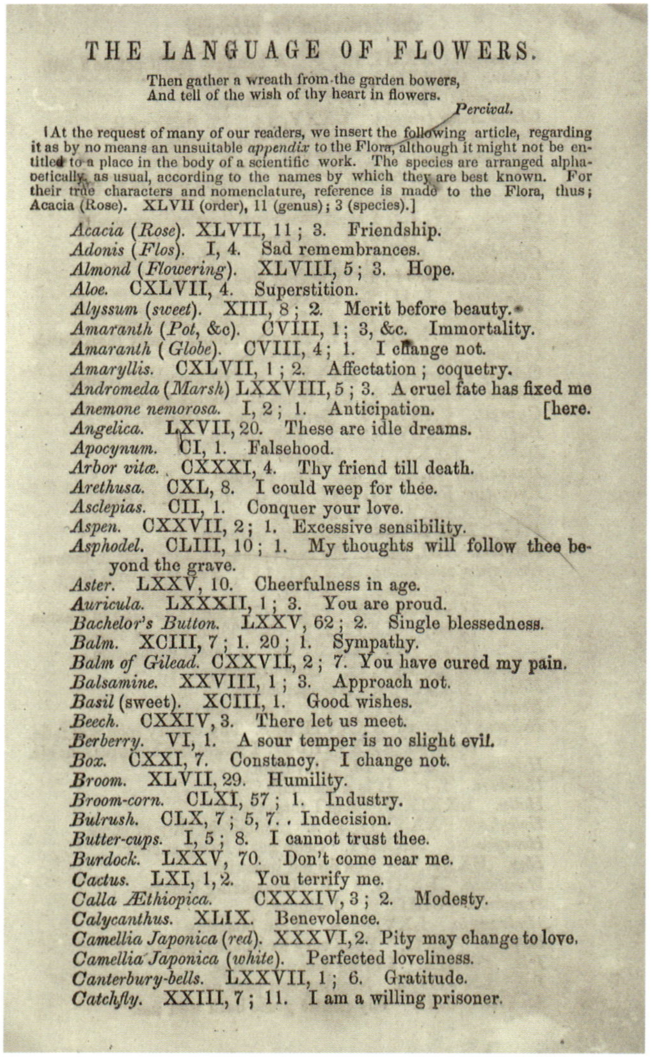

THE LANGUAGE OF FLOWERS.

Then gather a wreath from the garden bowers,
And tell of the wish of thy heart in flowers.
Percival.

[At the request of many of our readers, we insert the following article, regarding it as by no means an unsuitable *appendix* to the Flora, although it might not be entitled to a place in the body of a scientific work. The species are arranged alphabetically, as usual, according to the names by which they are best known. For their true characters and nomenclature, reference is made to the Flora, thus; Acacia (Rose). XLVII (order), 11 (genus); 3 (species).]

Acacia (*Rose*). XLVII, 11; 3. Friendship.
Adonis (*Flos*). I, 4. Sad remembrances.
Almond (*Flowering*). XLVIII, 5; 3. Hope.
Aloe. CXLVII, 4. Superstition.
Alyssum (*sweet*). XIII, 8; 2. Merit before beauty.
Amaranth (*Pot*, &c). CVIII, 1; 3, &c. Immortality.
Amaranth (*Globe*). CVIII, 4; 1. I change not.
Amaryllis. CXLVII, 1; 2. Affectation; coquetry.
Andromeda (*Marsh*) LXXVIII, 5; 3. A cruel fate has fixed me [here.
Anemone nemorosa. I, 2; 1. Anticipation.
Angelica. LXVII, 20. These are idle dreams.
Apocynum. CI, 1. Falsehood.
Arbor vitæ. CXXXI, 4. Thy friend till death.
Arethusa. CXL, 8. I could weep for thee.
Asclepias. CII, 1. Conquer your love.
Aspen. CXXVII, 2; 1. Excessive sensibility.
Asphodel. CLIII, 10; 1. My thoughts will follow thee beyond the grave.
Aster. LXXV, 10. Cheerfulness in age.
Auricula. LXXXII, 1; 3. You are proud.
Bachelor's Button. LXXV, 62; 2. Single blessedness.
Balm. XCIII, 7; 1. 20; 1. Sympathy.
Balm of Gilead. CXXVII, 2; 7. You have cured my pain.
Balsamine. XXVIII, 1; 3. Approach not.
Basil (sweet). XCIII, 1. Good wishes.
Beech. CXXIV, 3. There let us meet.
Berberry. VI, 1. A sour temper is no slight evil.
Box. CXXI, 7. Constancy. I change not.
Broom. XLVII, 29. Humility.
Broom-corn. CLXI, 57; 1. Industry.
Bulrush. CLX, 7; 5, 7. . Indecision.
Butter-cups. I, 5; 8. I cannot trust thee.
Burdock. LXXV, 70. Don't come near me.
Cactus. LXI, 1, 2. You terrify me.
Calla Æthiopica. CXXXIV, 3; 2. Modesty.
Calycanthus. XLIX. Benevolence.
Camellia Japonica (*red*). XXXVI, 2. Pity may change to love.
Camellia Japonica (*white*). Perfected loveliness.
Canterbury-bells. LXXVII, 1; 6. Gratitude.
Catchfly. XXIII, 7; 11. I am a willing prisoner.

FIGURE 5.21. Appendix page i, following page 645, Wood, *Class-Book of Botany* (1855)

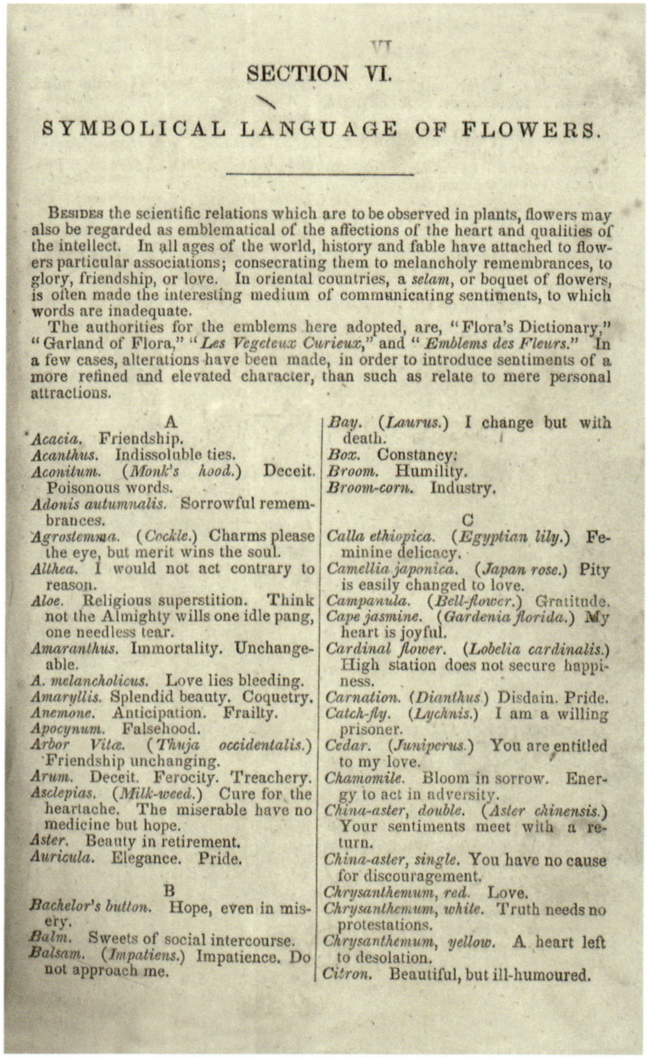

SECTION VI.

SYMBOLICAL LANGUAGE OF FLOWERS.

BESIDES the scientific relations which are to be observed in plants, flowers may also be regarded as emblematical of the affections of the heart and qualities of the intellect. In all ages of the world, history and fable have attached to flowers particular associations; consecrating them to melancholy remembrances, to glory, friendship, or love. In oriental countries, a *selam*, or boquet of flowers, is often made the interesting medium of communicating sentiments, to which words are inadequate.

The authorities for the emblems here adopted, are, "Flora's Dictionary," "Garland of Flora," "*Les Vegeteux Curieux*," and "*Emblems des Fleurs*." In a few cases, alterations have been made, in order to introduce sentiments of a more refined and elevated character, than such as relate to mere personal attractions.

A

Acacia. Friendship.
Acanthus. Indissoluble ties.
Aconitum. (*Monk's hood.*) Deceit. Poisonous words.
Adonis autumnalis. Sorrowful remembrances.
Agrostemma. (*Cockle.*) Charms please the eye, but merit wins the soul.
Althea. I would not act contrary to reason.
Aloe. Religious superstition. Think not the Almighty wills one idle pang, one needless tear.
Amaranthus. Immortality. Unchangeable.
A. melancholicus. Love lies bleeding.
Amaryllis. Splendid beauty. Coquetry.
Anemone. Anticipation. Frailty.
Apocynum. Falsehood.
Arbor Vitæ. (*Thuja occidentalis.*) Friendship unchanging.
Arum. Deceit. Ferocity. Treachery.
Asclepias. (*Milk-weed.*) Cure for the heartache. The miserable have no medicine but hope.
Aster. Beauty in retirement.
Auricula. Elegance. Pride.

B

Bachelor's button. Hope, even in misery.
Balm. Sweets of social intercourse.
Balsam. (*Impatiens.*) Impatience. Do not approach me.
Bay. (*Laurus.*) I change but with death.
Box. Constancy.
Broom. Humility.
Broom-corn. Industry.

C

Calla ethiopica. (*Egyptian lily.*) Feminine delicacy.
Camellia japonica. (*Japan rose.*) Pity is easily changed to love.
Campanula. (*Bell-flower.*) Gratitude.
Cape jasmine. (*Gardenia florida.*) My heart is joyful.
Cardinal flower. (*Lobelia cardinalis.*) High station does not secure happiness.
Carnation. (*Dianthus.*) Disdain. Pride.
Catch-fly. (*Lychnis.*) I am a willing prisoner.
Cedar. (*Juniperus.*) You are entitled to my love.
Chamomile. Bloom in sorrow. Energy to act in adversity.
China-aster, double. (*Aster chinensis.*) Your sentiments meet with a return.
China-aster, single. You have no cause for discouragement.
Chrysanthemum, red. Love.
Chrysanthemum, white. Truth needs no protestations.
Chrysanthemum, yellow. A heart left to desolation.
Citron. Beautiful, but ill-humoured.

FIGURE 5.22. Appendix page 171, Phelps, *Familiar Lectures on Botany* (1836)

Flower language dictionaries cultivated scientific literacy in their sections on botanical instruction and scientific information elsewhere in their contents. Tina Gianquitto argues that "science and sentiment uneasily shared the stage in early-nineteenth-century culture" and that flower language dictionaries were "geared toward a consumer culture whose interest in them had been primed by the very botanical education that Phelps advocated." But "for those who have not had the advantage of the study in their education,"

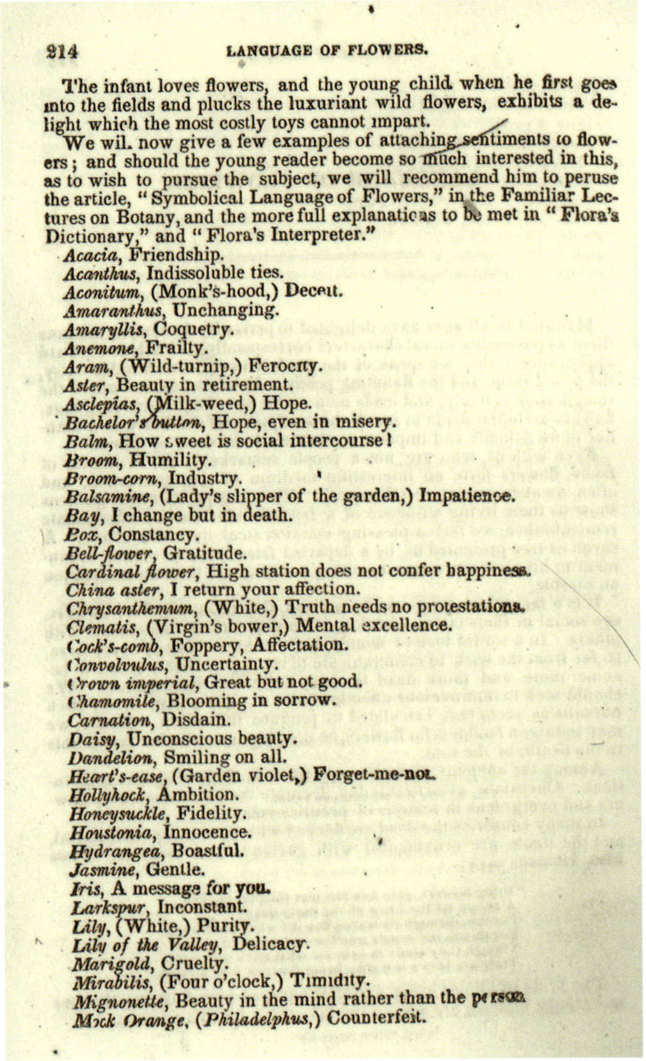

214 LANGUAGE OF FLOWERS.

The infant loves flowers, and the young child when he first goes into the fields and plucks the luxuriant wild flowers, exhibits a delight which the most costly toys cannot impart.

We will now give a few examples of attaching sentiments to flowers; and should the young reader become so much interested in this, as to wish to pursue the subject, we will recommend him to peruse the article, "Symbolical Language of Flowers," in the Familiar Lectures on Botany, and the more full explanations to be met in "Flora's Dictionary," and "Flora's Interpreter."

Acacia, Friendship.
Acanthus, Indissoluble ties.
Aconitum, (Monk's-hood,) Deceit.
Amaranthus, Unchanging.
Amaryllis, Coquetry.
Anemone, Frailty.
Aram, (Wild-turnip,) Ferocity.
Aster, Beauty in retirement.
Asclepias, (Milk-weed,) Hope.
Bachelor's button, Hope, even in misery.
Balm, How sweet is social intercourse!
Broom, Humility.
Broom-corn, Industry.
Balsamine, (Lady's slipper of the garden,) Impatience.
Bay, I change but in death.
Box, Constancy.
Bell-flower, Gratitude.
Cardinal flower, High station does not confer happiness.
China aster, I return your affection.
Chrysanthemum, (White,) Truth needs no protestations.
Clematis, (Virgin's bower,) Mental excellence.
Cock's-comb, Foppery, Affectation.
Convolvulus, Uncertainty.
Crown imperial, Great but not good.
Chamomile, Blooming in sorrow.
Carnation, Disdain.
Daisy, Unconscious beauty.
Dandelion, Smiling on all.
Heart's-ease, (Garden violet,) Forget-me-not.
Hollyhock, Ambition.
Honeysuckle, Fidelity.
Houstonia, Innocence.
Hydrangea, Boastful.
Jasmine, Gentle.
Iris, A message for you.
Larkspur, Inconstant.
Lily, (White,) Purity.
Lily of the Valley, Delicacy.
Marigold, Cruelty.
Mirabilis, (Four o'clock,) Timidity.
Mignonette, Beauty in the mind rather than the person.
Mock Orange, (*Philadelphus*,) Counterfeit.

FIGURE 5.23. Page 214, Phelps, *Botany for Beginners* (1867)

Hale, Waterman, and Wirt included legitimate botany instruction along with poetry and emblems.[26]

as Osgood attests in *The Poetry of Flowers and Flowers of Poetry*, "the chapter on botany . . . will serve as an interesting introduction." Rather than uneasily sharing science and sentiment in their flower language dictionaries, authors such as Osgood, Newman, Dix, Hooper, Hale, Waterman, and Wirt included legitimate botany instruction along with poetry and emblems.[26]

They also promoted a *specifics* of place rooted in the *species* they catalogued. These particulars included descriptions of habitat, both native and cultivated, and narratives of places and relationships. Although not as extensive, entries in flower language books are analogous to the "life histories" format popular in nineteenth-century natural history writing. As Margaret Welch explains, these were "virtual biographies containing every scrap of available evidence on appearance, reproduction, communication, habitat, life stages and geographical distribution of a species." Some specifics of place are geographical, as when Wirt notes that "the Rose Acacia, *R. Hispida*, or Hairy Robinia, is a native of America, particularly of the mountains of Virginia and Carolina" and "the *Pseudo Acacia*, or common locust, is also a native of North America, from Canada to Carolina," or when Hale states that the yellow acacia (*A. farnesiana*) is "common around New Orleans." Others focus on habitat: Osgood, ever the poet, describes how acacia "is found in the most retired places, and it blooms the fairest in the closeness of its own foliage. It loves the mossy rock and the solitary grove, and pines away in the gay garden and crowded parterre," using familiar terms to convey details of the acacia's preferred environment. Not all flower language dictionaries included Latin binomials, but those that did enabled readers to distinguish among different species (*R. hispida, A. farnesiana*) with the same common name (acacia).[27]

In addition to providing geography and habitat information, flower language dictionaries placed species in their cultural environments. When Hale discusses magnolia in *Flora's Interpreter*, she cites an American poem, in keeping with her stated purpose of foregrounding American authors. She excerpts James Wallis Eastburn's *Yamoyden: A Tale of the Wars of King Philip, in Six Cantos* (1820), an American epic poem about the late seventeenth-century wars among northeastern Indigenous nations and English colonists. The excerpt reads,

> Immortal in bloom,
> Soft waves the *Magnolia* its groves of perfume.
> (Emphasis in the original)

Osgood's entry on magnolia notes an interesting tradition in the tree's native southern habitat: "The large white leaf of the flower often serves the romantic southern youth for paper. He pricks upon it with a needle or pin the passionate thoughts of his heart, and commits his perfumed billetdoux to the care of zephyr to be wafted to the feet of his ladye-love." Like birch bark used as paper, the "leaf" or petal of the magnolia flower has material as well as symbolic uses, and the chivalry implied in terms such as "ladye-love" was an acknowledged—and controversial—cultural ideal of the American South.[28]

Stories that are repeated in multiple flower language books reveal what authors and readers found interesting and important. One anecdote that frequently appeared involved the Virginian (or Carolina) jessamine, or jasmine. The entry on the Carolina yellow jasmine (*Gelsemium nitidum*) in Waterman's *Flora's Lexicon* notes the hummingbird's preference for its nectar: "This beautiful flower grows wild in great abundance in our southern states, spreading over the hedges and trees and shedding a most delicious fragrance. It is a favourite of the Humming-Bird and a British writer regrets that in transferring it to England it is necessary to separate it from the sprightly and elegant little bird, who finds sustenance in the nectareous vessels of its flowers." Osgood similarly mentions hummingbirds in her entry: "Behold the Virginian jessamine, with its beautiful verdure and purple flowers! The humming-bird makes its nest in one of the leaves, which it rolls into the shape of a horn; it finds its sustenance in the nectareous vessels of its red flowers, which are similar to those of the foxglove: and its little body, when resting on the jasmine-flower, appears like an emerald set in coral. . . . This little being is the soul and the life of the plant that cherishes it."[29] A nearly identical passage, also noting the value of nectar and leaves for the hummingbird, appears in John Newman's *Illustrated Botany*, in the entry on trumpet flower, also native to Virginia and the Carolinas. Waterman has "separation" as the emblem, echoing the British writer who "regrets that in transferring it to England it is necessary to separate it from the sprightly and elegant little bird." Osgood, however, chooses "soul of my soul" as the emblem for Virginian jessamine and includes her poem:

> Oh! the pet and the darling of nature—whose plume,
> Through the woods, like a sunbeam, doth playfully dart—
> The humming-bird! bathing in dew and perfume,
> That hide in the crimson jasmine's heart,
> While he blends, with the blossom, his own rich bloom,
> Till you hardly can tell them apart.

Both Waterman and Osgood note the symbiosis of the flower and its pollinator. But whereas Waterman's entry and emblem emphasize the negative consequences of separating the symbionts, Osgood's highlights their union: "He blends, with the blossom . . . till you hardly can tell them apart." In both books, the location of the species is foregrounded: In Waterman, it's "our southern states"; in Osgood, it's the *Virginian* jessamine. Their narratives focus on this native American flower and its special relationship with another uniquely New World species, the hummingbird. With unfortunate consequences if separated or transplanted, jessamine and hummingbird *must* go together, as these U.S. authors are proud to emphasize.[30]

There are many more examples of stories like the Carolina or Virginian jessamine and its hummingbird, the magnolia flower as a vehicle for billet-doux, and the acacia, which Native Americans reportedly used for bows and arrows.[31] The impact of these repeated "plant biographies" becomes significant when multiplied across the hundreds of species in scores of flower language books. The sheer number and diversity of plants and environments in the United States in turn inspired flower language dictionary authors to compile multiple literary sources to demonstrate this impressive plenty. The next section will show how the visual aspects of flower language dictionaries also represented American abundance.

The Aesthetics of Abundance: Bindings and Bouquets

For gift books such as the flower language dictionary, the outside cover or binding was the first indicator of its value and appeal. Flower language books from the 1840s and 1850s typically featured stamped cloth bindings, often with a central image stamped in gold. Book historian Sue Allen writes that by the mid-1840s, publishers "wanted to draw attention to their covers. They could use images as topical and timely as the wood engravings in illustrated newspapers and magazines. Publishers needed stamps that were pictorial, specific, expressive of the nature of the book." The binding of Edgarton's *Fables of Flora*, a combination of flower language and flower fables, has a central image of the goddess Flora, typically depicted in a flowing dress and surrounded by flowers (figure 5.24). This image dynamically displays a female body in motion in the wild, walking barefoot and wearing a comfortably airy dress that does not impede her movement. Anticipating the dress reform movement of the early 1850s, Flora's free-flowing clothing might have inspired the book's readers to their own sartorial liberation.[32]

FIGURE 5.24. Cover, Edgarton, *Fables of Flora* (1844)

FIGURE 5.25. (*left*) Cover, Hanson, *Flora's Dial* (1846)

FIGURE 5.26. (*right*) Cover, Griffin, *Drops from Flora's Cup* (1846)

Pocket-sized flower language books such as *Fables of Flora* used a variety of central stamps. J. Wesley Hanson's *Flora's Dial* (figure 5.25) features a central bouquet of rose, tulip, forget-me-not, and snowdrop, with botanically accurate details of petals and leaves, in a loose arrangement that is more natural than flowers contained in vases. The image in Mary M. Griffin's *Drops from Flora's Cup* (figure 5.26) shows a lush blooming tree growing from an urn-shaped planter, surrounded by grass and other vegetation.

Octavo-sized flower language books from the 1840s and 1850s used different colors of cloth bindings (figures 5.27–5.30), reflecting publishing trends before the 1860s, when "it was typical for a single work to be offered for sale in a multitude of colors and grains of cloth and often in several binding designs or styles," an excellent marketing strategy for this popular genre, and a corollary to the many colors of flowers the books portrayed.[33] All four bindings pictured here have a combination of straight-ruled outer borders and gold-stamped

FIGURE 5.27. Cover, Osgood, *The Poetry of Flowers and Flowers of Poetry* (1841)

FIGURE 5.28. Cover, Dinnies, *The Floral Year* (1847)

FIGURE 5.29. Cover, Hale, *Flora's Interpreter, and Fortuna Flora* (1850)

FIGURE 5.30. Cover, Hooper, *The Lady's Book of Flowers and Poetry* (1858)

central floral images. *The Floral Year* (figure 5.28) and *Flora's Interpreter, and Fortuna Flora* (figure 5.29) additionally feature intricately detailed gold-stamped scrolling middle borders; *The Lady's Book of Flowers and Poetry* (figure 5.30) has its scrolling middle border in blind stamping. The imagery on all four covers was standard in these decades—cherubs, bouquets, generic five-petaled blossoms—yet other details show the same botanical accuracy seen in figure 5.25. *The Poetry of Flowers and Flowers of Poetry* (figure 5.27) has a central-image border of wooden branches, reminding us of paper's material origin from trees, and the flowers in the four corners of this twiggy border are all different. The oval border around the gold-stamped central title of *The Lady's Book of Flowers and Poetry* (figure 5.30) includes four passionflowers, a native of the Americas that fascinated readers of flower language books.

Postbellum innovations in cloth binding technologies enabled even more lavish and richly detailed covers, as in Cordelia Harris Turner's *The Floral Kingdom* (figure 5.31). It shows "the characteristic pictorial and 'eclectic' binding designs of the 1870s and 1880s."[34] An intricate Islamic-influenced border echoes the Alhambra-style palace at the center right margin, the ornate font of the title is further decorated, and lots of botanical imagery fills the space: water lilies and rushes, feathery fronds, a morning-glory vine climbing a tree whose branch shades the palace, and a trellis covered in a grape vine. The built environment is overpowered by the profusion of natural images: both palace and trellis are much smaller in scale by comparison, as befits a book on the "floral kingdom." Even the mountain peaks at the center fade into the distance, so as not to distract from the lush verdure of the foreground. *The Floral Kingdom*'s contents are extensive; as the title page notes, it is "a dictionary of more than three hundred plants, with the genera and families to which they belong, and the language of each illustrated with appropriate gems of poetry," plus "an autograph letter and introductory poem by William Cullen Bryant and a practical treatise for amateurs on the cultivation and analysis of plants." Repeating the ornate profusion of the binding, the illustrated title page abounds with stylized lettering, trailing ivy, a drooping fuchsia, bouquets, and even a couple of winged fairies (figure 5.32). Turner's preface asserts that her book includes "the general history of plants, their nativity, uses, the chief events in the history of each species, its cultivation and introduction into America"—in short, a wealth of information.[35]

Two different publishers' editions of C. M. Kirtland's *Poetry of the Flowers* exhibit the patterned zones characteristic of American cloth bindings in the 1880s (figures 5.33 and 5.34). Instead of a single central image, these books have titles at the top and multiple bordered sections in a variety of patterns. In the green binding of the Crowell edition (figure 5.33), leaves and flowers proliferate: a bottom border of repeating daisies, ivy leaves in the center left zone, gold-stamped flowers in the top section, and another repeating floral border just above the title. Even the embellished capital "P" has leaf and flower motifs. In the lower center section the lyre, symbol of poetry, is surrounded by leaves and tendrils. The red binding of the Lovell edition (figure 5.34) also uses repeating floral borders, an embellished capital, leaves "growing" around the title section, and a botanically accurate clover blossom as the central image of the lower section. By this point later in the century, flower language dictionaries included hundreds of species amassed from decades of predecessors, and the many different botanical details of bindings such as these announced that variety before the book was even opened.

FIGURE 5.31. Cover, C. Turner, *The Floral Kingdom* (1877)

FIGURE 5.32. Illustrated title page, C. Turner, *The Floral Kingdom* (1877)

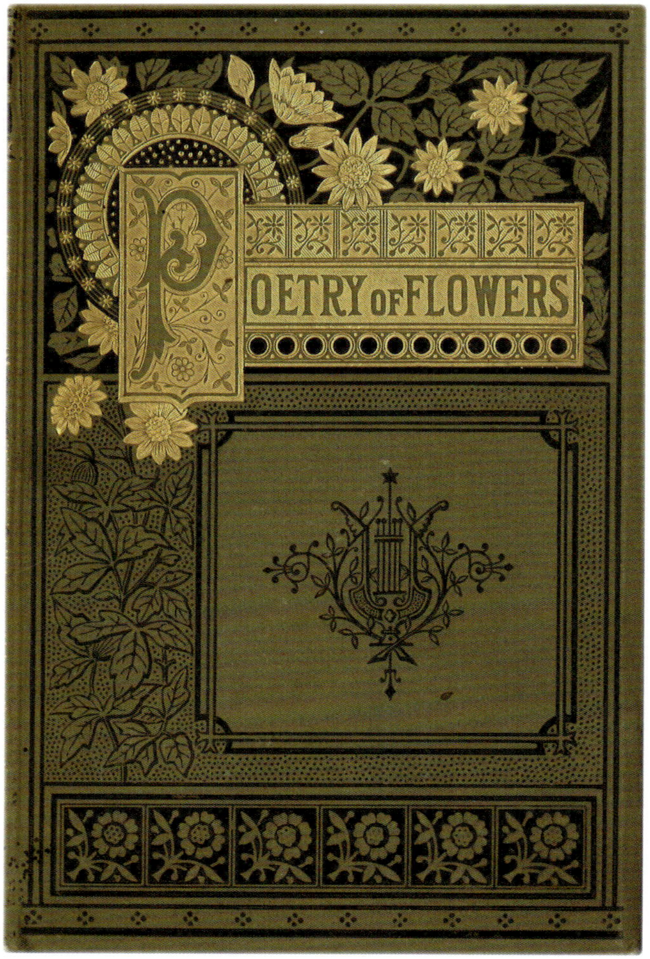

FIGURE 5.33. Cover, Kirtland, *Poetry of the Flowers* (Crowell edition)

Inside these books, beautiful illustrations added to their aesthetic appeal. Colored frontispieces appeared in all sizes of flower language books, including double frontispieces, as seen in Hanson's *Flora's Dial* and Dinnies's *The Floral Year* (figures 5.35 and 5.36). In both, the left-hand page shows a bouquet of multiple flowers, and the right-hand page a highly ornate version of the title. Viewed from left to right, as one reads in English, the first page shows only flowers, privileging the botanical; the words of the title, with their more constructed decoration of straight-lined borders, come second.

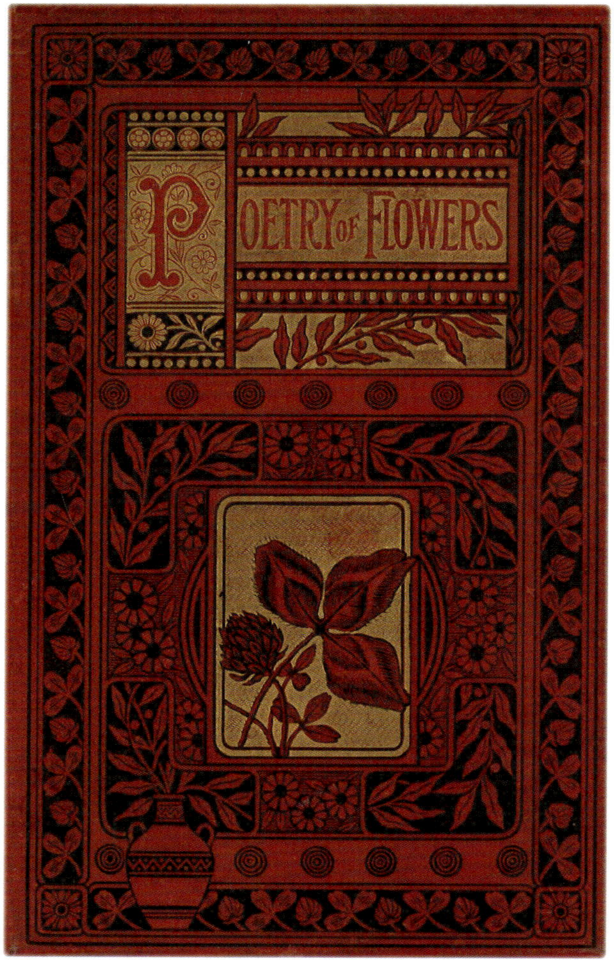

FIGURE 5.34. Cover, Kirtland, *Poetry of the Flowers* (Lovell edition)

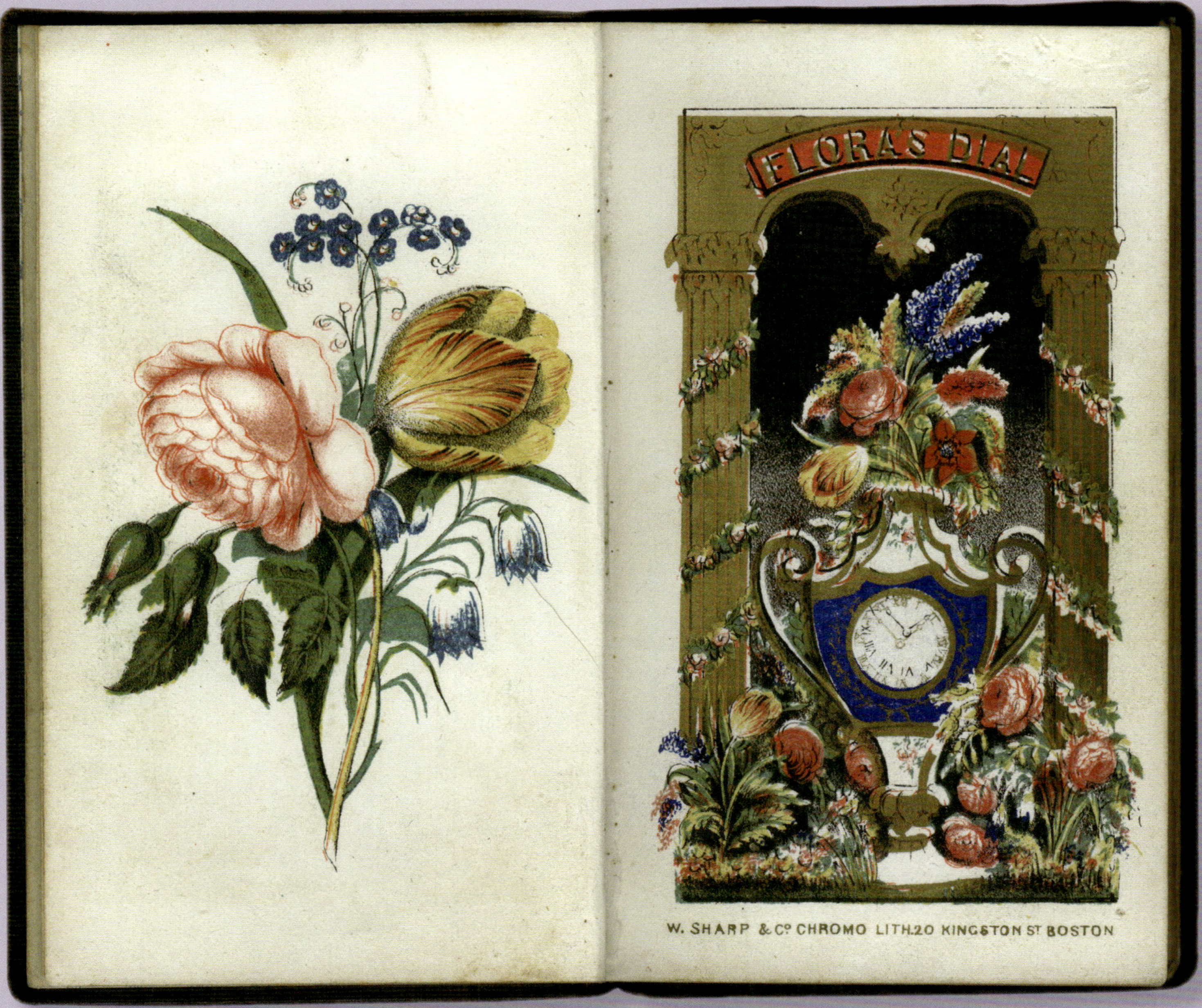

FIGURE 5.35. Double frontispiece, Hanson, *Flora's Dial* (1846)

FIGURE 5.36. Double frontispiece, Dinnies, *The Floral Year* (1847)

FIGURE 5.37. Illustrated title page, Hooper, *The Lady's Book of Flowers and Poetry* (1841), courtesy of HathiTrust

The illustrated title page for the first edition of Hooper's *The Lady's Book of Flowers and Poetry* (figure 5.37) combines flowers of different colors, shapes, and sizes with leaf motifs and a flower-garlanded woman framed in a woven basket. The 1858 edition's frontispiece (figure 5.38) features only flowers, rendered in more detail than those of the 1841 image, again in a variety of shapes and colors for both flowers and leaves. These different images would have been provided by the publisher and suggest an increasing attention to botanical accuracy.

The bouquet is the most prevalent arrangement in flower language books. One piece of flower language's alleged origin was the *sélam* or "talking bouquet."[36] As intentionally constructed displays of multiple flowers, bouquets emphasize diversity of color, shape, size, and even scent. They were also the preferred way of giving

FIGURE 5.38. Frontispiece, Hooper, *The Lady's Book of Flowers and Poetry* (1858)

FIGURE 5.39. Page 65, Osgood, *Poetry of Flowers and Flowers of Poetry* (1841)

FIGURE 5.40. Between pages 144 and 145, Dinnies, *The Floral Year* (1847)

flowers. In the language of flowers, the bouquet adds syntax to semantics, enabling a sentence rather than a single word. Osgood's *The Poetry of Flowers and Flowers of Poetry*, Dinnies's *The Floral Year*, and Dumont's *The Floral Offering* have multiple plates that show groups of flowers. Osgood constructs sentences from her bouquets; one with primrose, honeysuckle, and marigold translates to "Be not too early entangled in the chains of Love, or yours will be a life of inquietude," derived from the flowers' respective emblems of early youth, bond of love, and inquietude (figure 5.39). Dinnies's bouquet for August (figure 5.40) is composed of seasonally appropriate oats, passionflower, love-lies-bleeding, cypress vine, columbine, verbena, amaranth, periwinkle, and althea frutex (*Hibiscus syriacus*), whose emblems are included just below the list of flowers (as seen in

FIGURE 5.41. Between pages 198 and 199, Dumont, *The Floral Offering* (1851)

FIGURE 5.42. Between pages 252 and 253, Hale, *Flora's Interpreter, and Fortuna Flora* (1850)

figure 5.2). Dumont's less crowded bouquet (figure 5.41) contains only sweet pea and pansy, signifying "I depart - Think on me." The pansies and sweet peas in Dumont's colored plate are shown in two forms, bud and full bloom, with correctly depicted leaves. Osgood's arrangement (figure 5.39) similarly includes accurately rendered leaves for each of the three flowers, in a balanced composition of size, shape, and color. Dinnies's collection foregrounds the passionflower, whose details are not only botanically correct but also significant for their Christian iconography.[37]

Full-page color plates in Hale and Wirt support each author's stated goals for her book. A bouquet in Hale's *Flora's Interpreter, and Fortuna Flora* (1850) includes red, white, and blue flowers, supporting Hale's mission to prioritize *American* productions (figure 5.42). The plates in Wirt's *Flora's Dictionary* (ca. 1837) depict bouquets composed of alphabetically adjacent flowers, such as those from the section for the letter "C": clover, cockscomb, purple columbine, red columbine, and convolvulus (figure 5.43). The exquisite detail shows accurate leaves for each flower; columbine, convolvulus, and cockscomb in both bud and bloom; the twining tendrils of the convolvulus vine; and even the tiny black seeds that emerge from cockscomb. This artful composition has several shades of green for leaves and a combination of red, purple, and blue flowers. The large cockscomb and convolvulus anchor the center of the image, while columbine, clover, and a smaller bit of cockscomb fill the remaining space. The preference for multiple over individual flowers counters the single-specimen form of scientific illustration and the typical herbarium page, instead portraying flowers as ornament and gift. As accurate in detail as scientific botanical art, these plates are also worthy of display in one's home, the book left open to show off their beauty.

Many forms of abundance appear in nineteenth-century American flower language dictionaries: hundreds of species, multiple meanings, lots of literary extracts, place-based knowledge, botany instruction, ornate bindings, and beautiful illustrations. These books evoked expansive figurative possibilities without damaging actual flowers and promoted a practice of botanizing that was interpretive rather than empirical. Their frequent borrowing from other sources can be dismissed as a lack of originality—or viewed as a form of recycling that made productive use of materials already in circulation. American flower language books were not exclusively or exceptionally American in their subject matter; even Hale, champion of American poetry, included nonnative plant species and non-U.S. authors in *Flora's Interpreter*. As a phenomenon, the language of flowers was vibrantly generative, allowing new meanings for familiar flowers and adding new flowers as they were discovered in both the United States and elsewhere.

By the time *The Floral Kingdom* was published in 1877, Turner had benefited from the compilations of so many predecessors that her book included a whopping 326 species and literary extracts from 440 authors. The extravagant binding and ornate title page (figures 5.31 and 5.32) invite the reader into a wonderland of discoveries, and each page in the book's main flower language section includes an ornate capital, Latin binomial and taxonomic classification, an emblem, and multiple quotations, as seen in the examples of cranberry and rhodora (figures 5.44 and 5.45). "Nearly every one is familiar with" the cranberry, and its emblem "hardihood" inspired Turner to include a quatrain from Longfellow's "The Village Blacksmith" (1841) that commends the smith's "large and sinewy hands" and brawny arm muscles; by 1877, its opening line "under the spreading chestnut tree" was familiar to schoolchildren across the country. Turner also quotes from Sarah J. Hale's "The Storm" (1848), where hardihood is based instead on a soul "nerved to bear the blow"—and nods to Hale's work as an earlier floriographer. The rhodora entry includes Ralph Waldo Emerson's poem "The Rhodora" (1834), with its message that "beauty is its own excuse for being," a sentiment that aptly characterizes flower language books.

As a form of popular botany, the flower language genre is far less scientific than botany instruction books or herbaria, even as it includes scientific information

FIGURE 5.43. Between pages 41 and 42, Wirt, *Flora's Dictionary* (ca. 1837)

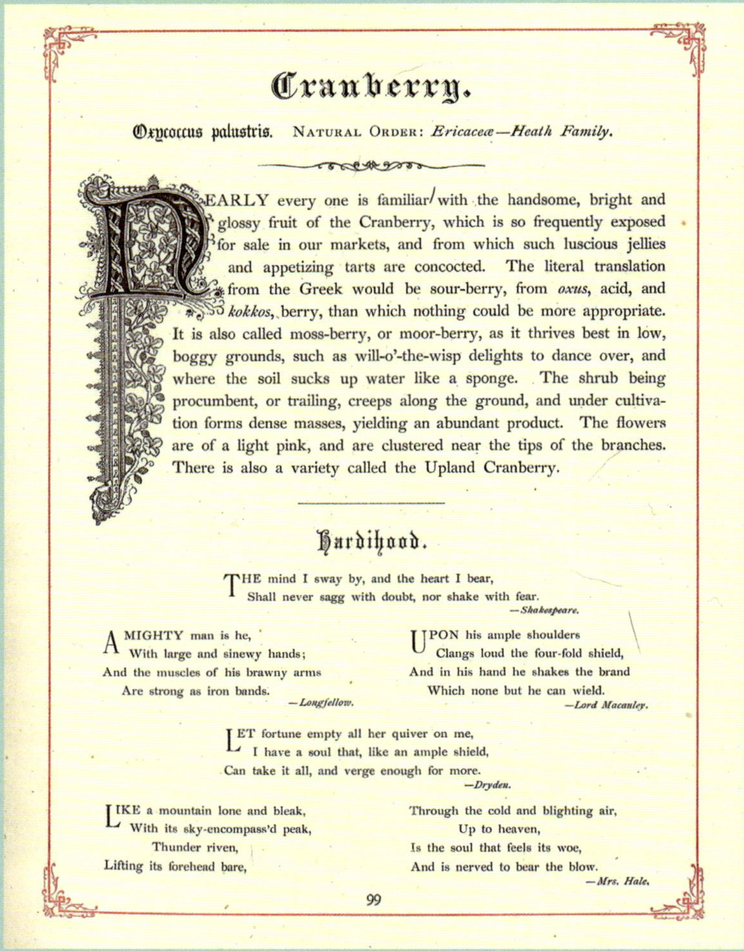

Cranberry.

Oxycoccus palustris. NATURAL ORDER: *Ericaceæ—Heath Family.*

NEARLY every one is familiar with the handsome, bright and glossy fruit of the Cranberry, which is so frequently exposed for sale in our markets, and from which such luscious jellies and appetizing tarts are concocted. The literal translation from the Greek would be sour-berry, from *oxus*, acid, and *kokkos*, berry, than which nothing could be more appropriate. It is also called moss-berry, or moor-berry, as it thrives best in low, boggy grounds, such as will-o'-the-wisp delights to dance over, and where the soil sucks up water like a sponge. The shrub being procumbent, or trailing, creeps along the ground, and under cultivation forms dense masses, yielding an abundant product. The flowers are of a light pink, and are clustered near the tips of the branches. There is also a variety called the Upland Cranberry.

Hardihood.

THE mind I sway by, and the heart I bear,
Shall never sagg with doubt, nor shake with fear.
—*Shakespeare.*

A MIGHTY man is he,
With large and sinewy hands;
And the muscles of his brawny arms
Are strong as iron bands.
—*Longfellow.*

UPON his ample shoulders
Clangs loud the four-fold shield,
And in his hand he shakes the brand
Which none but he can wield.
—*Lord Macaulay.*

LET fortune empty all her quiver on me,
I have a soul that, like an ample shield,
Can take it all, and verge enough for more.
—*Dryden.*

LIKE a mountain lone and bleak,
With its sky-encompass'd peak,
Thunder riven,
Lifting its forehead bare,
Through the cold and blighting air,
Up to heaven,
Is the soul that feels its woe,
And is nerved to bear the blow.
—*Mrs. Hale.*

99

FIGURE 5.44. "Cranberry," page 99, C. Turner, *The Floral Kingdom* (1877)

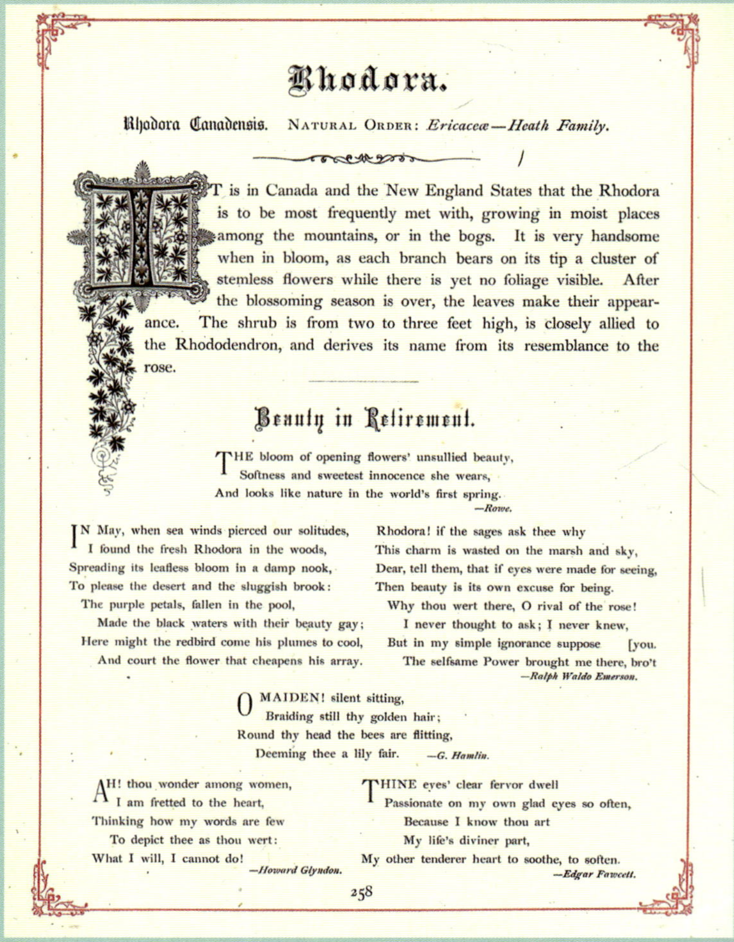

Rhodora.

Rhodora Canadensis. NATURAL ORDER: *Ericaceæ—Heath Family.*

IT is in Canada and the New England States that the Rhodora is to be most frequently met with, growing in moist places among the mountains, or in the bogs. It is very handsome when in bloom, as each branch bears on its tip a cluster of stemless flowers while there is yet no foliage visible. After the blossoming season is over, the leaves make their appearance. The shrub is from two to three feet high, is closely allied to the Rhododendron, and derives its name from its resemblance to the rose.

Beauty in Retirement.

THE bloom of opening flowers' unsullied beauty,
Softness and sweetest innocence she wears,
And looks like nature in the world's first spring.
—*Rowe.*

IN May, when sea winds pierced our solitudes,
I found the fresh Rhodora in the woods,
Spreading its leafless bloom in a damp nook,
To please the desert and the sluggish brook:
The purple petals, fallen in the pool,
Made the black waters with their beauty gay;
Here might the redbird come his plumes to cool,
And court the flower that cheapens his array.
Rhodora! if the sages ask thee why
This charm is wasted on the marsh and sky,
Dear, tell them, that if eyes were made for seeing,
Then beauty is its own excuse for being.
Why thou wert there, O rival of the rose!
I never thought to ask; I never knew,
But in my simple ignorance suppose [you.
The selfsame Power brought me there, bro't
—*Ralph Waldo Emerson.*

O MAIDEN! silent sitting,
Braiding still thy golden hair;
Round thy head the bees are flitting,
Deeming thee a lily fair.
—*G. Hamlin.*

AH! thou wonder among women,
I am fretted to the heart,
Thinking how my words are few
To depict thee as thou wert:
What I will, I cannot do!
—*Howard Glyndon.*

THINE eyes' clear fervor dwell
Passionate on my own glad eyes so often,
Because I know thou art
My life's diviner part,
My other tenderer heart to soothe, to soften.
—*Edgar Fawcett.*

258

FIGURE 5.45. "Rhodora," page 258, C. Turner, *The Floral Kingdom* (1877) and botany instruction. As a complement to these other genres in their nineteenth-century context, flower language books were significant, both conveying meaning and conferring importance. Early in the century, when natural history writers paid more attention to fauna than flora, Wirt's *Flora's Dictionary* gave readers several hundred pages of information about flowers. By the end of the century, James Terry White's *Captive Memories: Commemorative Verses Interwoven with California Flowers* (1897) expanded the floral lexicon to "the Sierras" and included species such as manzanita. In the decades between, flower language books made abundantly clear that American flora were well worth getting to know for their botanical, scientific, cultural, figurative, and aesthetic significance.

The exuberant abundance of flower language books included the discourses of their verbal texts—dictionary entries, literary excerpts, botany instruction, taxonomic information, folklore, cultivation information, and symbolism—as well as the number and diversity of species and corresponding emblems. These were in turn reinforced by the visual abundance of ornate bindings and beautiful illustrations. They contained copious resources for their readers to use in whatever ways they liked. But unlike nineteenth-century America's natural resources of water, soil, timber, and grasslands, the abundant materials of the flower language dictionary were figurative and representational and thus could not be exhausted in ways detrimental to the environment.

The next chapter focuses on a third archival object, *The Scriptural Album with Floral Illustrations*. It uses the language of flowers in interesting ways, and its floral illustrations rival those of flower language dictionaries for their vivid beauty. These albums featured their own forms of abundance, as friends filled the pages with their well-wishes, poetry, sketches, and more.

FIGURE 6.1. Title page, Louisa Jane Page, *The Scriptural Album with Floral Illustrations*, Ms. Harris Codex 1010, John Hay Library, Brown University

6

FROM THE ARCHIVES

"Friendship Is a Holy Thing"

The Scriptural Album with Floral Illustrations

Dear girl for thee I dedicate this Book
In which thine Eyes will often fondly look;
To read the wishes for thy happiness here penned,
By many a kind and ever constant friend.
And recollections of Youth's happy days,
Will gild the future with Hope's brightest rays
And like a spell, in many a lonesome hour,
Will cheer the Dearest with its nameless power.

—Mary W. Laughton, *The Scriptural Album with Floral Illustrations*

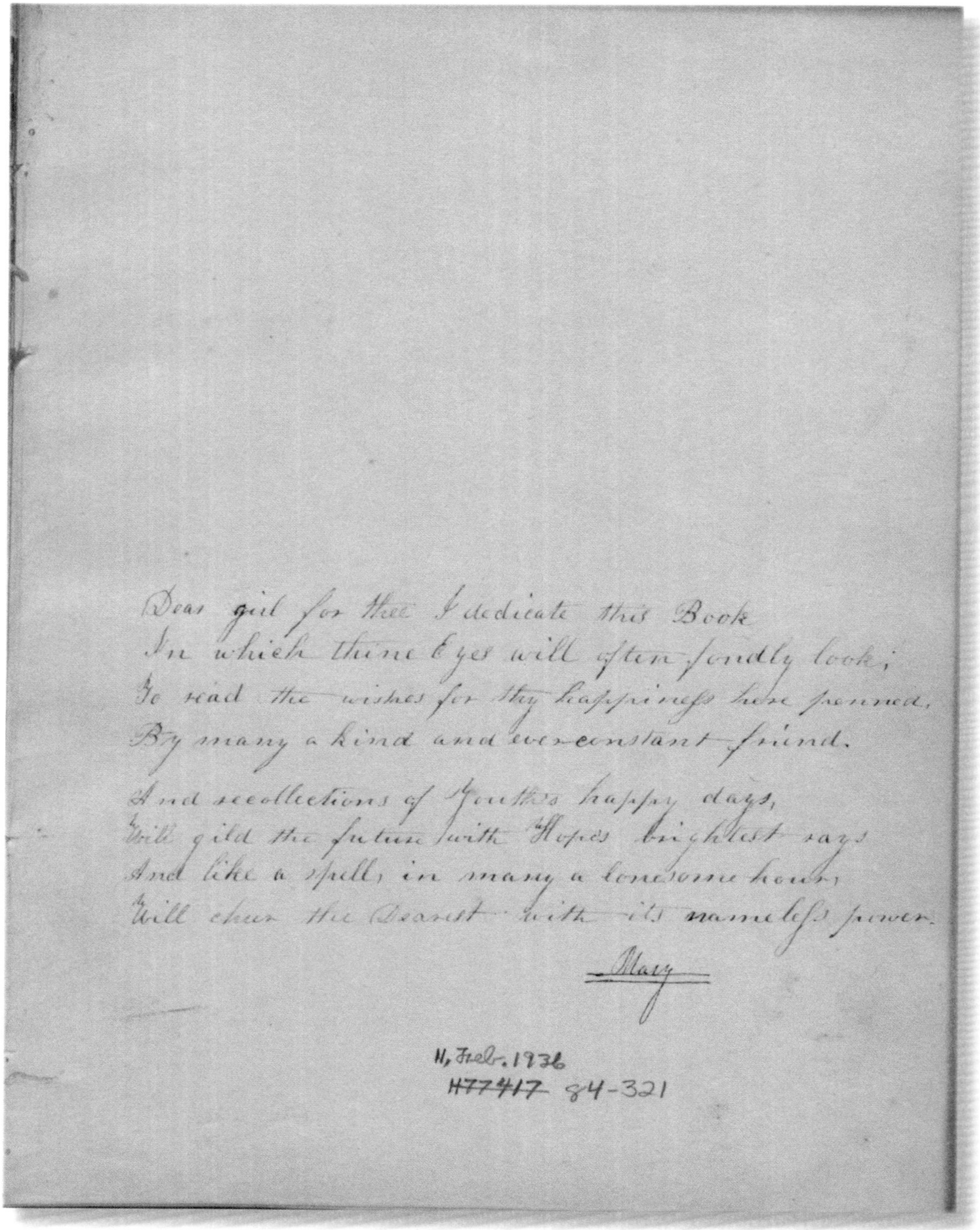

Dear girl for thee I dedicate this Book
In which thine Eyes will often fondly look;
To read the wishes for thy happiness here penned,
By many a kind and ever constant friend.

And recollections of Youth's happy days,
Will gild the future with Hope's brightest rays
And like a spell, in many a lonesome hour,
Will cheer the Dearest with its nameless power.

Mary

H, Feb. 1936
~~H77417~~ 84-321

FIGURE 6.2. Inscription, Mary W. Laughton, *The Scriptural Album with Floral Illustrations*, Ms. Harris Codex 1165, John Hay Library, Brown University

The opening inscription in the friendship album of Mary W. Laughton neatly summarizes a central purpose of such albums: for "many a kind and ever constant friend" to write "wishes for thy happiness" (figure 6.2). Preserved within the pages of "this Book," those "recollections of Youth's happy days" would later be certain to "cheer" "many a lonesome hour" of the album's owner, who might be far distant from the inscriber in time or space.

The friendship album, also known as a token album, was a staple of mid-nineteenth-century U.S. sentimental culture, in which earnest expressions of emotion, religious piety, and virtuous living were valued in men and expected of women. One facet of this culture was the sentimental botany of the 1830s to 1860s, as discussed in chapter 5. The middle decades of the nineteenth century in the United States also encompassed the social and political upheavals of slavery and abolitionism, imperialism and territorial expansion, rising capitalism and industrialization. While the prevalence of sentimental culture has sometimes been framed as a conservative response to so much change, it also inspired activism, particularly by women who promoted their causes in writing for the public, including in literary genres such as fiction and poetry. In turn, they were able to reach large audiences because book production technologies and distribution networks made print material more widely available than ever before, in newspapers, periodicals, and books.

Mid-century American women also wrote for more intimate and private audiences in their letters and diaries. The friendship album occupied a middle ground between fully public and entirely private; as Laura Zebuhr argues, "Friendship albums have a decidedly public genesis; they are 'made' through circulation and collaboration. Nineteenth-century albums might be exhibited in the areas of a house open to guests so that visitors could look through their pages and add their own contributions, and some collectors traveled with their albums and shared them with new acquaintances."[1] Like its cousins the commonplace book and the scrapbook, the friendship album was a bound, blank book for collecting memories. But whereas the commonplace book usually preserved quotations, poems, and passages of its owner's choosing, and the scrapbook archived clippings from newspapers and magazines, the friendship album was filled with other people's handwritten messages. These albums were often given on the occasion of a major life event: graduation from school, moving to a new place, or marriage. Inscriptions wished the owner well, gave pious advice, reflected on shared happy times, and promised remembrance for years to come.

Friendship albums flourished in the United States in the middle decades of the nineteenth century and remained popular well into the twentieth century. Like scrapbooks and commonplace books, they often included material copied from other sources. Historian Jenifer Blouin explains that "inscriptions within the albums were generally not original compositions. Many of the entries came from verses that had already been published in women's magazines and literary publications such as *Godey's Lady's Book*, *The Ladies' Repository*, and *Ladies' Cabinet of Fashion, Music, and Romance*. The transcribed verses dealt with a wide range of issues including death, nature, religion, family, and, of course, the joy of friendship."[2] These "not original compositions" can sound trite or overly sentimental to the modern ear. They might be impassioned expressions of religious faith, or metrically clunky stanzas of verse. After reading hundreds of album entries, I have not encountered any that would strike today's reader as informal or spontaneous: They were carefully chosen, serious, formal in tone, beautiful in their penmanship, and meant to be savored by the recipient over a lifetime.

I've pored through scores of friendship albums in archives. The books themselves are usually finished in beautiful bindings, like that of the *Leaves of Friendship* album by New York publisher Leavitt and Allen in the mid-1850s (figure 6.3). These intricate, luxury bindings indicate the preciousness of the friendship album, something worth keeping for years, a treasured item and frequent family heirloom. Publishers such as Leavitt and Allen or J. C. Riker offered multiple titles of friendship albums. Each title was generally available in several colors of cloth bindings, with varying gold-stamped images on their covers and spines. Riker published books such as *The Token Album* and *The Religious Album* in brown cloth, often a more affordable option than other colors of bindings (figures 6.4 and 6.5). These albums had fifty to sixty pages, most of them blank; some included a few full-page illustrations. In most of the friendship albums I've seen, such illustrations are rendered in black-and-white mezzotint and feature nature scenes or religious themes (figures 6.6 and 6.7). These albums' illustrations rarely went beyond duotone engravings in dark ink on a blank page. Their somber tones and themes support the gravity of the friendship album. No mere trifle, it was a place to reflect thoughtfully, in one's best penmanship. Album inscribers knew they were writing for posterity.

FIGURE 6.3. Cover, *Leaves of Friendship* album, courtesy of the American Antiquarian Society

FIGURE 6.4. (*left*) Cover, *The Token Album*

FIGURE 6.5. (*right*) Cover, *The Religious Album*

FIGURE 6.6. (*left*) "The Summer Stream," *The Token Album*

FIGURE 6.7. (*right*) "Moses on Mt. Horeb," The *Religious Album*

Imagine my delight, then, when I discovered *The Scriptural Album with Floral Illustrations*, first published by J. C. Riker in 1842. Like other friendship albums, it featured beautiful gold-stamped bindings, mostly blank pages, and a few illustrations interspersed throughout the book. Unlike the images in other albums, however, the ones in *The Scriptural Album* fairly leap off the page with their vibrant colors and mix of sacred and secular iconography. These vividly colored illustrations would have been painted by hand; chromolithography, which enabled the mass production of colored pages, became prevalent in the United States only later in the century.[3]

Figure 6.8 depicts the biblical story of Jeptha [*sic*], visually narrated in a black-and-white drawing along the bottom and right margin of the page. The figures are small and some fade into seemingly unfinished line drawings at the top of the stairs. What immediately catches the eye are the flowers, which are much larger than the human figures. Rich in accurate botanical detail, the flowers are numbered and linked to a key at the bottom of the page that reveals their meanings. In this version of the language of flowers, sweet pea and Tangier pea signify "An Appointed meeting," spiderwort "Transient Happiness," and crown imperial "Power. Majesty." All of these emblems are relevant to the story in Judges 11: Jephthah has "power and majesty" when he defeats the Ammonites but then experiences excruciatingly "transient happiness" when he realizes that his rash vow to God to ensure his victory will result in the sacrifice of his only child, a daughter. Her happiness is similarly transient when she accepts her fate, asking only for a two-month reprieve to go into the wilderness with her friends. She then dutifully keeps the "appointed meeting," and her father kills her.

FIGURE 6.8. "Jeptha" page, Augusta, *The Scriptural Album with Floral Illustrations*, Ms. Harris Codex 2063, John Hay Library, Brown University

This combination of sacred narrative and secular imagery distinguishes *The Scriptural Album with Floral Illustrations* from other friendship albums. Riker also published *The Religious Album* and *The Sacred Album*, with illustrations depicting scenes from both the

FIGURE 6.9. Title page, *The Floral Album*, courtesy of the American Antiquarian Society

FIGURE 6.10. Frontispiece, *The Boquet* [sic] *Album*, courtesy of the American Antiquarian Society

Hebrew Bible and the New Testament in duotone engravings, as seen in figure 6.7. Riker's *Floral Album* and *Boquet* [sic] *Album* included hand-colored, full-page illustrations of flowers, but these decorous arrangements have nothing to do with the language of flowers (figures 6.9 and 6.10). Christian piety, central to nineteenth-century U.S. sentimental culture, emphasizes morality and spiritual devotion. The language of flowers, by contrast, conveys a wide range of attributes, including those antithetical to the ideals of devout Christians, such as deceit, hatred, ingratitude, perfidy, revenge, rudeness, and treachery.[4] The distinctive element of *The Scriptural Album* is the language of flowers, and its use to further illuminate biblical scenes shows how its popularity extended beyond the flower language dictionaries discussed in the previous chapter.

I have researched sixteen examples of *The Scriptural Album with Floral Illustrations*. All were owned by women; the dates of their inscriptions range from 1843 to 1883. When locations were noted, most were in New England or the mid-Atlantic, with one album from the Midwest and another that extends to India.[5] The least inscribed has only one entry; the most, seventy-seven, filling every page. Their contents are handwritten, almost always in ink, with an occasional pressed flower, sketch, calling card, newspaper clipping, lock of hair, or piece of ribbon.

The Scriptural Album included six illustrated pages: the title page with Rebecca (figure 6.1), then Hagar, "Jeptha," Ruth, "The Adoration of the Wise Men," and "The Annunciation." The order of the illustrations follows their appearance in the Bible, from Genesis to Luke. These stories all foreground women, who were also the target market for friendship albums. *The Scriptural Album* is typical of mid-nineteenth-century friendship albums in many ways: The gold-stamped binding and gilt-edged pages made it a worthy gift book; its owners cherished their friends' and families' inscriptions for years; and these precious objects were kept long enough to land in special collections and archives many decades later. The scriptural subject matter reflects the centrality of Protestant piety to sentimental culture, but the images of large, brightly colored flowers that overwhelm the biblical scenes suggest a more subversive element. In each of the full-page illustrations, flowers dominate: They are vibrant, large, and exuberant, while the human figures are faint, small, and contained.

For example, the illustrated page titled "Hagar" depicts her sitting on a rock with her young son Ishmael's head in her lap in a parched desert environment (figure 6.11). The flowers, from bottom to top, are garden anemone, great flowered heath, and honeywort flowered heath, as indicated at the bottom of the page. As with the Jeptha image, the brightly colored flowers take over the page; Hagar and Ishmael, in black and white, fade into the background by comparison. Hagar and Ishmael are further contained within a scrolled border, while the flowers run riot in a dynamic arrangement of blooms, leaves, and tendrils. Consistent with the emblems provided for the garden anemone and great flowered heath, Hagar was indeed "forsaken" and endured "solitude," as told in Genesis 16:1–16 and 21:8–21. She was an enslaved woman in the household of Abraham, who impregnated her because his wife, Sarai, could not conceive. When Sarai mistreated the pregnant Hagar, she fled. She later returned to Abraham's household and bore him a son, Ishmael. When Sarai then had a son of her own (Isaac), she told Abraham to send Hagar and Ishmael away. They wandered in the desert of Beersheba, until God answered Hagar's prayers and they were saved.

The story of Hagar particularly resonated in the mid-nineteenth century, including for fiction writers such as the best-selling E. D. E. N. Southworth—with a protagonist named Hagar in *The Deserted Wife* (1850) and a version of the story in a later novel, *Ishmael* (1876)—and H. Marion Stephens, in *Hagar the Martyr* (1855). The biblical Hagar's enslaved status and Egyptian origins signified in divergent ways, often depending on the race of the author and the extent to which Hagar is portrayed as tragic or triumphant.[6] Women of the Bible inspired a number of books at mid-century, such as Eliza R. Steele's *Heroines of Sacred History* (1841) and Frances M. Caulkins's *Eve and Her Daughters of Holy Writ, Or, Women Of The Bible* (1850), which capitalized on the popularity of collective biography in America at this time.[7]

FIGURE 6.11. "Hagar" page, Augusta, *The Scriptural Album with Floral Illustrations*, Ms. Harris Codex 2063, John Hay Library, Brown University

The Hagar page in figure 6.11 is blank, but in other albums, the illustrated pages are often inscribed, as in a page showing "The Adoration of the Wise Men" (figure 6.12).

To Miss Dickinson

There is a flower of heavenly worth
That blooms within the mind
Its origin is not of earth
Tis heavenly and divine

It sheds a lustre round the heart
It lifts the thoughts on high
It bids us cling to friendships gem
For purest sympathy

This amaranthine flower is love
For this the Saviour died
For this he left the courts above
For this was crucified

Sarah A Bumstead

Chicopee Mass. March 11th - 50
(emphasis in the original)

The verse inscription features a "flower of heavenly worth," the "amaranthine flower" of love. In the language of flowers, the emblem for amaranth is "immortality," likely because, as Catharine Waterman notes

FIGURE 6.12. "The Adoration of the Wise Men" page, Mary Ann Dickinson, *The Scriptural Album with Floral Illustrations,* courtesy of the American Antiquarian Society

in *Flora's Lexicon*, "when dead its flowers retain their rich scarlet colour."[8] The flowers for the Adoration of the Wise Men page are, appropriately, *Lychnis coronata* ("religious enthusiasm") and *Convolvulus purpurea* ("night"). The inscriber Sarah Bumstead may have chosen this page specifically for its connection to the birth of the Savior mentioned in the verse she inscribed, or perhaps because, like the wise men, she had a gift for Miss Dickinson: "friendships gem."

Some inscribers refer specifically to the narrative depicted in the illustration, as in a page from Tacy Anna Walker's album (figure 6.13):

> *She fell upon her face, and bowed her head*
> *Low to the ground, and modestly replied*
> *"O what am I, that thou shoulds't knowledge take,*
> *Of one so poor—a stranger too, to thee;*
> *Why in thine eyes have I found grace?"*
> *Oh sweet humility, a precious grace,*
> *May we profess it—even make it ours*
> *And may it ever in our actions shine.*
> *S. S. W.*

The inscriber's quotation is from Ruth 2:10, where she is gleaning in the fields belonging to Boaz and he comes to greet her. The flowers and their listed meanings are purple rudbeckia (justice), Persian cyclamen (diffidence), coreopsis (love at first sight), and scarlet azalea (your blush has won me). Rather than focusing on the well-known story of Ruth's devotion to her mother-in-law Naomi ("Whither thou goest I will go," Ruth 1:16), this scene depicts the first meeting of Ruth and Boaz, who eventually marry. Ruth's agricultural labor is centered at the bottom of the page, with Boaz at the right. To the left, the leaves of the Persian cyclamen run over into the ground where Ruth is kneeling, an insistent reminder

FIGURE 6.13. "Ruth" page, Tacy Anna Walker, *The Scriptural Album with Floral Illustrations*, Ms. Harris Codex 1128, John Hay Library, Brown University

that in *The Scriptural Album with Floral Illustrations*, the flowers are the main attraction, with the scriptural figures in a supporting role, much like the vivid flowers and shadowy landscapes in Emma Embury's *American Wild Flowers in Their Native Haunts* (see introduction).

FIGURE 6.14. "The Annunciation" page, Elizabeth Grier, *The Scriptural Album with Floral Illustrations*, courtesy of the American Antiquarian Society

Despite the centrality of floriography in the illustrated pages, inscribers in *The Scriptural Album* examples that I have seen do not refer to the language of flowers. They do, however, mention flowers more generally, as in this example from "The Annunciation" page in Elizabeth Grier's album (figure 6.14):

Through storm & sunshine,
Flower and thorn,
Pursue thine own way—
Nor let thy better hopes be born
Of things that must decay.
Rejoice with trembling,
Mourn with hope,
Take life as life was given,
Its rough ascent
Its flowery slope
May lead alike to Heaven.

That thus may your way be
and thus life's close
is the sincere wish
of your friend & teacher

Fannie Henry

Oakland, Feb. 20th
1847.
(emphasis in the original)

In Fannie Henry's verse inscription—the widely reprinted poem "Counsels" by Bernard Barton—the flower/thorn dichotomy parallels that of storm/sunshine, suggesting that the pursuit of "thine own way" should occur on scales both grand (meteorological) and small (floral).[9] The "flowery slope" and "rough ascent" similarly provide two equally likely paths "to Heaven." The choice of the Annunciation page for this inscription is not clear, but the meaning of two of its featured flowers—"a message" (*Iris versicolor* and Persian iris)—is certainly relevant to "the sincere wish" of inscriber Fannie Henry, who taught music at the Oakland Female Institute in Norristown, Pennsylvania.[10]

Elsewhere in Grier's album, I. H. Hartville copied "The Rose Without a Thorn" and Mary Elizabeth shared Daniel Clement Colesworthy's "A Little Word"; the first stanzas of each poem appear below.

The Rose Without a Thorn

The flower, in all its sweetness,
Must wither and decay,
And soon, my child, time's fleetness
Will bear thy frame away.

A Little Word

A little word in kindness spoken,
A motion or a tear,
Has often healed a heart that's broken,
And made a friend sincere.[11]

Grier's friend Jane copied Charles Force Deems's "Lines Written in the Album of a Little Stranger" (1839). As late as 1880, Deems's poem appeared in Marie Tomisetta's *Poetical Selections from Celebrated Authors, Suitable for Inscription in Autograph Albums, Comprising a Choice Collection of Humorous, Friendly, Affectionate and Miscellaneous Verses,* showing that copying poems into autograph albums remained a popular pastime throughout the nineteenth century.[12]

In the albums I have studied, verse inscriptions far outnumber prose inscriptions. In nineteenth-century America, poetry was a familiar part of people's daily lives: They memorized the verses found in newspapers, magazines, and anthologies; they exchanged poems in letters and gift messages; they read poetry aloud to one another, for entertainment and edification; and they frequently recorded poems in friendship albums. Tacy Anna Walker's album included "The Dial of Flowers," a poem by Felicia Hemans, based on Linnaeus's floral horologe that charted the hours of the day according to when various flowers opened.[13] The inscriber, who signed only the initials M. E. S., provided the title of the poem but not its original author; typically, inscribers supplied neither title nor author for what they copied (figure 6.15). While Linnaeus used particular species for his dial, Hemans's poem more broadly addresses the concept of time. Like much of the sentimental poetry valued by nineteenth-century Americans, "The Dial of Flowers" is both melancholy and hopeful, ranging in subject from prelapsarian Eden to Arcadian shepherds to real life on earth, as the final two stanzas show. These lines might have cheered Tacy Anna

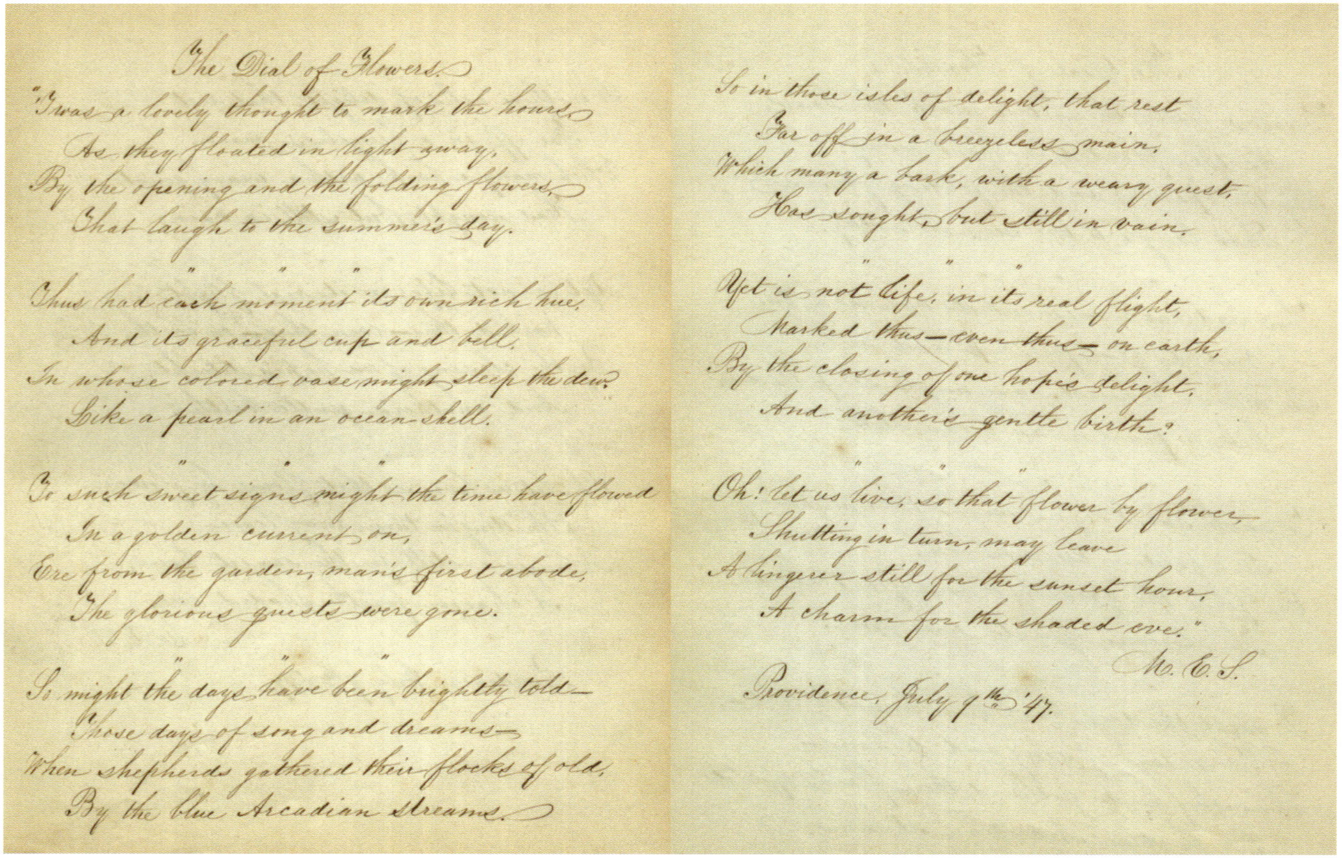
The Dial of Flowers.

'Twas a lovely thought to mark the hours,
As they floated in light away,
By the opening and the folding flowers,
That laugh to the summer's day.

Thus had each moment its own rich hue,
And its graceful cup and bell,
In whose colored vase might sleep the dew,
Like a pearl in an ocean-shell.

To such sweet signs might the time have flowed
In a golden current on,
Ere from the garden, man's first abode,
The glorious guests were gone.

So might the days have been brightly told—
Those days of song and dreams—
When shepherds gathered their flocks of old,
By the blue Arcadian streams.

So in those isles of delight, that rest
Far off in a breezeless main,
Which many a bark, with a weary quest,
Has sought, but still in vain.

Yet is not life, in its real flight,
Marked thus—even thus—on earth,
By the closing of one hope's delight,
And another's gentle birth?

Oh! let us live, so that flower by flower,
Shutting in turn, may leave
A lingerer still for the sunset hour,
A charm for the shaded eve."

M. E. S.

Providence, July 7th '47.

FIGURE 6.15. "The Dial of Flowers," Tacy Anna Walker, *The Scriptural Album with Floral Illustrations*, Ms. Harris Codex 1128, John Hay Library, Brown University

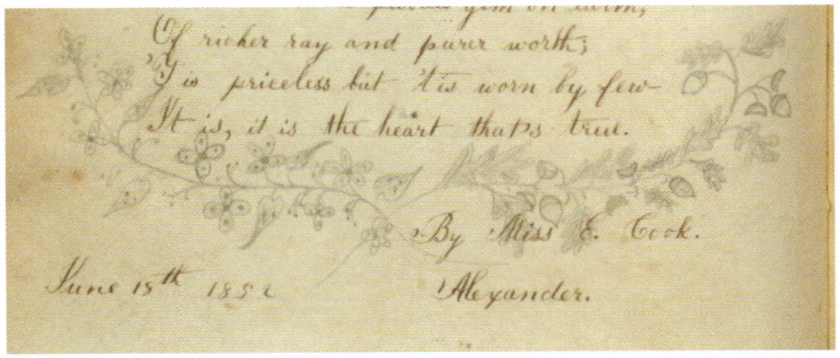

FIGURE 6.16. (*left*) Full-page drawing signed "Isabell Pinkham"; FIGURE 6.17. (*top right*) Header drawing; FIGURE 6.18. (*above*) Footer drawing. Lucy Pinkham, *The Scriptural Album with Floral Illustrations*, Ms. Harris Codex 1120, John Hay Library, Brown University

Walker, with the reminder that "the closing of one hope's delight" gives birth to new ones, and the exhortation to live knowing that even as flowers shut "in turn," there is always something that lingers to "charm . . . the shaded eve."

Although these albums were intended primarily for people to *write* messages to the albums' owners, sometimes friends and family left drawings and other visual matter. Lucy Pinkham's album includes several sketches of flowers, all done in pencil (figures 6.16–6.18). Figure 6.16 is signed by Isabell Pinkham, Lucy's sister. Figures 6.17 and 6.18 are on different pages in the album and might have been done by someone other than the inscriber, as the drawings are in pencil and the inscriptions in ink. These smaller drawings might also have been done by Isabell Pinkham, whose style they resemble, and perhaps at a later date. These decorative touches reflect how flower drawing and painting were part of a young lady's artistic repertoire. Books such as *A Series of Progressive Lessons, Intended to Elucidate the Art of Flower Painting in Water Colours* (1820) and Thomas Addison Richards's *The American Artist; or, Young Ladies' Instructor, in the Art of Flower Painting, in Water Colours* (1838) would have been available to middle-class families and were also used in schools that taught these "finishing" accomplishments.

J. C. Riker, the publisher of *The Scriptural Album, The Token Album, The Religious Album, The Floral Album,* and *The Boquet Album,* also offered *The Drawing Album.* Like friendship albums, drawing albums were beautifully bound with gold-stamped covers and had blank pages for the owner's sketches. The Winterthur Library has two drawing albums with flower paintings, one owned by Susan Nichols and the other by Eleuthera du Pont (1806–1876), daughter of gunpowder manufacturer Eleuthère du Pont, whose family later established the Winterthur estate. Art historian Joy Peterson Heyrman explains that "drawings shared space with the sentimental fiction and illustrated gift books that both pictured and shaped American middle-class culture. . . . Drawing instruction was widespread and drawing was an expected accomplishment of an educated person. Drawing albums served as mementi of personal association and as entertainment during social visiting, while they assembled and commemorated social contact, and promulgated a particular form of sentimental culture."[14]

FIGURE 6.19. Back cover, Martha B. Morrill, *The Scriptural Album with Floral Illustrations,* Ms. Harris Codex 1031, John Hay Library, Brown University

FIGURE 6.20. Back cover, Augusta, *The Scriptural Album with Floral Illustrations,* Ms. Harris Codex 2063, John Hay Library, Brown University

Isabell Pinkham's penciled bouquet (figure 6.16) recalls those in flower language dictionaries, and other floral embellishments in Lucy Pinkham's album resemble the decorative fonts and patterns used in flower books more generally.

The Scriptural Album's cover designs also used floral imagery; the examples I have seen are bound in red or brown cloth, with gold-stamped images on the front, back, and spine (figures 6.19 and 6.20). The cover images vary from album to album, but one popular stamp features the image of Flora, also seen on the covers of flower language books such as Sarah Edgarton's *Fables of Flora* (figure 5.24). In figure 6.19, the back cover of Martha Morrill's album has the central image stamped in gold, with the surrounding border blind-stamped; the back cover of the "Augusta" album (figure 6.20) has the reverse: gold stamping for the border, with the central image blind-stamped. These options for gold and blind stamping enabled publishers to sell a variety of album covers, and as book historian Sue Allen notes, "in this period [1840s–50s] the canons of cover design virtually demanded a blind-stamped border, ranging in complexity from a plain rule to ornate scrolling cornerpieces," as well as the center pictorial images.[15] While many bindings of *The Scriptural Album* used the image of Flora, others featured different pictorial central stamps, such as a scene of Jesus and the woman at the well (figures 6.21; see also figure 6.5) and the ornate decorative borders that Allen references (figure 6.22). Overall, the combination of flowers in *The Scriptural Album*'s cover images, hand-colored printed illustrations keyed to floriographic meaning, and individuals' drawings creates a layered, multiply connotative floral display of popular botany that was material, aesthetic, and personal, like the herbaria and their offshoots discussed in chapter 3.

FIGURE 6.21. Back cover detail, Mary W. Laughton, *The Scriptural Album with Floral Illustrations*, Ms. Harris Codex 1165, John Hay Library, Brown University

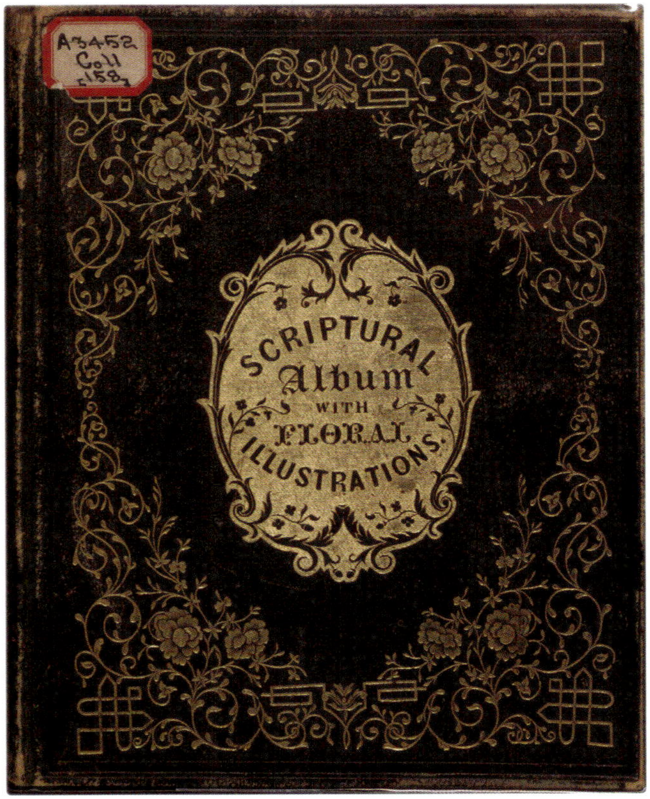

FIGURE 6.22. Front cover, Martha Downing, *The Scriptural Album with Floral Illustrations*, Ms. Harris Codex 1172, John Hay Library, Brown University

Occasionally an album might include verbal and visual examples of flowers in the same inscription, as in two facing pages of an entry titled "Copied by Maria P. Davies for her friend Mrs. Martha Downing" (figure 6.23). Rarely do inscribers provide the source of copied material, but Davies does: "from Willis's Magazine Dec. 1830." The inscribed poem, "Flowers," appeared in *The American Monthly Magazine*, edited by Nathaniel P. Willis. The poet was identified as J. H. S., likely Julia H. Scott, the "Poet of Sheshequin" (Pennsylvania).[16] The epigraph to the published poem,

No marvel woman should love flowers, they bear
So much of fanciful similitude
To her own history

aligns with *The Scriptural Album*'s combination of flowers and the biblical women Rebecca, Hagar, Ruth, Jephthah's daughter, and Mary, mother of Jesus. The poem has a melancholy tone, detailing the sadness of various flowers from being

by some reckless wanderer torn
. . . and borne
To some far clime.

The forty-two-line poem's final stanza concludes:

Flowers are like fond hearts—for they
When beauty shall have passed away,
Breathe the same fragrance to the air
As when the bloom was lingering there
So hearts, though by unkindness broken,
Will still retain some cherish'd token.

The "lingering" materiality of flowers can be found elsewhere in Downing's album. A pressed specimen (figure 6.24) appears on a blank page, after several blank

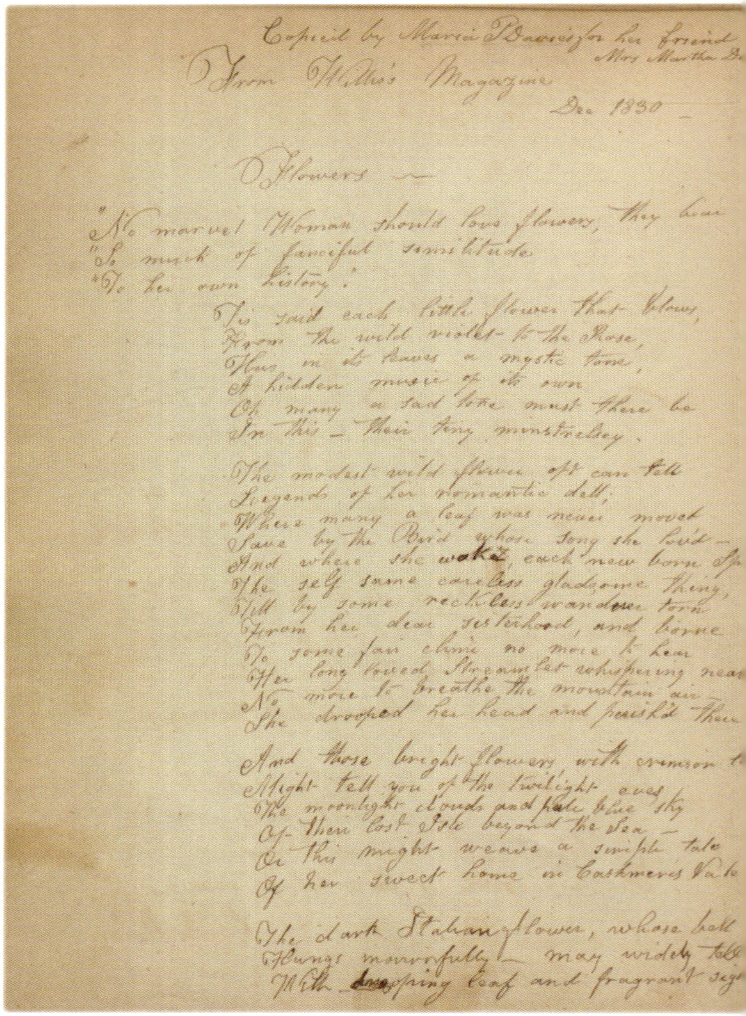
Copied by Maria P Davies for her friend Mrs Martha D
From Willis's Magazine
Dec 1830
Flowers ~
"No marvel Woman should love flowers, they bear
"So much of fanciful similitude
"To her own history."
'Tis said each little flower that blows,
From the wild violet to the Rose,
Has in its leaves a mystic tone,
A hidden music of its own
Of many a sad tale must there be
In this – their tiny minstrelsy.
The modest wild flower oft can tell
Legends of her romantic dell,
Where many a leaf was never moved
Save by the Bird whose song she lov'd –
And where she woke, each new born sp
The self same careless gladsome thing,
Till by some reckless wanderer torn
From her dear sisterhood, and borne
To some fair clime no more to hear
Her long loved Streamlet whispering nea
No more to breathe the mountain air –
She drooped her head and perish'd there.
And those bright flowers with crimson t
Might tell you of the twilight eves
The moonlight clouds and pale blue sky
Of their lost Isle beyond the Sea –
Or this might weave a simple tale
Of her sweet home in Cashmere's Vale
The dark Italian flower, whose bell
Hangs mournfully – may widely tell
With drooping leaf and fragrant sig

pages, with no indication of who preserved it there or why. It looks like lily of the valley (*Convallaria majalis*), known for its lovely fragrance. Its shape echoes that of the drawing in Maria Davies's inscription (detail, rotated, figure 6.25; full page, figure 6.23). Across several pages in this one album, a preserved specimen, sketch, and poem collaborate in a multimedia homage linking memory and flowers, "beauty" and "fragrance," "fond

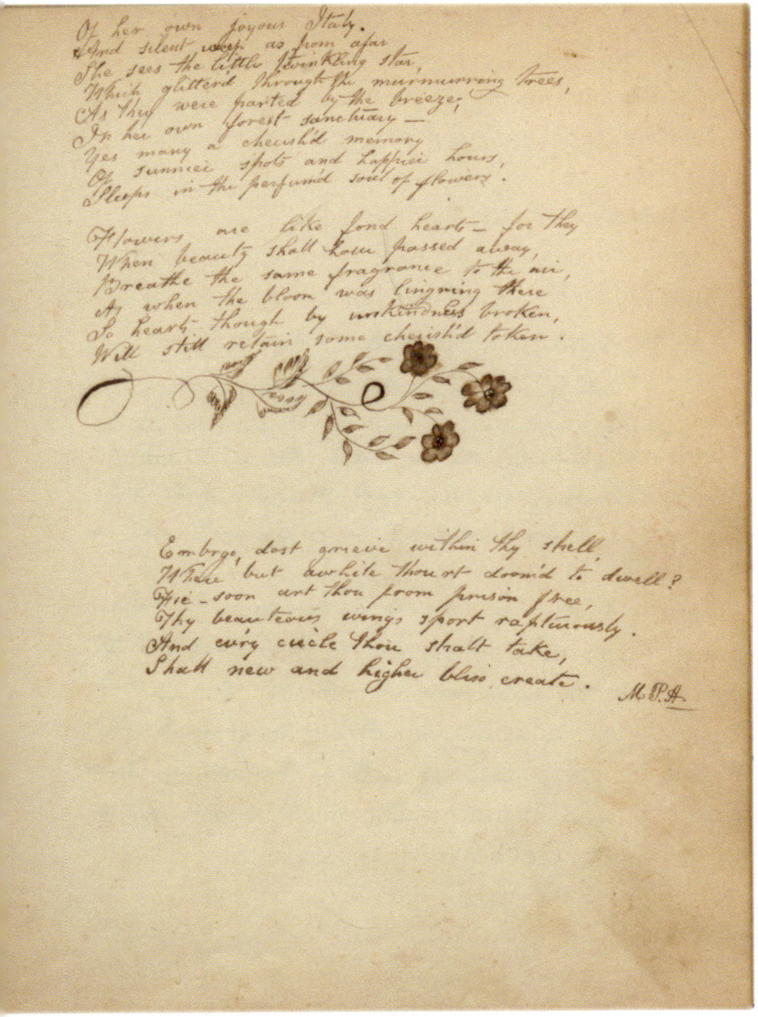

Of her own joyous Italy.
And silent weep as from afar
She sees the little twinkling star,
Which glitter'd through the murmuring trees,
As they were parted by the breeze;
In her own forest sanctuary—
Yes many a cherish'd memory
Of sunnier spots and happier hours,
Sleeps in the perfum'd soul of flowers.

Flowers are like fond hearts—for they
When beauty shall have passed away,
Breathe the same fragrance to the air,
As when the bloom was lingering there
So hearts though by unkindness broken,
Will still retain some cherish'd token.

Embryo, dost grieve within thy shell,
Where but awhile thou art doom'd to dwell?
Ere—soon art thou from prison free,
Thy beauteous wings sport rapturously.
And every circle thou shalt take,
Shall new and higher bliss create.
M.P.A.

FIGURE 6.23. (*left*) Maria Davies inscriptions;
FIGURE 6.24. (*above left*) Pressed specimen;
FIGURE 6.25. (*above right*) Drawing detail (rotated).
Martha Downing, *The Scriptural Album with Floral Illustrations*, Ms. Harris Codex 1172, John Hay Library, Brown University

hearts," and the "cherish'd token" of this inscription and of the album as a whole.

An album can tell a captivatingly incomplete story of its owner's life, as does Mary W. Laughton's *Scriptural Album*. It contains only nine inscriptions, two of which are just names and dates; the first inscription appears in figure 6.2. The inscribers' well-wishes indicate that Mary Laughton was leaving home: "I would not say farewell," writes S. A. Ballard in the second inscription. "It is a melancholy word." Mary Laughton seems to be journeying far away, for as Ballard continues, "We meet no more till many months are past." The third inscription, by Mary A. Warren, is dated October 12, 1846, in Lancaster (Massachusetts) and expresses friendship by copying a poem published as "To Emeline."

Midst all the clear, and valued ones who claim
A kind remembrance in thy youthful breast
Respected friend, I fain would have my name,
By memory's hand inscribed among the rest—
And when in future years, though far away
And many kindly whispered words of praise,
Or when the untiring powers of memory,
Love to recall the scenes of by gone days—
Deign to bestow a passing thought on me,
However bless'd may be thy happy lot,
And if thou art clouded in adversity;
Midst thoughts of dearer friends, forget me not.[17]

Through the quoted poem, Warren posits a time "in future years" when her friend, "though far away," will still "deign to bestow a passing thought" on her. Given the uncertainties of life, particularly for those traveling far from home, Warren notes the equal possibilities of a "happy lot" or being "clouded in adversity." Another inscriber wrote these couplets:

Friendship is a holy thing,
Surer, than a ripling [sic] stream,
On it goes, and on it flows:
Forever stronger, as it goes.
When this you see, remember me,
Though many miles we distant be.

The first line, "friendship is a holy thing," inspired this chapter's title, and the inscribed verse is another contribution to the album's themes of friendship, religion, and being "many miles" distant.

The fourth inscription, titled "To My Dear Only Sister" and dated October 15, 1846, in Boston, is by Caroline, who clearly grieves her sister's departure:

Am I called upon so suddenly to bid you farewell? My bosom heaves. My heart is filled to overflowing for thee. . . . If it is our fate never more to meet in this world, may we both be reconciled with submission, and think it is all for the best . . .

Though thousands of miles may separate us, yet we may have sweet communion with each other. Live to do good if you wish for happiness. May we walk in the paths of righteousness and lead such lives that we may hope to meet in Heaven.

It seems certain that Mary Laughton is indeed traveling far away, "thousands of miles," perhaps even somewhere dangerous. Caroline considers the very real possibility that it may be the sisters' "fate never more to meet in this world."

The fifth inscriber, Susan Delia Hamblet, signs herself "your affectionate cousin" and reiterates both the good wishes and probable sorrows shared by other friends. In a lengthy prose inscription, Hamblet notes that instead of asking "that your lot might be free from sorrow . . . I would rather ask that you might always love the hand that leads you, though your path be dark and beset with dangers. I would ask that the peace of believing and of trust in God might be your portion, though each earthly prospect should be blighted and pass away." Hamblet's closing wishes continue in the fervently religious mode of the entire inscription: "When life is done with us and we with it, we with Christ and our beloved friends and the seraphic hosts shall meet where parting will never come, No never Never Never come."

The ninth and final inscription in Laughton's album comes after a number of blank pages. It fills two facing pages and is ominously titled "To Mary = in Heaven" (figure 6.26).

If I had thought thou coud'st have died,
I might not weep for thee;
But I forgot, when by my side,
That thou coud'st mortal be:
It never through my mind had past,
The time would e'er be o'er,
And I on thee should look my last,
And thou should'st smile no more.

And still upon that face I look,
And think twill smile again;
And still the thought I will not brook,
That I must look in vain!
But when I speak—thou dost not say
What thou ne'er left'st unsaid;
And now I feel, as well I may
Sweet Mary, thou art dead!

If thou would'st stay e'en as thou art,
All cold, and all serene—
I still might press thy silent heart,
And where thy smiles have been!
While e'en thy chill, bleak corse I have,
Thou seemest still my own;
But there I lay thee in thy grave—
And I am all alone!

I do not think, where'er thou art,
Thou hast forgotten me;
And I, perhaps may sooth this heart,
In thinking too of thee:
Yet there was round thee such a dawn
Of light ne'er seen before,
As fancy never could have drawn,
And never can restore!

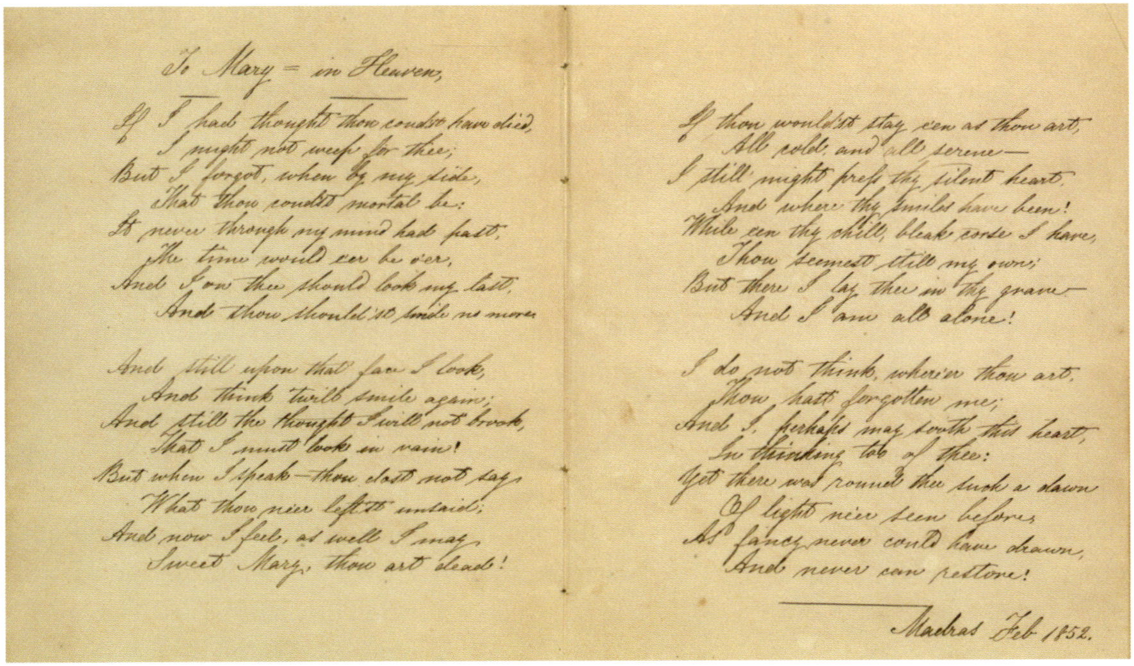
To Mary — in Heaven,

If I had thought thou coudst have died,
I might not weep for thee;
But I forgot, when by my side,
That thou coudst mortal be:
It never through my mind had past,
The time would e'er be o'er,
And I on thee should look my last,
And thou should'st smile no more.

And still upon that face I look,
And think twill smile again;
And still the thought I will not brook,
That I must look in vain!
But when I speak—thou dost not say
What thou ne'er left'st unsaid;
And now I feel, as well I may,
Sweet Mary, thou art dead!

If thou would'st stay een as thou art,
All cold, and all serene—
I still might press thy silent heart,
And where thy smiles have been!
While een thy chill, bleak corse I have,
Thou seemest still my own;
But there I lay thee in thy grave—
And I am all alone!

I do not think, where'er thou art,
Thou hast forgotten me;
And I, perhaps may sooth this heart,
In thinking too of thee:
Yet there was round thee such a dawn
Of light ne'er seen before,
As fancy never could have drawn,
And never can restore!

Madras Feb 1852.

FIGURE 6.26. Final inscription, Mary W. Laughton, *The Scriptural Album with Floral Illustrations*, Ms. Harris Codex 1165, John Hay Library, Brown University

"Sweet Mary, thou art dead!" The shock of these words still registers 170 years later. The inscription is dated "Feb. 1852" in Madras. The inscribed poem is "To Mary" by Charles Wolfe, an Irish poet who died in 1823; the poem, like so many others in these albums, was frequently reprinted and even quoted in other albums.[18]

The neat penmanship and ordered arrangement of alternately indented lines suggest that the bereaved inscriber carefully copied them into Mary's album. But why would this person directly address a dead woman, in *her* album? She did not live to read this sorrowful inscription; who, then, was its intended audience? The suggestive clues of Mary's album—her voyage far away, its danger, the possibility that she might not return, the highly religious inscriptions by friends and family members—lead to a probable identification. A genealogical record states that "Andrew Bancroft married Mary Laughton (daughter of Stephen Laughton) in 1847; had three children. Both Andrew and Mary died overseas. Andrew died (age 45) in Hong Kong on 8 May 1866; Mary died in Madras, India on 5 February 1852 (age 25)."[19] Given the ardent religious language of some of Laughton's inscribers, a likely reason that she and Andrew went abroad was to serve as missionaries: Both India (where Mary died, in Madras) and Hong Kong (where Andrew died) had active American missions at this time.

As missionaries abroad from the late 1840s to 1860s, they might have gone under the auspices of the American Board of Commissioners for Foreign Missions (ABCFM, or American Board). Founded in 1810, it was "the oldest and largest foreign missionary organization in the United States [and] the first of the voluntary societies produced by the aggressive evangelical activism of the period."[20] By 1850, around the time the Bancrofts were in India, "the Board was in charge of 40 percent of all American missionary personnel."[21] Like Laughton and Bancroft, most of the ABCFM's missionaries were from the Northeast. While Mary might have traveled as a single woman, it's far more likely that she accompanied her husband, especially if they were missionaries with the American Board: By the 1830s, the ABCFM "prohibited unmarried persons from entering the mission field" and "required young men to be engaged at least two months before entering the mission field. . . . Often these missionary couples would leave for foreign lands within a week after their marriage." Mary Laughton and Andrew Bancroft were married in 1847 and may have left for India shortly thereafter, especially given the dates in her album from October 1846. Many missionaries at this time expected to serve the rest of their lives away from the United States and die abroad. Perhaps they hoped for ripe old age at their posts, but far more likely was death from diseases such as malaria or cholera or, for young women like Mary Laughton, in childbirth.[22]

I have discussed Mary Laughton at length to give a sense of the roles that friendship albums played. Laughton's album conveys the despair at her death, the deep faith of her well-wishers, and the danger that her journey entailed—all in only nine inscriptions. Albums like Mary's were carefully kept for years, perhaps even across oceans. As treasured memorial objects, they provided a place to record significant events in the life of the album owner, sometimes even including her death. Mary Laughton may have taken the album abroad with her, and Andrew Bancroft may have passed it along to

FIGURE 6.27. Adoration of the Wise Men page, Augusta, *The Scriptural Album with Floral Illustrations*, Ms. Harris Codex 2063, John Hay Library, Brown University

FIGURE 6.28. The Annunciation page, Augusta, *The Scriptural Album with Floral Illustrations*, Ms. Harris Codex 2063, John Hay Library, Brown University

one of their children. Or, given the small number of inscriptions, the album might have stayed in the United States, and a friend or relative memorialized Laughton with the poem "To Mary." Someone kept it, and eventually it ended up at the John Hay Library at Brown University.

The inscriptions in these albums are decorous and serious, as befits the scriptural content of the illustrated pages. Several of the figures—the wise men, Mary, Hagar and Ishmael—are depicted within scrolling borders, confining them in appropriately decorative frames (figures 6.27–6.29; red outlines added). Similarly,

FIGURE 6.29. (*above left*) Hagar page; FIGURE 6.30. (*above*) "Jeptha" page; FIGURE 6.31. (*left*) Rebecca (title page). Augusta, *The Scriptural Album with Floral Illustrations*, Ms. Harris Codex 2063, John Hay Library, Brown University

in figure 6.30, Jephthah and his daughter are placed outside a straight line at the right side of the page, while the iris at the lower left trails into the black-and-white image at the bottom of the page, much like the leaves of the cyclamen on the page featuring Ruth (figure 6.13). The title page of the album shows Rebecca demurely kneeling within a rectangular proscenium at the bottom of the page (figure 6.31). By contrast, the vivid red and pink hibiscus at the top left is boisterously budding *and* blooming, with its showy, phallic stamen reaching upward, in stark contrast to Rebecca's downcast gaze.

As I have shown in previous chapters, the forms of nineteenth-century American flower books often exceeded the boundaries of their genres: Botany textbooks were used as field guides, impromptu herbaria, and even excursion journals. The herbarium evolved from loose pages used for scientific classification to ready-made collections of pressed flowers as travel souvenirs. The language of flowers existed not only in book-length dictionaries but also as appendices in botany textbooks—and in *The Scriptural Album with Floral Illustrations*. The bright, botanically accurate, hand-colored flowers in these archival treasures burst with life beyond the constructed borders enclosing faintly drawn people, showcasing their beauty and ability to signify.

The conclusion begins with flower fables: In this iteration of popular botany, the floral subject transcends botanical taxonomy, becoming personified and sometimes even hybridized as fairies, whose wings liken them to insects and provide a range of motion unavailable to rooted plants. As these personified flowers take flight, they instill a sense of freedom, while also conveying important messages of valuing diversity and appreciating differences.

FIGURE C.1. "Tiger Lily," Clarkson, *The Gathering of the Lilies* (1877)

Tiger Lily.

Surely, friends, I bear your name;
Do not look at me with blame,
And so sadly put to shame
My yellow face.
I am lily-born like you;
But the morning bath in dew
Will not make me fair to view
As my race."

Conclusion

Critical Plants and Fables of Transformation

The first object that met her vision was a tulip superbly arrayed in scarlet and gold. When she was aware of the intention of her neighbor, her cheeks flamed with anger, her eyes flashed indignantly, and she haughtily refused to change her proud robes for the garb the rose tree had prepared for her. . . . The snow drop drew around her robes of virgin white; she would not willingly exchange them for the most brilliant attire that ever decked a flower's form; to her they were the emblems of purity and innocence.

—Frances Ellen Watkins Harper, "The Mission of the Flowers" (1869)

"I propose our stately cousin, Violet Aster, for queen this year. Whoever agrees with me, say Aye."

Quite a shout went up from all the Asters; and the late Clovers and Buttercups joined in it, for they were honest, sensible flowers, and liked fair play. To their great delight the Pitcher-plant, or Forefathers' Cup, said "Aye" most decidedly, and that impressed all the other plants; for this fine family came over in the "Mayflower," and was much honored every where.

But the proud Cardinals by the brook blushed with shame at the idea of a queen; the Fringed Gentians shut their blue eyes that they might not see the bold Asters; and Clematis fainted away in the grass, she was so shocked.

—Louisa May Alcott, "A Flower Fable" (1887)

FIGURE C.2. Trade card designed by Charlotte Perkins Gilman, early 1880s, Jay T. Last Collection of Horticulture Prints and Ephemera, 1840–1933, series 1, binder 1, The Huntington Library, San Marino, California

The previous chapters have delineated one trajectory of popular botany in nineteenth-century America, from the concrete and scientific to the symbolic and metaphorical. Through the specific genres of botany instruction books, herbaria, and flower language dictionaries, popular botany both mirrors and shapes larger issues of American citizen science, preservation, and abundance. This conclusion begins by analyzing flower fables and personified flowers, which extend plants into agents of reform and transformation. With their ability to act, feel, and communicate, the personified flowers of nineteenth-century fables anticipate twenty-first concepts of plant agency and sentience, providing a useful bridge to this book's concluding arguments about the continued relevance of popular botany in the twenty-first century.

The epigraphs to this conclusion show literary and visual examples of floral characters that promote ideals of freedom, diversity, and equality in the latter half of the nineteenth century. Frances E. W. Harper's fable "The Mission of the Flowers" was published with the long poem *Moses: A Story of the Nile* in 1869. "Tiger Lily" (figure C.1) appears in Lida Clarkson's *The Gathering of the Lilies* (1877), a collection of drawings and poems. The writer and graphic artist Charlotte Perkins Gilman designed a series of trade cards featuring personified flowers for a soap company in the early 1880s (figure C.2). "A Flower Fable" (reprinted as "Queen Aster") was first published in 1887 and is one of the many flower fables that Louisa May Alcott wrote over the course of three decades.

Fables of Transformation

Frances Ellen Watkins Harper (1825–1911), a prolific African American poet, essayist, and lecturer, used the popular form of the flower fable for a cautionary tale about the perils of enforcing conformity. In this brief allegory, Harper personifies a number of floral species:

> In a lovely garden, filled with fair and blooming flowers, stood a beautiful rose tree. It was the center of attraction, and won the admiration of every eye. . . . Now the rose was very kind and generous hearted, and, seeing how much joy she dispensed, wished that every flower could only be a rose, and like herself have the privilege of giving joy to the children of men; and while she thus mused, a bright and lovely spirit approached her and said, "I know thy wishes and will grant thy desires. Thou shalt have power to change every flower in the garden to thine own likeness." . . . The first object that met her vision was a tulip superbly arrayed in scarlet and gold. . . . She haughtily refused to change her proud robes for the garb the rose tree had prepared for her; but she could not resist the spell that was upon her, and she passively submitted the garments of the rose to enfold her yielding limbs. The verbenas . . . felt a change pass through their sensitive frames, and instead of gentle verbenas they were blushing roses. . . . The sleeping poppies . . . too had changed to bright and beautiful roses. The heliotrope . . . gracefully submitted to her unwelcome destiny. The violets . . . were averse to losing their identity. "Surely," said they, "we have a mission as well as the rose;" but with heavy hearts they saw themselves changed like their sister plants.

In like fashion, the pure and innocent snowdrop, the proud dahlias, the "modest lily," "marygolds," garden pinks, and other flowers meet the same fate. The rose beholds a world "become a perfect wilderness of roses" but then mourns the loss of "her sister plants." Fortunately for the rose, she wakes and realizes the vision was only a dream; "but an important lesson had been taught; she had learned to respect the individuality of her sister flowers, and began to see that they, as well as herself, had their own missions . . . and of those whose mission she did not understand, she wisely concluded there must be some object in their creation, and resolved to be true to her own earth-mission."[1] Harper had a strong sense of her own mission as an abolitionist and champion of equality for women and African Americans. She was an activist who felt the prejudice of her white sister feminists and deeply understood the need for the movement's metaphorical gardens to feature more than just roses.

As literary scholar Alice Rutkowski argues, "The Mission of the Flowers" is "a very public rejection of the benevolent maternalism of white, middle-class feminism and a powerful assertion of black self-determination" that "represents homogeneity as coercive, debilitating, and undesirable."[2] In her book on Harper, *Discarded Legacy*, Melba Joyce Boyd similarly notes that "the lesson imparted in this story applies both to white feminists who are ignorant of the especial problem of racism that women of color face, and to elite black women who are ignorant of the cultural difference and intelligence of freedwomen." Rutkowski states that the 1869 publication of "The Mission of the Flowers" situates it not in the context of slavery and abolition but of Reconstruction, when "for newly freed slaves, a crucial question was whether their quest for civil, political, and social equality would receive white assistance."[3] As Harper's flower fable suggests, even the well-meaning attempts of "kind and generous hearted" folks like the rose could be seriously misguided. This flower fable resonates not only within its Reconstruction-era context of advocating for the benefits of a racially plural society but also with known botanical principles regarding the dangers of plant monocultures.

In another example from the Reconstruction period, Lida Clarkson's "Tiger Lily" racializes this flower (figure C.1). *The Gathering of the Lilies* features Clarkson's poems about and illustrations of various kinds of lilies, including Japan lily, meadow lily, lily of the valley, daylily, and water lily—all of which are white.[4] The poem for the tiger lily, however, is the only one that mentions race. Predating J. M. Barrie's Tiger Lily character from *Peter Pan* by several decades, Clarkson's has a "yellow face," similar to the actual tiger lily flower's orange coloring:

> Surely, friends, I bear your name;
> Do not look at me with blame,
> And so sadly put to shame
> My yellow face.
> I am lily-born like you;
> But the morning bath in dew
> Will not make me fair to view
> As my race.

It's not clear whether the racialized yellow here is meant to be Asian or light-skinned African American; the color was used in the latter part of the nineteenth century for both.[5] Visually, however, Clarkson's rendering of Tiger Lily's face is not distinctly racialized (figure C.3); she resembles other dark-haired personified lilies in the book (figures C.4 and C.5). What is clear, however, is that textually her "yellow" coloring is associated with "blame" and "shame," as replicated visually by the

FIGURE C.3. Detail, "Tiger Lily," Clarkson, *The Gathering of the Lilies* (1877)

FIGURE C.4. Detail, "Garden Lily," Clarkson, *The Gathering of the Lilies* (1877)

FIGURE C.5. Detail, "Water Lily," Clarkson, *The Gathering of the Lilies* (1877)

Figure C.6. Frontispiece, Clarkson, *The Gathering of the Lilies* (1877)

postures of the tiger lily personifications in the book's frontispiece (figure C.6), with their downcast gazes and hands covering the face (figure C.7). When Clarkson's Tiger Lily gently chides her white lily friends and asserts her equality—"I am lily-born like you"—she explicitly legitimates her place in the lily family, both metaphorically and botanically: The *Liliaceae* family includes *Lilium lancifolium* and *Lilium tigrinum*, both varieties of tiger lily. Clarkson's message of diversity within this "gathering" of lilies shows another postbellum poet and artist using personified flowers as vehicles of reform.

FIGURE C.7. Detail, frontispiece, Clarkson, *The Gathering of the Lilies* (1877)

A third example comes from the visual work of Charlotte Perkins Gilman (1860–1935), a writer who promoted feminist ideals in her short story "The Yellow Wallpaper" (1892), novel *Herland* (1915), and nonfiction work such as *Women and Economics* (1898). Her first career was in graphic art, and in the early 1880s, she supported herself by designing advertising cards for several different companies and their household products, including the soap company Soapine. One set featured six related images in a series that collector Kevin MacDonnell has labeled the "three ages of woman": babies, young girls, and young women, each in a religious and a secular version. With their female subjects depicted as flowers, Gilman's cards use compelling iconography, including that of the language of flowers. The religious version of the baby shows a personified calla lily with iconography of the baby Moses (figure C.2). Gilman portrays the appropriate botanical habitat: Calla lilies are water-loving plants, and in the wild, they grow near streams and marshes—just as the calla lily baby here has her feet in the water next to the bulrushes. MacDonnell identifies the flower language symbolism of the calla lily as "youthful innocence, purity, and chastity" and explains that the baby Moses is "sometimes shown wrapped in a white sheet that closely resembles the lily that wraps tightly around this baby."[6] In Gilman's depiction, the calla stares directly at the viewer, bare feet firmly planted in the flowing water, and competently handles an oar-like branch. Neither swaddled nor contained in a basket, this confident, feminist Moses is ready to lead her people to freedom.

Louisa May Alcott (1832–1888) wrote about personified flowers for decades, starting in 1854 with *Flower Fables*, her first published book. These stories, originally written for Ralph Waldo Emerson's daughter, anticipate Alcott's later writing on slavery and abolitionism. Elizabeth Keyser argues that "for all their cloying sweetness, *Flower Fables* contains nascent subversive elements." When read in the context of mid-nineteenth-century social and political issues, one of her flower fables' "subversive elements" is a condemnation of slavery and prejudice. For example, in "Little Annie's

Dream, or The Fairy Flower" (1855), the child Annie confesses to being "not happy, for I am not good; how shall I learn to be a patient, gentle child." A fairy flower gives her a magic flower that will reward the child for goodness by emitting a sweet fragrance—and will note her anger, selfishness, or unkind words with a "soft, low chime of the flower-bell." Annie is initially delighted but soon slides back into her willful ways and wants to be rid of the annoying, constantly chiming flower. While in search of the fairy flower who bestowed it, the child falls asleep and dreams that her angry thoughts and feelings take form and gain "a strange power over her." A "high, dark wall" encloses her within a garden where the flowers near her all die; "then the spirits gathered round her, whispering strange things in her ear, bidding her obey, for by her own will she had yielded up her heart to be their home, and she was now their slave." The fairy intervenes and saves the child from her bondage. As Annie vows to reform and be a better child, the garden flourishes: "flowering vines crept up the high wall, and . . . wherever green leaves and flowers bloomed, the wall beneath grew weak, and fell apart." In this fable, botanical profusion conquers barriers to free movement, and a single flower can effect the child's release from slavery.[7]

Three decades later, "A Flower Fable" focuses on the rivalry of the Golden-rods and Asters; the former have long ruled over the meadow as kings, but the Asters campaign to put a queen on the throne. A variety of personified flowers drive the narrative: honest clovers, cheerful buttercups, proud and haughty cardinal flowers, demure fringed gentians, even a "delicate, high-born" clematis. In this protodemocratic meadow, Aster is elected queen and soon implements a series of reforms. The defeated Prince Golden-rod eventually acknowledges that Queen Aster "has done more than ever we did to make the kingdom beautiful and safe and happy" and offers his allegiance, and the two flowers decide to rule jointly. The story ends, "This is as it should be: love and strength going hand in hand, and justice making the earth glad." Throughout the story, Queen Aster makes difficult and often unpopular decisions for the good of the meadow, without capitulating to flowers often critical of her reforms. Keyser notes that "A Flower Fable" was first published in *The Woman's Journal*, "an organ of the women's rights movement"; she states that Alcott's story "clearly can be read as a feminist fable, offering a utopian vision of social harmony and justice under matriarchal rule."[8]

The genre of flower fables remained popular throughout the nineteenth century and was part of a larger phenomenon of socially and politically engaged floral discourse.[9] These simple didactic tales portray a world in which creatures of all species and varieties interact and can coexist both peacefully and productively. Human children, personified flowers, and insects comingle with fanciful, hybrid creatures such as fairies, whose winged bodies and human faces bridge the gap between animal and human. Floral tropes were ubiquitous in nineteenth-century America and are analogous to the equally widespread tropes of white/black, dark/light that signaled preoccupations with race and slavery. In their range of colors, forms, blooming seasons, habitats, scents, and healing properties, flowers are the perfect emblem for a society seeking to embrace diversity and flourish according to ideals of universal equality. In the value they place on species diversity and plant particularity, flower fables also resist what Michael Marder terms "a signature gesture of metaphysical violence seeking to eliminate differences" by reducing them to the single "conceptual unity 'plant.'"[10]

Critical Plants and Flower Books' Legacy

The personified plants of these flower fables were critical of society's ills, decrying greed, racism, and selfishness. In the present, plants are critical in other senses: crucial to ecosystems' healthy functioning, but also in critical condition, threatened by climate change, habitat destruction, commercial exploitation, and competition from invasive species. The *State of the World's Plants and Fungi 2023* report is a collaboration among two hundred scientists representing 102 institutions in thirty countries worldwide, spearheaded by the Royal Botanic Gardens at Kew, England. The 2023 report states that "45% of flowering plant species are potentially threatened with extinction," up from an estimated 40 percent of plants threatened with extinction in the 2020 report, and a startling increase over the 2016 report's finding of 21 percent then threatened. While the 2020 and 2023 figures are estimates, the reports' analyses by top plant scientists nonetheless convey the enormity of plants' critical status.[11]

The introduction to the *State of the World's Plants and Fungi 2023* closes by asserting, "Just as our early ancestors needed to know what grows where for their own survival, so plants and fungi need us to know where they grow—to enable us to safeguard their continued existence for generations to come." The popular botany of the past has important lessons for us today: Interactions with plants by nineteenth-century American writers, visual artists, collectors, students, and everyday citizens combined a commitment to knowledge with an ethics of care and emotions of wonder and pride. As Mary Kuhn states, "Nineteenth-century Americans' immediate, intimate experiences cultivating plants brought them into a caretaking relationship with forms of life that were organized differently from their own. They reflected seriously on the significance of plant life for their own lives, and in paying close attention, many saw models of collective and of dynamic multispecies interactions that escaped taxonomy's ready vocabulary."[12] Having firsthand experiences with and seriously reflecting on plants might be unfamiliar to many of us today, making our predecessors' examples all the more relevant.

The particular genres of botany instruction books, herbaria, and flower language dictionaries have diverse afterlives that indicate continued interest in plants. Nineteenth-century American citizen science was typically small in scale, and its practitioners valued their personal learning as much as their contributions to communal knowledge. Botanizers brought books by Amos Eaton, Asa Gray, Alphonso Wood, and John Newman into the field and used them to identify and record their findings; today we use applications such as Seek and iNaturalist for similar purposes. A cell phone is smaller and lighter than a seven-hundred-page book, and its immediate connection to millions of crowdsourced data points would have astounded people two centuries ago. The landing page of the iNaturalist website states, "Every observation can contribute to biodiversity science, from the rarest butterfly to the most common backyard weed. We share your findings with scientific data repositories like the Global Biodiversity Information Facility to help scientists find and use your data. All you have to do is observe." Although much of plant science has changed, the basic instruction to *observe* remains constant. In 1824, Eaton encouraged students to "search for all they [can] find" because botany "depends on continued research, and the accurate examination of such a multitude of individuals."[13] The worldwide reach of today's iNaturalists has resulted in these contributions:

487,517 species observed
7,995,976 people signed up
208,157,149 observations to date.

The individual observations and local efforts of amateur botanists have become communal and global, and the technology of the internet has superseded the technology of the book. The *State of the World's Plants and Fungi 2023* accordingly notes that "while remote habitats are often the source of new plant species, novel additions also arise from eureka moments in herbaria or even on social media."[14]

Scientists continue to create and learn from herbaria in both physical and digital formats. Herbarium specimens of the past contain valuable evidence for conservation, including for "identifying critical biodiversity regions."[15] In her chapter on online herbaria, Maura Flannery describes the "massive efforts to image specimens and enter their labels into databases, making them available to a broader community of users via the internet." Botanical data can take several forms: a live plant in its native habitat, a live plant extracted from its habitat, a physical plant specimen in an herbarium, the digitized image of a physical plant specimen with its label data, or a photograph of a plant uploaded to an online herbarium, global biodiversity database, or crowdsourced repository such as iNaturalist. Each form has advantages and disadvantages. Researching biota *in situ* can harm fragile environments, and extracting live plants can lead to species endangerment or even extinction. There are limitations to using the physical plant specimens in herbarium collections; as W. John Kress and Vicki A. Funk point out, they can be "geographically biased, favoring more easily accessed areas; taxonomically incomplete, including only easy-to-study species [. . . and] temporally biased, based on a single survey in a single season." Historical specimens are not only scientifically limited but can also be culturally and ethically problematic, given the damages of colonialism and imperialism on plant collecting. Online images of physical specimens and photographs of live plants are much more accessible, but a researcher cannot (yet) extract DNA from those images. Despite these limitations, data from physical herbarium specimens remain significant to scientists. In a North American example, the biologist Richard Primack has used Henry David Thoreau's herbarium, journal, field observations, and published works such as *Walden* to track the impacts of climate change on the flora of Concord, Massachusetts. Erin Zimmerman calls herbaria "simple in concept, but powerful in that each specimen is a data point: This species lived in this place at this time. Many data points together can be used to test all manner of hypotheses. . . . Dusty old cabinets of brittle brown plants have never been so cutting edge."[16]

The persistent fascination with the language of flowers would not have surprised nineteenth-century floriographers. The combination of flowers' beauty and their ability to convey meaning in "hidden" messages makes their symbolism perpetually enticing. Two recent novels use the language of flowers as their central conceit and as agents of healing: Vanessa Diffenbaugh's *The Language of Flowers: A Novel* (2012) and Holly Ringland's *The Lost Flowers of Alice Hart* (2018). Present-day flower language dictionaries abound, as do other formats such as coloring books, affirmation cards, guides to floriographic arrangements, online courses, YouTube videos, and more. Beyond these pop culture phenomena is the environmental legacy of nineteenth-century American flower language books' celebration of abundance. The ecological concept of species abundance is one of the primary indicators of biodiversity and a valuable tool for measuring other interactions such as competition and predation.[17] Sustainability advocate Jay Friedlander argues that "abundance turns scarcity on its head and reorients toward finding new possibilities," as did the small island community of Samsø, Denmark, in becoming the world's first carbon-negative community. When local leader Søren Hermansen adopted an abundant perspective, "the nature of the conversation changed from tradeoffs to benefits" and emphasized thriving community, a robust economy, and energy independence—all "without sacrificing comfort or quality of life."[18] As environmental humanities scholars attest, scientific solutions alone will not solve environmental problems; the ability to reorient perspectives, including through language, is also necessary.

The focus of *Flora's Interpreters* on print and manuscript narrows the representation of both producers and consumers to those who can read and write. Literacy rates for whites of all social classes were much higher than for people of color throughout the nineteenth century. For example, in 1870 the proportion of illiterate white people aged ten or older was 11 percent; for African Americans, it was nearly 80 percent. African Americans and Native Americans in the nineteenth century had extensive plant knowledge that typically was not conveyed in writing or print, either within their own communities or more broadly through publication. When white authors of flower books shared "Indian lore," they did so at several layers of remove from authentic and reliable sources, depending instead on accounts that usually originated in early European colonists' misinformed and incorrect assumptions. Excellent current scholarship has recovered African Americans' botanical and natural history knowledge, including the medicinal plant knowledge of enslaved people on Southern plantations and in the eighteenth-century Atlantic world. Native American ethnobotany is another burgeoning field, with databases created by Indigenous scientists and sourced from generations of Native peoples' knowledge, encyclopedic books, scholarly articles, and resources for practitioners.[19]

Several prominent plant scientists working today have made their research accessible to wider audiences in books that counter early American popular botany's legacy of excluding the perspectives of people of color. Potawatomi botanist Robin Wall Kimmerer's *Braiding Sweetgrass: Indigenous Wisdom, Scientific Knowledge, and the Teachings of Plants* (2013) combines the three strands of her subtitle with memoir and science in a beautifully written homage to her Indigenous culture, scientific training as a bryologist (specialist on mosses), and family. She tells stories of plant agency and sentience by focusing on the sweetgrass plant, known as *Hierochloe odorata* ("fragrant, holy grass") according to its Latin binomial, and *wiingaashk* ("sweet-smelling hair of Mother Earth") in Potawatomi.[20] The Rarámuri ethnobotanist Enrique Salmón developed the term "kincentric ecology" to describe "an awareness that life in any environment is viable only when humans view the life surrounding them as kin."[21] His book *Iwígara:*

The Kinship of People and Plants (2020) is specifically written for laypeople and focuses on eighty plants "that are the most culturally relevant to North American native peoples." He compiled the information for *Iwígara* by consulting with "native plant practitioners, [his] professional ethnobotanical network, and close friends," thus dismantling a hierarchy that valorizes trained scientists over so-called nonscientists.[22] A third book, Beronda Montgomery's *Lessons from Plants* (2021), addresses "how plants' individual and collective strategies and behaviors result in adaptable and productive living, and how we can learn from them." Montgomery's research focuses on "the responses of photosynthetic organisms (i.e., plants and cyanobacteria) to external light cues," and in *Lessons from Plants* and her wider work, she applies this knowledge in promoting strategies for mentoring, leadership, and success. In one chapter, she explains how plants distinguish "friend or foe" using examples of plants' ability to perceive proximity of neighbors, employ "shade avoidance behavior" to compete for light resources, and "recruit 'friendly' microorganisms" to access nutrients from soil. From these examples, she draws lessons for humans on how to "avoid expending excessive energy in antagonistic behaviors," "forge beneficial relationships," and establish "an ecosystem of support, collegiality, and community." As an African American woman scientist, Montgomery learned early on "how important mentoring and leading were to [her] pursuit of success."[23] All three of these plant scientists—Kimmerer, Salmón, and Montgomery—are contributing to popular botany by using their research and personal experiences to reach audiences of everyday people interested in plants and their teachings.

I, too, have learned from plants as I have researched and written *Flora's Interpreters*. A decade ago, I started work on my yard in the Piedmont of North Carolina. There were a few mature rhododendrons, some azaleas, and lots of trees, including native willow oak, persimmon, sycamore, sassafras, and dogwood. Previous residents had presumably given up growing grass in this shade-filled space and instead planted English ivy (*Hedera helix*), which completely covered the ground, climbed up the trees, and took over other parts of the neighborhood. Step one: Eradicate the ivy—literally, by pulling it up by the roots. Step two: Enlist the expertise of a landscape designer for a plan to fill this newly blank canvas of red clay. Many steps and many years followed, as I built winding paths and planted hundreds of shade-loving, deer-resistant, heat-tolerant, mostly native plants. I immersed myself in nineteenth-century American popular botany and saw plant metaphors everywhere; my writing and gardening went hand in hand. Transplanting from one part of the yard to another was like cutting and pasting a section of prose into a different chapter where it made much more sense. I pruned a large lady fern and discovered a Distylium shrub that had languished; even though I struggled to get rid of those big, beautiful fronds, it significantly improved that corner of the yard. I promptly went inside and hacked my way through pages of rambling thought; several thousand words fewer, the draft became much clearer and I could see the point I was trying to make. I planted for pollinators and bird habitat, learned from locals with deep roots in the area, and continued to weed my prose and revise the garden.

I also thought a lot about the environmental issues of today, which are huge, dire, global, and unjustly disproportionate in their impacts. Biodiversity loss is irreversible when species go extinct. The window for keeping levels of atmospheric carbon dioxide below 350 parts per million—what scientists deem necessary for a healthily functioning planet—closed in 1988, and only large-scale, radical change will get us back there. Massive wildfires, extreme storms, severe flooding, and record temperatures are no longer the projections into a dystopian future but our daily reality. How can we possibly respond to all of this productively? Perhaps paradoxically, we might learn from nineteenth-century amateur botanists, who knew the pleasures, rewards, and importance of small-scale environmentalism.

To validate the small scale in a U.S. context is to resist discourse about the large, epic, vast, and exceptional—discourse that thrived in the nineteenth century and persists today. (For example, millions of voters would not have been persuaded by a slogan such as "Make America Manageable Again.") The primary materials of *Flora's Interpreters* provide a useful starting point for rethinking the large. A book is sized for human hands. It's portable and durable and doesn't take up a lot of room on a table or shelf. Books last, even when they've been through the wars—as did Hezekiah Marbourg's botany textbook. Inscriptions, annotations, and marginalia are also small in scale. People write them in the space available. Mattie Snowden reduced her literary quotations to the few lines that would fit in the margins of her copy of *Illustrated Botany*. Diaries and friendship albums record the personal; they are one-of-a-kind creations whose worth comes from their individuality and irreproducibility.

Flowers are small in size and brief in their lifespan. Their ephemerality teaches us to appreciate them in the moment. They are beautiful in their variety of colors, patterns, and shapes, as artists such as Lydia Bustill, Childe Hassam, and Isaac Sprague conveyed. Pressing and drying a single flower is a deliberate act of preservation; as herbarium instructions detailed, this process took days and required careful attention to accurate placement and levels of moisture. Annie Fields pressed flowers given to her by poets such as Tennyson when she was traveling in Europe; years later, she placed these mementos in the diary she titled "Fragrant Memories." Botanizing required patience, developed powers of observation, and led to deeper appreciation of the natural world.

Botany textbooks, herbaria, flower language dictionaries: These disparate genres of flower books contributed to nineteenth-century American environmentalism in their commitment to place-based knowledge and detailed attention to plants. Their creators and consumers valued these books' combinations of scientific information, beautiful design, and compelling narratives, as do today's plant advocates in digital formats. The Herbaria 3.0 website extends this invitation: "Everyone has a story to tell about a plant. What's yours?" This project "emerged as a collaborative effort of environmental humanities scholars, scholars of science pedagogy, and plant biologists to contemplate the role and meaning of plants in an era of rapid climate change and species displacement." By sharing stories about all kinds of plants—wild, cultivated, medicinal, flowering, edible, carnivorous, and more—contributors honor their relationships with these extraordinary beings. The Plant Humanities Lab's collection of plant narratives combines scientific, historical, visual, and cultural perspectives on crop species (maize, quinoa, rice, soybean), herbs and spices (cinnamon, mint, turmeric), fruits and vegetables (banana, breadfruit, cassava, okra, tamarind), flowers (heliconia, opium poppy, peony, sunflower) and more in fascinating stories that uncover details such as Amazonian *ayahuasca* rituals, herbal abortifacients, and transoceanic migrations.[24] I imagine that my delight in discovering these websites might be like that of a nineteenth-century reader delving into a book like Sophie Herrick's *Wonders of Plant Life Under the Microscope* or Cordelia Harris Turner's *The Floral Kingdom* or Isaac Sprague and A. B. Hervey's *Beautiful Wild Flowers of America*.

Plants matter. Telling their stories, sharing images of their beauty, observing them in their native habitats, and making scientists' botanical research accessible are critical to saving the world's remaining and yet to be discovered species. Flora's interpreters still exist, and we still have much to learn from them.

Notes

Introduction. Flower Books, Popular Botany, and American Environmentalism

Epigraph sources: Phelps, *Botany for Beginners* (1833), 22; Beecher, "Discourse on Flowers" (1853), 98; "Introduction," *Wild Flowers of America*, n.p.

1. Sarah Alden Bradford Ripley "was a classical scholar of note, knew many languages as well as mathematics, chemistry, astronomy, and botany. . . . This was in addition to raising seven children of her own and an adopted niece while helping her Unitarian minister husband run a boarding school to prepare boys for Harvard." Goodwin, *Remarkable Mrs. Ripley*, xvii.

2. *Wild Flowers of America* (unpaginated), title page, plate 191, plate 164; F. Turner, "Significance of the Frontier," 227, 200.

3. For more on the cultural applications of flowers in the nineteenth century, see Seaton, *Language of Flowers*, 4–15.

4. Welch, *Book of Nature*, 68.

5. Others include Lydia Maria Child, Annie Adams Fields, and Sarah Josepha Hale.

6. Embury, *American Wild Flowers in Their Native Haunts*, iii, 10.

7. Embury, *American Wild Flowers in Their Native Haunts*, 42, 43; Welch, *Book of Nature*, 5.

8. Winship, *American Literary Publishing in the Mid-Nineteenth Century*; Senchyne, *Intimacy of Paper*; Moylan and Stiles, *Reading Books*; Casper, Chaison, and Groves, *Perspectives on American Book History*; Gross and Kelley, *Extensive Republic*; Casper et al., *Industrial Book*.

9. Senchyne, "Vibrant Material Textuality," 72.

10. Multigraph Collective, *Interacting with Print*, 50.

11. Multigraph Collective, *Interacting with Print*, 205; Iovino, "Stories from the Thick of Things," 450.

12. Keeney, *Botanizers*, 1. See Stefaniak, "Botanical Gleanings," for a discussion of the contrast between Catharine Parr Traill's utilitarian and Susan Fenimore Cooper's aestheticized responses to their local flora, and for its analysis of botanizing as a gendered practice. See also Kuhn, *Garden Politic*.

13. Goodwin, *Remarkable Mrs. Ripley*, 28, 30; E. Dickinson, *Letters*, 35; Godfrey, Diary.

14. M. Treat, "Botany for Invalids," 39. For more on Treat, see Gianquitto, *"Good Observers of Nature,"* 32, 136–176.

15. DeJong, introduction to *Sentimentalism in Nineteenth-Century America*, 9, 10; Stefaniak, "Botanical Gleanings," 237; Kuhn, *Garden Politic*, 4. Chapter 3 of Keeney, *Botanizers*, is about botanizing and self-improvement.

16. Lawrence, "Listening to Plants," 633; Gibson and Sandilands, "Introduction: Plant Performance," 4.

17. Sandilands, "Plants," 157; Gibson and Sandilands, "Introduction: Plant Performance," 3, 7; Gutierrez, "From Objects of Study to Worldmaking Beings," 7, 8. An incomplete list of other excellent sources in critical plant studies, particularly from humanistic perspectives, includes Gaard, "Critical Ecofeminism"; Gagliano, Ryan, and Vieira, *Language of Plants*; Hall, *Plants as Persons*; Laist, *Plants and Literature*; Marder, *Philosopher's Plant*; and Vieira, Gagliano, and Ryan, *Green Thread*.

18. "Plant Humanities Initiative"; Batsaki and Fine, "Cultivating History."

19. Cultural work is the reciprocal influence and mutually constitutive practices of culture and cultural production; see Tompkins, *Sensational Designs*.

20. R. Johnson, *Passions for Nature*, 4. For brief examples of how environmental humanities scholars of nineteenth-century America articulate the relevance of their work in the

present, see R. Johnson, *Passions for Nature*, 21–22; Kuhn, *Garden Politic*, 24; and Kilcup, *Fallen Forests*, 3.

21. Lewis, *Democracy of Facts*, 2; Welch, *Book of Nature*, 3.

22. Emerson, *Nature*, 33. For an overview of Swedenborg's influence on Emerson's thinking, see Branch, *"Best Read Naturalist,"* xvi–xxi.

23. A *very* partial (and predominantly North American) list of influential ecocritics—many of whom are also leading environmental humanities scholars—includes Joni Adamson, Stacy Alaimo, Michael P. Branch, Lawrence Buell, Elizabeth DeLoughrey, Greta Gaard, Greg Garrard, Cheryll Glotfelty, Ursula Heise, Serenella Iovino, Stephanie LeMenager, Timothy Morton, Rob Nixon, Serpil Oppermann, Catriona Sandilands, Nirmal Selvamony, Scott Slovic, and Julie Sze.

24. HathiTrust and the Biodiversity Heritage Library are excellent online sources for almost all of the pre-twentieth-century books I discuss, as is the online antiquarian bookseller abebooks.com.

25. Through the generosity of an annual research fund, I've been able to visit archives and purchase nineteenth-century American flower books for my own study; these will be housed in Special Collections at the E. H. Little Library at Davidson College. The bibliography indicates works in this collection with an asterisk.

Chapter 1. Botany Instruction Books

Epigraph sources: Phelps, *Familiar Lectures on Botany* (1829), 12; Wood, *Class-Book of Botany* (1855), 14; Green and Congdon, *Analytical Class-Book of Botany*, iv.

1. Keeney, *Botanizers*, 51; Kohlstedt, "Parlors, Primers, and Public Schooling," 428.

2. Phelps, *Familiar Lectures on Botany* (1836), 14.

3. Crowdsourcing and Citizen Science Act of 2016, (c) Definitions. "Although citizen science has been around for centuries, the term citizen science was coined in the 1990s." Vohland et al., "Editorial," 1–2. The terms for participants and activities appear in a compilation of twenty-four definitions of citizen science in Table 2.1 in Haklay et al., "What Is Citizen Science?," 15–18.

4. Lewis, *Democracy of Facts*, 11.

5. Irving, *Catechism of Botany*, 5; Phelps, *Botany for Beginners* (1833), 22–23.

6. Welsh, *Botanical Catechism*, iii–iv; Phelps, *Botany for Beginners* (1833), 144; Comstock, *Introduction to the Study of Botany*, 76.

7. Davis and Tucker, "Impact of the Christian Faith on Books," 115, 116; Little, "Brief History of the American Sunday School Union." Little notes that the American Sunday School Union "played a role in shaping the direction of 19th century children's literature in America." Edwin Rice explains that the American Sunday School Union "proposed to make definite advances along three related lines," the first of which was "educational: by providing a system of lessons, a decidedly juvenile literature, a complete equipment for the school, and definite information upon principles and methods of teaching." Rice, *Sunday-School Movement*, 90.

8. Wood, *First Lessons in Botany* (1851), 6; Gray, *Botany for Young People*, part 1, 2. Other botany instruction books that cited the lilies of the field include Bourne, *Flores Poetici*, title page; Green and Congdon, *Analytical Class-Book of Botany*, 1; Wood, *Class-Book of Botany* (1855), title page; and Baily, *Trees, Plants, and Flowers*, vi.

9. Welsh, *Botanical Catechism*, iii. Welsh asserts the importance of working with actual specimens: "The Teacher will find it expedient to have an example of some perfect and complete flower, for the purpose of pointing out the elementary organs as the answers are given. Specimens dried with care, may be used with advantage, in attending to the elementary principles of this study in the winter season" (5).

10. Irving, *Catechism of Botany*, 34, 38, 69, 73; Goodrich, *Child's Botany*, 26–27, 28.

11. Youmans, *First Book of Botany* (1870), 4; R. Johnson, *Passions for Nature*, 13.

12. Herrick, *Wonders of Plant Life*, 3; Morley, *Few Familiar Flowers*, xi, xii.

13. Comstock, *Young Botanist*, 1; Gray, *Botany for Young People*, part 1, 4.

14. Marshall, *Arbustrum Americanum*, v; Lewis, *Democracy of Facts*, 55.

15. Sumner, *Compendium of Physiological and Systematic Botany*, vi; Eaton, *Manual of Botany* (1824), v; Phelps, *Familiar Lectures on Botany* (1832), vi; Comstock, *Introduction to the Study of Botany*, vi; Beck, *Botany of the United States North of Virginia*, iii–iv; Gray, *Manual of the Botany of the Northern United States*, vi.

16. Keeney, *Botanizers*, 52.

17. Sundberg, "Botanical Education in the United States," 106; Wood, *Class-Book of Botany* (1845), 3, 4.

18. Gray, *Manual of the Botany of the Northern United States*, v; Angelo, "Thoreau as Botanist," 17. See also Moore, Macklin, and DeCesare, "Brief History of Asa Gray's 'Manual of Botany.'"

19. Gray, *Manual of the Botany of the Northern United States*, ix.

20. Wood, *Class-Book of Botany* (1845); Wood, *Class-Book of Botany* (1855), title page, 4. Wood identifies his sources as "the botanical works of Bigelow, Eaton, Wright, Pursh, Michaux, Smith, Nuttall, Torrey, Beck, Loudon, Elliott, Darlington, Dewey, Barton, Hooker, Decandolle, and Torrey and Gray." *Class-Book of Botany* (1855), 4.

21. Sophocles's annotated copy of Gray is at Harvard and has been digitized at HathiTrust. "Evangelinos Apostolides Sophocles."

22. Dupree, *Asa Gray*, 119.

23. For botany as an appropriate science for women, see Baym, *American Women of Letters*; Gianquitto, *"Good Observers of Nature"*; and Keeney, *Botanizers*.

24. Quoted in the front matter of Wood, *First Lessons in Botany*.

25. Warner, "Science Education for Women in Antebellum America," 58; Baym, *American Women of Letters*, 4.

26. Baym, *American Women of Letters*, 36; "Preservation of Wild Flowers."

27. Eaton, *Botanical Exercises*, 33; Eaton, *Manual of Botany*, 11.

28. Eliza Roberts Nye was born in 1811 in Sandwich, Massachusetts, the ninth of eleven children of Allen and Rebecca Nye. After her husband's death, Eliza Nye Fitz "founded and directed the Brookline Seminary for young ladies," a boarding school in Brookline, Massachusetts. Nye and Best, *Genealogy of the Nye Family*, 151–153. See also "79 Marlborough" and Woods, *Historical Sketches of Brookline*, 127.

29. Kelley, *Learning to Stand and Speak*, 67–68.

30. Green and Congdon, *Analytical Class-Book of Botany*, iv. Biographical information on Stearns is from Bond, *Genealogies of the Families*, 470, and New Ipswich Academy record books, 1854–1856; on Lee, from Chandler, *History of New Ipswich*, 514, and Student Name Index Project, Bradford Academy.

31. Youmans, *First Book of Botany* (1870), 4, 8, 11; "Newberry Street School."

32. Noyes, "From Lover's Lane to Sparks Street," 157, 160. Noyes also remembered visiting "Mrs. Asa Gray" and that her "mother's edition of the Botany is that of 1875" (164).

33. Angelo, "Thoreau as Botanist," 13; Sattelmeyer, *Thoreau's Reading*, 116–296.

34. Angelo, "Thoreau as Botanist," 14, 15; Dean, *Wild Fruits*, ix; Sattelmeyer, *Thoreau's Reading*, 79.

35. Dean, "Possible Source," 1; Dean, *Wild Fruits*, x–xi. For use of Thoreau's journals and other materials to track climate change, see Primack, *Walden Warming*.

36. Corinne Smith, *Westward I Go Free*, 100; Sattelmeyer, *Thoreau's Reading*, 191.

37. Thoreau, "Flowering of Plants"; Warren, *Thoreau's Botany*, 4, 9.

38. Biographical information about Marbourg comes from Pennsylvania College Alumni Association, *General Report*, 361; Storey, *History of Cambria County*, 564–566; *Biographical and Portrait Cyclopedia of Cambria County*, 142; and *Portrait and Biographical Record of the State of Colorado*, 150.

39. Goldsboro, Raleigh, New Bern, Morehead City, and Roanoke Island all had hospitals during the Civil War. See Pollitt, *African American Hospitals*, 7–9.

40. Maura Flannery, *In the Herbarium*, 138–139; LaFauci, "Peculiar Nature," 28–29.

41. Storey, *History of Cambria County*, 566; Fee et al., "Medical Education for Women."

42. Mohapatra et al., "Benefits of Black Cohosh"; DeSoto, "What Is Black Cohosh?"; "Women's Medical College of Pennsylvania Student Theses"; Wall, "Medical Ethics of Dr. J. Marion Sims," 346.

43. Keeney, *Botanizers*, 11. Fortunately, Marbourg's *Class-Book of Botany* will now be available to other researchers at the Special Collections of the Davidson College Library.

Chapter 2. "You Old Botany"

1. This copy of John Newman's *Illustrated Botany* is housed at the Harvard University Botany Libraries, Arnold Arboretum.

2. Karen Kilcup argues that nineteenth-century American women's environmental writing was characterized by this kind of "hybridity," and while she is discussing published work, the argument also applies to Snowden's private writing. Kilcup, *Fallen Forests*, 3–7.

3. Newman, *Illustrated Botany* (1850), iii.

4. *Alpine Heritage*, 6.

5. Karen Dustman, email to author, July 8, 2022. Dustman cites the *Alpine Chronicle* from June 22, 1867, for the Woodfords event.

6. Cummins, *Lamplighter*, 12, 60; Fern, *Ruth Hall and Other Writings*, 150.

7. The quoted marginalia appear on pages 78, 118, 194, 122, and 168 and the last page (recto). Elsewhere in the text of *Illustrated Botany*, "indigenous" appears only twice and "endogen" not at all.

8. Drug Enforcement Administration, "Jimson Weed (Datura stramonium)." See also Hasegawa, "Quinine Substitutes in the Confederate Army," and Michael Flannery, *Civil War Pharmacy*, 202, 217.

9. Moerman, *Native American Ethnobotany*, 510. This entry on *Salvia dorrii* ssp. *dorrii* var. *incana* notes, "Washo *Cold Remedy* Infusion or decoction of leaves, and sometimes stems, taken for colds. *Respiratory Aid* Dried leaves smoked in a pipe to clear congested nasal passages." Snowden marginalia, *Illustrated Botany*, 144.

10. Dustman, *Silver Mountain City*, 70. When the California Gold Rush and subsequent Nevada silver rush brought settlers to the area, "thousands of miners and immigrants flooded Wá·šiw lands in a relatively short time period and they stayed here, disrupting the balance; changing the Wá·šiw world forever in only a few short years." "Changes to Washoe Life." For more on white women's participation in the Indian reform movement, see Mathes, *Women's National Indian Association*; Rhea, *Field of Their Own*; and Mathes, *Gender, Race, and Power in the Indian Reform Movement*.

11. Webb, "Chemist as Consultant," 11. From here on I use "Silliman" to refer to Benjamin Silliman Jr.

12. Newman, *Illustrated Botany* (1850), 11; Webb, "Chemist as Consultant," 10.

13. Newman, *Illustrated Botany* (1850), 129.

14. See, for example, Kemble, "Panamá Route." Mattie Snowden mentions an uncle in at least one journal entry; he may have been Henry R. Snowden, who "left New York for California by the Opposition steam *Arago*, January 14th, 1868" with his wife. "California Bound Passengers."

15. The same information about Arethusa appears nearly verbatim in I. Sprague and Hervey, *Flowers of the Field and Forest*, 46–47. While this might indicate that Snowden was still using her *Illustrated Botany* years later, it is more likely that Snowden and Hervey borrowed from the same source.

16. I am indebted to Tina Gianquitto for these insights.

17. See Cordell and Mullen, "'Fugitive Verses.'"

18. Longfellow, *Courtship of Miles Standish and Other Poems*, 29–30.

19. According to Open Source Shakespeare, "violet" or "violets" appear in *Cymbeline, Hamlet, Henry V, King John, Love's Labour's Lost, Measure for Measure, A Midsummer Night's Dream, Pericles, Richard II*, the *Sonnets, Twelfth Night, Venus and Adonis*, and *The Winter's Tale*.

20. Sources for this "Eastern tale" include Wilkinson, *Weeds and Wild Flowers*, 363; Thiselton-Dyer, *Folk-lore of Plants*, 311; "Wild Geranium"; and Tuttle, *Book of Rural Life*, 2346. Tuttle notes that mallow and geranium are not botanically related.

21. Page numbers in *Illustrated Botany* for poems quoted by Snowden: Herbert, "The Flower," 4; Coolbrith, "Among the Daisies," 151; Cary, "The Thistle Flower," 197; Wordsworth, *Peter Bell*, 86; Longfellow, "A Day of Sunshine," 91; and Thaxter, "Before Sunrise," 160.

22. Cohen, "Album Verse and the Poetics of Scribal Circulation," 72; Garvey, *Writing with Scissors*, 15.

23. Fay, "When the Autumn Comes Again."

24. "Lewis Chalmers also sent 'kind regards to a Miss Snowden' in March 1869 in one of his letters; it appears she may have boarded with Barney Pilkington. (1868–69 Chalmers Letters, p. 327). So it does sound like 'Lewis C.' would be Lewis Chalmers." Dustman, email to author, July 8, 2022.

25. Multigraph Collective, *Interacting with Print*, 207.

26. United States Census, 1870.

Chapter 3. Herbaria

Epigraph source: E. Dickinson, *Letters of Emily Dickinson*, 35.

1. Sophia Thoreau's herbarium is at the Concord Free Public Library, Concord, Massachusetts.

2. Ulrich et al., *Tangible Things*, 162–163.

3. "About," Harvard University Herbaria website, https://www.huh.harvard.edu/pages/about.

4. Ulrich et al., *Tangible Things*, 163. The authors are specifically writing about the herbaria collections at Harvard University, but the analysis applies to nineteenth-century American herbaria more generally.

5. "About," Harvard University Herbaria website.

6. From the *Oxford English Dictionary* entry for "paper": "Anglo-Norman *papir*, *paper*, *papere*, *papire* and Middle French, French *papier* paper, written document [. . .], ultimately classical Latin *papȳrus* the papyrus or paper-reed of the Nile." *OED* entry for "book": "A long-standing and still widely accepted etymology assumes that the Germanic base of book *n.* is related to that of beech *n.*" However, this entry also notes that "objections to this etymology have been made on several grounds." For more on the etymological debate, see https://www.oed.com/dictionary/book_n?tab=etymology#16619117. Nineteenth-century writers on plants noted the book/beech connection: Wood writes, "To the German name of the Beech (buch) we owe our English word book, the sides of thick books having formerly been made of beech boards." Wood, *How to Study Plants*, 213. See also Pratt, *Flowering Plants*, 179. Mattie Snowden, subject of chapter 3, quotes the same passage on page 89 of her *Illustrated Botany*.

7. Dickinson's herbarium is at the Houghton Library at Harvard University, and although its fragility prohibits access in person, digital facsimiles are available, and individual species are searchable in the library's online catalog. See also E. Dickinson, *Emily Dickinson's Herbarium*. Most of the Lewis and Clark herbarium is at the Academy of Natural Sciences at Drexel University in Philadelphia; see also Moulton, *Journals of Lewis and Clark*. Thoreau's herbarium is at the Gray Herbarium at Harvard; on the 200th anniversary of Thoreau's birth, the Harvard University Herbaria provided open access to digital images of Thoreau's herbarium. Herbarium pages by Vasey, "the future Chief Botanist of the Department of Agriculture," are in various locations; see "J. W. Powell Colorado Exploring Expedition, 1868," Smithsonian's National Museum of Natural History, https://naturalhistory.si.edu/research/botany/about/historical-expeditions/j-w-powell-colorado-exploring-expedition-1868. Muir's herbarium is at the John Muir National Historic Site in Martinez, California; see also Gisel, *Nature's Beloved Son*. Addams's herbarium is available on microfilm in the Jane Addams Collection, Swarthmore College Peace Collection. In *The Selected Papers of Jane Addams*, Bryan, Bair, and de Angury note that "part of the botany class that JA took in the second semester of her first year entailed the making of a herbarium" (1:233–234).

8. For details on the taxonomic history of the Oregon grape herbarium page (figure 3.6), see "Oregon Grapes," Discover Lewis & Clark website, https://lewis-clark.org/sciences/plants/oregon-grape/; and Maura Flannery, *In the Herbarium*, 2, x. Thoreau labeled the plant in figure 3.7 as *Rosa carolina*, but Ray Angelo has also identified *Rosa virginiana* and *R. palustris* specimens on this page; for details, see "Index of Botanical Specimens," Harvard University Herbaria & Libraries, https://kiki.huh.harvard.edu/databases/specimen_search.php?barcode=872506. For an explanation of his meticulous photo restoration process, see Joseph, "Technical and Aesthetic Application."

9. Maura Flannery, *In the Herbarium*, 5, 7.

10. Phelps, *Botany for Beginners* (1833), 21–22; L. Johnson, *Botanical Teacher for North America*, 14–15; Wood, *Class-Book of Botany* (1861), 15.

11. Gray, *Botanical Text-Book*, 506.

12. Wood, *First Lessons in Botany*, 87.

13. Thoreau quoted in Warren, *Thoreau's Botany*, 5; Phelps, *Botany for Beginners* (1833), 22; Phelps, *Familiar Lectures on Botany* (1849), 30.

Bradley Dean further notes that Thoreau "built a 'scaffold' inside the crown of his hat to hold plant specimens." Dean, *Wild Fruits*, ix.

14. *Leaf and Flower Pictures*, 15, 38–39.

15. Phelps, *Familiar Lectures on Botany* (1829), 44; L. Johnson, *Botanical Teacher for North America*, 15; Maura Flannery, *In the Herbarium*, 146–147.

16. Phelps, *Botany for Beginners* (1833), 22; Welsh, *Pastime of Learning*, 11; Irving, *Catechism of Botany* (1852), 71; L. Johnson, *Botanical Teacher for North America*, 15; Coxe, *The Young Lady's Companion and Token of Affection* (1846), 165.

17. Eaton, *Manual of Botany* (1829), 6–7, 5. Other botanists concur on the importance of exchanging specimens: Laura Johnson, Eaton's student, similarly recommends that "the students of a class should go out, in small parties, in different directions, and each should collect more than he needs, in order to exchange with other students." L. Johnson, *Botanical Teacher for North America*, 14. Nearly thirty years later, Gray was still advocating for sharing plants with others, for one's herbarium would then comprise "not only the specimens which the proprietor has himself collected, but those which he acquires through friendly exchanges with distant botanists, or in other ways." Gray, *First Lessons in Botany and Vegetable Physiology*, 201.

18. Quoted in LaMond, "History Reclaimed," 6.

19. Britton, "Preservation of Native Plants," 1; Smithsonian Institution, "Spare the Flowers." See also Armitage, *Nature Study Movement*, 124.

20. Wood, *First Lessons in Botany*, 86–87.

21. Phelps, *Familiar Lectures on Botany* (1849), 31–32.

22. See, for example, Curtiss, "Hints on Herborizing." *The Bulletin of the Torrey Botanical Club* ran a regular column called "Herbarium Suggestions." See also "Forming and Preserving Herbariums or Dried Plants"; "Preserving Plants, and Method of Making an Herbarium," 40; "The Herbarium," 255; Coultas, "On the Formation of an Herbarium," 317–318; [Ida], "My Herbarium"; "The Ruralist."

23. "Wild Flowers of Early Spring-Time," 343; "Collection and Preservation of Plants," 352.

24. "Drying and Preserving Flowers"; "The Wild Flowers of Early Spring-Time," 343. The unattributed "valuable little work" was Oliver, *Lessons in Elementary Botany*.

25. Keeney, *Botanizers*, 13; Dupuy, "Father's Revenge," 176, 177.

26. Harland, "Rizpah's Idols" (August 1866), 116.

27. Harland, "Bred in the Bone," 214, 215.

28. Bailey, *Botanizing*, 5, 9; Manton, *Field Botany*, 27, 28, 30.

29. Manton, *Field Botany*, 31, 36, 38, 40.

30. Soule's "Skeleton Leaves" series also featured Henry Ward Beecher and Charles Sumner. Later stereoscope cards by other photographers have skeleton-leaf-framed portraits of Presidents James Garfield, Grover Cleveland, William Henry Harrison, and William McKinley with their wives. The Library of Congress has a collection of more than fifty-two thousand stereograph cards; for information about the collection as well as more detail about the process of stereoscopy, see "Collection: Stereograph Cards, About This Collection,"

Library of Congress, https://www.loc.gov/collections/stereograph-cards/about-this-collection/.

31. Lightfoot, *Culture and Art of Death*, 211, 212,

32. See Lightfoot, *Culture and Art of Death*, chapter 8, "Photography and Death."

33. Parrish, *Phantom Bouquet*, 6, 8–9.

34. Parrish, *Phantom Bouquet*, 17–18, 39, 38.

35. *Phantom Flowers*, 11.

36. *Phantom Flowers*, 72–73, 79.

37. DiNoto and Winter, *Pressed Plant*, 28; Maura Flannery, *In the Herbarium*, 150.

38. Greene, *Wild Flowers from the Holy Land*, 10; Greene, *Wild Flowers from Palestine*, n.p.

39. Greene, *Wild Flowers from Palestine*, n.p.; Greene, *Wild Flowers from the Holy Land*, 19.

40. Musselman, *Dictionary of Bible Plants*, 4; Bielo, "Flower, Soil, Water, Stone," 372. See also Bielo, *Materializing the Bible*.

41. Ewing, *Bible Natural History*, 3–4.

42. Examples are available at the Chester County (Pa.) History Center Herbarium Collection, Lewis and Clark College, Mason Area (Mich.) Historical Society, Maynard (Mass.) Historical Society, and the Smithsonian Institution. *Herbarium and Plant Descriptions* examples have also been listed on AbeBooks, Etsy, and eBay.

43. Nelson, *Herbarium and Plant Descriptions*, 1.

44. R. Johnson, *Passions for Nature*, 10. Johnson uses the phrase in the context of her argument that the "passion for nature" of Susan Fenimore Cooper and Henry David Thoreau was more rooted in the physical world, in contrast to that of Thomas Cole, Andrew Jackson Downing, and Ralph Waldo Emerson, who expressed theirs primarily through metaphor.

45. I. Sprague and Hervey, *Flowers of the Field and Forest*, title page, 34–35, 59–60. Barbara Thiers supports Hervey's assertion: "The first herbarium specimens collected in North America are probably those of Michel Sarrazin, a physician living in Quebec. Beginning in 1697 he sent specimens annually to French botanist Joseph Pitton de Tournefort, including the first known specimens of *Aralia nudicaulis* (ginseng) and *Sarracenia purpurea* (common pitcherplant)." Thiers, *Herbarium*, 103.

46. An example that combines an actual herbarium page and a watercolor drawing of the same plant is discussed in Ulrich et al., *Tangible Things*, 38–45.

47. Marsh, *Address Delivered before the Agricultural Society of Rutland County*, 18; "Preservation of Wild Flowers"; Northrop, "Arbor Day," 410.

48. Crowe-Carraco, "'Perusin' the Pennyrile,'" 252–253; U.S. Fish and Wildlife Service, "Price's Potato-Bean (*Apios priceana*)," 4.

49. Miller-Rushing et al., "History of Citizen Science," 19.

50. For coverage of the closure, see Pennisi, "'Tragic Mistake,'" and Zimmer, "Duke Shuts Down Huge Plant Collection."

51. Maura Flannery, *In the Herbarium*, 9.

Chapter 4. "Diary Kept in Flowers"

1. Annie Adams Fields's "Fragrant Memories" diary is at the Huntington Library, San Marino, California.

2. There are two book-length biographies of Fields: Roman, *Annie Adams Fields: The Spirit of Charles Street*, and Gollin, *Annie Adams Fields: Woman of Letters*. Throughout this chapter, "Fields" refers to Annie Fields; James Fields is identified by first and last name.

3. Sixty-one volumes of Fields's diaries are at the Massachusetts Historical Society, Boston.

4. Deborah Evans notes that "while these diaries were personal and private, the information Fields records suggests that she was aware that they might serve a more public use at some time in the future." Evans, "Annie Adams Fields," 123.

5. As I did with Mattie Snowden's writings in chapter 2, I have retained Fields's line breaks in my transcriptions of the handwritten diary entries.

6. Gollin, *Annie Adams Fields*, 113. Houghton Mifflin published both the first edition of *An Island Garden* and a beautiful facsimile edition in 2001. The cover design is by Sarah Wyman Whitman (1842–1904), who also designed the covers of works by Sarah Orne Jewett.

7. Gollin, *Annie Adams Fields*, 151.

8. Roman, *Annie Adams Fields*, 14; Gollin, *Annie Adams Fields*, 26; Roman, *Annie Adams Fields*, 124–125.

9. Gollin, *Annie Adams Fields*, 13. Fields recorded visits to the "Pamfili Doria Villa" in her diary on February 28, 1860, and March 5, 1860; see transcripts of these entries at Fields, "Diary of Annie Adams Fields."

10. The ship first arrived in New York in November 1863. "Our Italian Visitors." When the ship landed, at the Western Islands (now Azores), "her masts were carried away, her bulwarks smashed, and all her guns thrown overboard, besides other heavy articles, but no lives were lost." "By Telegraph: The Missing Italian Frigate Heard From."

11. For more on Longfellow's as the "first American translation," see Roylance, "Longfellow's Dante"; for details about the dinner, see Dana and Dupont, "Longfellow and Dante," 257–258.

12. On March 20, 1864, Fields wrote, "From the church of the 'Immaculate Conception' where Mrs Stowe and I went on Palm Sunday—It was a refreshing season of devotion—"

13. Gollin, *Annie Adams Fields*, 79.

14. Scharnhorst, *Julian Hawthorne*, 19.

15. The handwriting in the inserted note strongly resembles that found in manuscript materials by Sophia Hawthorne, and Gollin identifies Sophia as the writer. Gollin, *Annie Adams Field*, 101. "Often Sophia [Hawthorne] addressed the much younger Annie as her 'dear Moonlight' or 'dearest Lily'" (92).

16. Fields, *James T. Fields*, 91.

17. Gollin, *Annie Adams Fields*, 38; "A Proclamation, for a Day of National Humiliation, Fasting and Prayer!"

18. "Reception and Departure of the Fifty-Fourth Regiment."

19. Fuller, *Good Thoughts in Bad Times*, v, x.

20. Giesburg, *Emilie Davis' Civil War*, 5.

21. Ticknor, Reed, and Fields first published *Angel Voices* in 1849; Ticknor and Fields republished it in multiple editions, including in 1855, 1858, 1861, 1864, and later. An advertisement from June 1864 announced two editions of *Angel Voices*: a "16mo. 75 cts." and "an entirely new and much enlarged edition, beautifully printed on tinted paper, and richly bound . . . small 4to. $2.50." "A List of Books Published by Messrs. Ticknor and Fields," 2. The size of "Fragrant Memories" and of my copy of *Angel Voices* is sextodecimo, or sixteenmo.

22. Multigraph Collective, *Interacting with Print*, 288, 289.

Chapter 5. Flower Language Dictionaries

Epigraph sources: Hale, *Flora's Interpreter, or, the American Book of Flowers and Sentiments* (1832), iv; Hooper, *Lady's Book of Flowers and Poetry* (1842), 8; *Flowers, Their Language, Poetry, and Sentiment*, ix.

1. Waterman, *Flora's Lexicon* (1840), 128.

2. Hooper, *Lady's Book of Flowers and Poetry (1842), 8.*

3. Osgood, *Flower Alphabet*, n.p.

4. Branch, *Reading the Roots*, 325. In his anthology of early American nature writing, Michael Branch includes excerpts from Lucy Hooper's *The Lady's Book of Flowers and Poetry* (1842), identifying her as "an accomplished botanist and purveyor of popular science" and further noting that flower language books "might be seen as a popular corollary to the Transcendentalists' more intellectualized assertion that, as Emerson wrote in *Nature* (1836), the natural world is a 'perpetual allegory' within which '[e]very natural fact is a symbol of some spiritual fact.'" Branch, *Reading the Roots*, 326.

5. Seaton, *Language of Flowers*, 69, 70, 80. For analysis of flower language's Orientalism, see Engelhardt, "Victorian Sélams and Talking Bouquets."

6. Wirt, preface, *Flora's Dictionary* (1832).

7. Table 5.1 provides the publication date of the first edition; figures 5.6–5.10 provide the publication date of the featured page(s).

8. Shinn, "Gift Books," 180; Urakova, "'Ghost Book' and the Gift Book," 44; Goodrich quoted in Lehuu, *Carnival on the Page*, 77.

9. Wirt, *Flora's Dictionary*; Waterman, *Flora's Lexicon*; Carter, *Lexicon of Ladies' Names with their Floral Emblems; Flora's Pocket Dictionary*; Dix, *Garland of Flora*; Edgarton, *Flower Vase*; *Bouquet*; J. Adams, *Flora's Album, Containing*

the Language of Flowers, Poetically Expressed; J. Adams, *Floral Wreath*; Anna Elizabeth, *Vase of Flowers*; Greenwood, *Rural Wreath*; Hooper, *Lady's Book of Flowers and Poetry*; Osgood, *Poetry of Flowers and Flowers of Poetry*; J. Adams, *Language of Flowers*; Frisbie, *Songs of the Flowers*; *Flowers, Their Language, Poetry, and Sentiment*; Dumont, *Floral Offering*; Chauncey, *Floral Gift, from Nature and the Heart*; Hanson, *Flora's Dial*; Dinnies, *Floral Year*; Hale, *Flora's Interpreter*.

10. Seaton, *Language of Flowers*, 108–110.

11. This copy of *The Bouquet* is at the University of Washington Library Special Collections. Mary H. Morrill was the sister of Martha B. Morrill, whose copy of *The Scriptural Album with Floral Illustrations* is discussed in chapter 6; for information about their parents, John Morrill and Nancy Jones, see "Morrill Online." The Morrill Act (1862), which created the system of land grant universities, is named for their relative Justin Smith Morrill, a congressman and senator from Vermont.

12. Bartlett, *Dictionary of Americanisms*, 138. Bartlett's definition traces the custom to Germany, based on the word *Vielliebchen*.

13. Edgarton, *Flower Vase*, iii; Cook, "'Oh Dear! How the Factory Girls Do Rig Up!,'" 223.

14. Adelaide, "Hill Side and the Fountain Rill," 20. Water lily has the emblem "purity" in Dix, *Garland of Flora* (1829), and "purity of heart" in Hale, *Flora's Interpreter or, the American Book of Flowers and Sentiments* (1832), both published before 1840.

15. J. Cobb, *Picture Freedom*, 92; Bondy, "Freedom Flora," 53.

16. Pratt, *Flora and Thalia*, 209.

17. Osgood, *Poetry of Flowers*, 244.

18. "[Assorted flowers] [graphic]."

19. J. Cobb, *Picture Freedom*, 67; Rusert, *Fugitive Science*, 206.

20. See, for example, Bennett, *Poets in the Public Sphere*, 159–180; Briggs, "Language of Flowers in *O Pioneers!*"; Downs, "Ellen Glasgow's Darwinian Language of Flowers"; Ellington, "Sentiment and Satire"; Marino, "Powerful Emblems of Protest"; Palleau-Papin, "Subversive Language of Flowers in *Sapphira and the Slave Girl*"; Rhyner, "Flowers of Rhetoric"; Rosenthal, "Floral Counterdiscourse"; and Sheley, "Language of Flowers in Harriet Beecher Stowe's *Uncle Tom's Cabin*." A recent exception to scholarship on the language of flowers for literary interpretation is Jasper, "Gathering Flowers."

21. T. Brown, *Dorothea Dix*, 41, 45, 48.

22. Osgood, *Poetry of Flowers and Flowers of Poetry*, "Botany," 245–256; Waterman, *Flora's Lexicon* (1840), "Botany," 233–244; Hale, *Flora's Interpreter or, the American Book of Flowers and Sentiments* (1832), iii; Wirt, *Flora's Dictionary* (1832), preface, n.p.

23. Waterman, *Flora's Lexicon* (1840), 3; Hale, *Flora's Interpreter, and Fortuna Flora* (1849), iv; Hooper, *Lady's Book of Flowers and Poetry* (1842), 256.

24. Dix, *Garland of Flora*, 22.

25. Wood, *Class-Book of Botany* (1855), p. i of appendix, following p. 645; Wood, *Leaves and Flowers*, "Appendix: The Language of Flowers," 341–346; Phelps, *Familiar Lectures on Botany* (1836), appendix, 171; Phelps, *Botany for Beginners* (1867), 214.

26. Gianquitto, *"Good Observers of Nature,"* 54, 49; Osgood, *Poetry of Flowers and Flowers of Poetry*, 4.

27. Welch, *Book of Nature*, 5; Wirt, *Flora's Dictionary* (1832), "Notes" section, 1; Hale, *Flora's Interpreter or, the American Book of Flowers and Sentiments* (1832), 13; Osgood, *Poetry of Flowers and Flowers of Poetry*, 177.

28. Hale, *Flora's Interpreter or, the American Book of Flowers and Sentiments* (1832), 103; Osgood, *Poetry of Flowers and Flowers of Poetry*, 186. Osgood's poem "The Southerner to a Yankee" is also about the magnolia billet-doux custom (186). I haven't located a definitive source for this custom, but later references to the magnolia billet-doux include "A Double Handful; or, The Heirs of Mariposa Joe" ("It is only Southern women who have that delicious creamy complexion—just like the magnolia-petals that they write their *billet-doux* upon" [450]); and Lägerlof, *The Miracles of Antichrist* ("In my time young girls used to prick love-letters on the

magnolia-blossoms" [118]). For a discussion of the literary and cultural foundations of Southern chivalry, see Genovese, "Chivalric Tradition in the Old South."

29. Waterman, *Flora's Lexicon* (1840), 114; Osgood, *Poetry of Flowers and Flowers of Poetry*, 205–206.

30. Newman, *Illustrated Botany* (1846), 96; Osgood, *Poetry of Flowers and Flowers of Poetry*, 206.

31. Waterman is one of several flower language dictionary authors who mentions this use: The entry on acacia (*Robinia pseudoacacia*) in *Flora's Lexicon* (1840) notes that Native Americans' "bows are made from the incorruptible wood of this tree, their arrows are armed with one of its thorns" (13). See also Osgood, *Poetry of Flowers and Flowers of Poetry*, 177, and Dumont, *Floral Offering* (1851), 123.

32. Allen, "Book-Cover Stamps," 13–14; Edgarton, *Fables of Flora*, 10. This stamp was also used for *Scriptural Album with Floral Illustrations*; see chapter 6.

33. Winship, "Manufacturing and Book Production," 60–61.

34. Morris and Levin, "Introduction," 15.

35. C. Turner, *Floral Kingdom*, iii.

36. See Engelhardt, "Victorian Sélams and Talking Bouquets."

37. Many nineteenth-century sources detailed the significance of the passionflower's iconography, including "Sacred Trees and Flowers," which quotes this description from an early seventeenth-century account: "The fringelike filaments above are blood-red; 'as though referring to the scourge with which Our Lord was beaten.' In the midst of the flower rises the column to which He was bound; and above are the nails, both of a 'clear green.' Above, again, is the crown of thorns, surrounded by a kind of veil of threads—seventy-two in number—(the traditional number of the thorns on Our Lord's crown) coloured like a peacock's feather ('di color pavonazzo'). In the centre of the flower, and under the column, are five marks or spots, of a blood colour, 'clearly representing the five chief wounds that Christ received on the cross.' The plant, he continues, is rich in leaves, which in shape resemble the iron of a pike or lance-head, and refer to that with which Our Lord's side was pierced" (248).

Chapter 6. "Friendship Is a Holy Thing"

1. Zebuhr, "Work of Friendship," 437.

2. Blouin, "Eternal Perspectives," 64.

3. See Reese, *Stamped with a National Character*, 9–22. Included in Reese's exhibition catalog are several works and authors I analyze: Wirt, *Flora's Dictionary* (69); Hooper, *The Lady's Book of Flowers and Poetry* (70); and Lida Clarkson (72).

4. "Deceit" corresponds to monkshood in Dorothea Dix, Sarah Hale, and Mary Chauncey; to Venus flytrap in Catharine Waterman, Lucy Hooper, C. M. Kirtland, and *The Language and Poetry of Flowers*; to dogbane in J. Wesley Hanson; and to apocynum (dogsbane) and geranium in Kirtland and *The Language and Poetry of Flowers*. "Hatred" corresponds to basil in Waterman, Hooper, Frances Osgood, Kirtland, and *The Language and Poetry of Flowers*. "Ingratitude" corresponds to thornless rose in Elizabeth Wirt and Hale; to buttercup in Waterman, Osgood, Hanson, Kirtland, and *The Language and Poetry of Flowers*; to yellow gentian in Hooper and Osgood; and to crowfoot in Kirtland and *The Language and Poetry of Flowers*. "Perfidy" corresponds to almond laurel or "common laurel in flower" in Hooper, Osgood, Kirtland, and *The Language and Poetry of Flowers*. "Revenge" corresponds to bird's-foot trefoil in Wirt, Waterman, Hooper, Osgood, Kirtland, and *The Language and Poetry of Flowers*. "Rudeness" corresponds to *Xanthium* (clotbur) in Hanson, Kirtland, and *The Language and Poetry of Flowers*. "Treachery" corresponds to laurel in Wirt; to bilberry in Osgood, Hanson, John Adams, Kirtland, and *The Language and Poetry of Flowers*; and to sweet william and whortleberry in Hanson.

5. The prevalence of northeastern locations could be attributed to the archives in which I found them: the American Antiquarian Society (Mass.), Brown University (R.I.), the New York Public Library, and Winterthur Library (Del.). The album owners are "Augusta," Ann Louise Berry, Marcella Brown, Mary Ann Dickinson, Martha Downing, Elizabeth Grier, Laughton, Clara Mitchell, Martha Morrill,

Louisa Jane Page, Lucy Pinkham, Jane Roberts, Susan Sherman, Catherine Smith, Sarah Ann Taber, and Tacy Anna Walker.

6. For fuller discussions of Hagar in nineteenth-century America, see Gabler-Hover, *Dreaming Black/Writing White*; Jefferson, "Emergence of a Pioneer"; and Junior, *Reimagining Hagar*.

7. Additional titles include C. Adams and Kidder, *Women of the Bible*; Aguilar, *Women of Israel*; Ashton, *Mothers of the Bible*; Baldwin, *Representative Women*; Cox, *Female Scripture Biography*; Weld, *Women of the Scriptures*; and, later in the century, Stowe, *Women in Sacred History*. For a detailed analysis of prosopography, or collective biography, see Booth, *How to Make It as a Woman*.

8. Waterman, *Flora's Lexicon* (1839), 18.

9. Barton's "Counsels" was reprinted in numerous sources, including *The Ladies' Miscellany*, *The Olive Wreath*, *The York Poetical Miscellany*, and *The Book of Notions*; the final stanza was reprinted in [Treat], *Angel Voices* (discussed in chapter 4), 46.

10. The Library Company of Philadelphia identifies Fannie Henry as "[Oakland] Institute music teacher Mrs. John Hunsicker, i.e., Fannie Henry Hunsicker" and notes that "Fannie Henry Hunsicker was a woman of Chinese descent. Official records are conflicting as to whether she was born in China, at sea, or in Connecticut." "Ralston Family Cased Photograph Collection." Elizabeth "Lizzie" Grier, this album's owner, was a student at Oakland and later a teacher there. "Oakland Female Institute."

11. "The Rose Without a Thorn"; Colesworthy, "A Little Word."

12. This poem was reprinted numerous times, including in Deems, *Triumph of Peace*, 40, and remains frequently quoted today.

13. Much has been written about the popularity of poetry in nineteenth-century America; two excellent starting points are Bennett, *Poets in the Public Sphere*, and Kilcup, *Who Killed American Poetry?*

14. Heyrman, *"'Signature Drawings,'"* 77–78.

15. Allen, "Book-Cover Stamps," 12.

16. Heverly, *Annual*, 129; "Flowers."

17. J. G. M., "To Emeline."

18. Gniadek identifies Wolfe as the author in her discussion of Elizabeth Craw's diary, which quotes from the poem in 1836. Gniadek, *Oceans at Home*, 84–86. "A Song—To Mary" appears in Charles Dexter Cleveland's edited volume *English Literature of the Nineteenth Century*. Wolfe, "A Song—To Mary," 133–134.

19. "Tarbel Bancroft."

20. Schneider, "Royal G. Wilder," 74.

21. Conroy-Krutz, *Christian Imperialism*, 3.

22. "Did You Know?," 3.

Conclusion. Critical Plants and Fables of Transformation

Epigraph sources: Harper, "Mission of the Flowers," "Flower Fable."

1. Harper, "Mission of the Flowers," 44–46.

2. Rutkowski, "Leaving the Good Mother," 85–86, 99.

3. Boyd, *Discarded Legacy*, 129; Rutkowski, "Leaving the Good Mother," 84.

4. The other lilies featured are garden lily, calla lily, Easter lily, and lily of St. Joseph, along with crocuses, which are of the Iris (*Iridaceae*) family, not Lily (*Liliaecae*).

5. Michael Keevak attributes yellow's racialized origin in part to Linnaeus's *Systema Naturae*: "*Homo asiaticus*, he said, was yellow." Keevak, *Becoming Yellow*, 12. "Yellow" was used in the nineteenth-century United States to refer to lighter-skinned African Americans.

6. MacDonnell, "Trade Cards by Charlotte Perkins Gilman," 23. The six trade cards in this series are labeled "Three Ages of Woman" and can be viewed online. See Gilman, "Three Ages of Woman."

7. Keyser, *Portable Louisa May Alcott*, 207; Alcott, "Little Annie's Dream," 59, 60, 62, 63.

8. Alcott, "Flower Fable"; Keyser, *Portable Louisa May Alcott*, 30.

9. Daniel Shealy's introduction to *Louisa May Alcott's Fairy Tales and Fantasy Stories* informs my reading of her flower fables as vehicles for reform. For African American women's politicized use of floral discourse and imagery, see J. Cobb, *Picture Freedom*; Bondy, "Freedom Flora"; and Rusert, *Fugitive Science*.

10. Marder, "Vegetal Anti-Metaphysics," 469.

11. A. Antonelli et al., *State of the World's Plants and Fungi 2023*, 68; A. Antonelli et al., *State of the World's Plants and Fungi 2020*, 11; RBG Kew, *The State of the World's Plants Report—2016*, 3.

12. Alexandre Antonelli, introduction; Kuhn, *Garden Politic*, 173.

13. iNaturalist, https://www.inaturalist.org, accessed September 4, 2024; Eaton, *Manual of Botany*, v.

14. iNaturalist; A. Antonelli et al., *State of the World's Plants and Fungi 2023*, 21.

15. Kress and Funk, "Herbarium Collections," 217.

16. Maura Flannery, *In the Herbarium*, 216; Kress and Funk, "Herbarium Collections," 210; Primack, *Walden Warming*; Zimmerman, *Unrooted*, 12.

17. Verberk, "Explaining General Patterns."

18. Friedlander, "From Sustainability to Abundance."

19. Mintz, "Statistics." Sources on early African American scientific knowledge include Fett, *Working Cures*; Schiebinger, *Secret Cures of Slaves*; and Rusert, *Fugitive Science*. The Native American Ethnobotany Database can be found at http://naeb.brit.org/.

20. Kimmerer, *Braiding Sweetgrass*, ix.

21. Salmón, "Kincentric Ecology," 1327.

22. Salmón, *Iwígara*, 11.

23. Montgomery, *Lessons from Plants*, xiv, 39, 45, 50–51, xii.

24. Herbaria 3.0 is located at www.herbaria3.org. The Plant Humanities Lab is at https://lab.plant-humanities.org.

Bibliography

(*) Available in Special Collections, Davidson College

"An Act to Set Apart a Certain Tract of Land Lying near the Head-waters of the Yellowstone River as a Public Park," Senate Bill S. 392 (1872), *U.S. Statutes at Large*, vol. 17, ch. 24, pp. 32–33.

Adams, Charles, and Daniel P. Kidder. *Women of the Bible*. New York: Lane and Scott, 1851.

Adams, John S. *The Floral Wreath: An Offering of Friendship*. Boston: J. Buffum, 1851.

———. *Flora's Album, Containing the Language of Flowers, Poetically Expressed*. New York: Leavitt, Trow, 1848.

*———. *The Language of Flowers, Poetically Expressed*. New York: Leavitt and Allen, 1852.

Adelaide. "The Hill Side and the Fountain Rill." *Lowell Offering* 2 (December 1840): 20–21.

Aguilar, Grace. *The Women of Israel*. New York: D. Appleton, 1853.

Alcott, Louisa May. "A Flower Fable." ["Queen Aster"]. *Woman's Journal* 18, no. 9 (February 26, 1887): 67.

———. "Little Annie's Dream." In Shealy, *Louisa May Alcott's Fairy Tales and Fantasy Stories*, 59–66.

———. "Queen Aster." In Shealy, *Louisa May Alcott's Fairy Tales and Fantasy Stories*, 334–338.

Allen, Sue. "Book-Cover Stamps Engraved by John Feely, 1842–1877." In *Decorated Cloth in America: Publishers' Bindings, 1840–1910*, edited by Sue Allen and Charles Gullan, 11–51. Los Angeles: UCLA Center for 17th- and 18th-Century Studies, William Andrews Clark Memorial Library, 1994.

Alpine Heritage: One Hundred Years of History-Recreation-Lore in Alpine County, California, 1864–1964. Alpine County: Centennial Book Committee, 1964.

Angelo, Ray. "Thoreau as Botanist: An Appreciation and a Critique." *Arnoldia* 45, no. 3 (Summer 1985): 13–23.

*Anna Elizabeth. *Vase of Flowers*. Boston: G. W. Cottrell, 1851.

Antonelli, A., C. Fry, R. J. Smith, J. Eden, R. H. A. Govaerts, P. Kersey, E. Nic Lughadha, R. E. Onstein, M. S. J. Simmonds, A. Zizka, et al. *State of the World's Plants and Fungi 2023*. Royal Botanic Gardens, Kew. https://doi.org/10.34885/wnwn-6s63.

Antonelli, A., C. Fry, R. J. Smith, M. S. J. Simmonds, P. J. Kersey, H. W. Pritchard, et al. *State of the World's Plants and Fungi 2020*. Royal Botanic Gardens, Kew. https://doi.org/10.34885/172.

Antonelli, Alexandre. Introduction to *State of the World's Plants and Fungi 2023*, by Antonelli et al., 2.

Armitage, Kevin C. *The Nature Study Movement: The Forgotten Popularizer of America's Conservation Ethic*. Lawrence: University Press of Kansas, 2009.

Ashton, S. G. *The Mothers of the Bible*. Boston: J. E. Tilton, 1855.

"[Assorted flowers] [graphic] / L.A.B." Amy Matilda Cassey Album webpage, Library Company of Philadelphia. https://digital.librarycompany.org/islandora/object/Islandora%3ACASS1?display=list.

Augusta, owner. *The Scriptural Album with Floral Illustrations*. New York: J. C. Riker, n.d. John Hay Library, Brown University.

*Bailey, William Whitman. *Botanizing: A Guide to Field-Collecting and Herbarium Work*. Providence, R.I.: Preston and Rounds, 1899.

*Baily, William L. *Trees, Plants, and Flowers: Where and How They Grow*. Philadelphia: J. B. Lippincott, 1870.

Baldwin, George C. *Representative Women: From Eve, the Wife of the First, to Mary, the Mother of the Second Adam*. New York: Sheldon, Lamport and Blakeman, 1856.

Bartlett, John Russell. *Dictionary of Americanisms: A Glossary of Words and Phrases, Usually Regarded as Peculiar to the United States.* New York: Bartlett and Welford, 1848.
Barton, Bernard. "Counsels." *Ladies' Miscellany* 2, no. 1 (April 1830): 4.
———. "Counsels." In *The Book of Notions*, edited by John Hayward, 131. Boston: Bela Marsh, 1850.
———. "Counsels." In *The Olive Wreath*, 86. Lowell, Mass.: Nathaniel Dayton, 1847.
———. "Counsels." In *The York Poetical Miscellany*, edited by William Hargrove, 205. York: W. and J. Hargrove, 1835.
Batsaki, Yota, and Julia Fine. "Cultivating History: The Plant Humanities at Dumbarton Oaks." *Perspectives on History* 60, no. 7 (October 2022). https://www.historians.org/perspectives-article/cultivating-history-the-plant-humanities-at-dumbarton-oaks-october-2022/.
Baym, Nina. *American Women of Letters and the Nineteenth-Century Sciences: Styles of Affiliation.* New Brunswick, N.J.: Rutgers University Press, 2001.
*Beck, Lewis C. *Botany of the United States North of Virginia.* New York: Harper and Brothers, 1848.
Beecher, Henry Ward. "A Discourse on Flowers." *Independent* 5, no. 238 (June 1853): 98.
*———. "A Discourse on Flowers." In *Star Papers; or, Experiences of Art and Nature*, 93–105. New York: J. C. Derby, 1855.
Bennett, Paula Bernat. *Poets in the Public Sphere: The Emancipatory Project of American Women's Poetry, 1800–1900.* Princeton: Princeton University Press, 2003.
*Berry, Ann Louise, owner. *The Scriptural Album with Floral Illustrations.* New York: J. C. Riker, n.d.
Bielo, James S. "Flower, Soil, Water, Stone: Biblical Landscape Items and Protestant Materiality." *Journal of Material Culture* 23, no. 3 (September 2018): 368–387.
———. *Materializing the Bible: Scripture, Sensation, Place.* New York: Bloomsbury, 2021.
Biographical and Portrait Cyclopedia of Cambria County, Pennsylvania. Philadelphia: Union Publishing, 1896.
Blouin, Jenifer. "Eternal Perspectives in Nineteenth-Century Friendship Albums." *Hilltop Review* 9, no. 1 (December 2016): article 7.
Bond, Henry. *Genealogies of the Families and Descendants of the Early Settlers of Watertown, Massachusetts, including Waltham and Weston.* Boston: Little, Brown, 1855.
Bondy, Katharine Isabel. "Freedom Flora: Botanical Revision and Community in African American Friendship Albums." *J19: The Journal of Nineteenth-Century Americanists* 10, no. 1 (Spring 2022): 49–76.
Booth, Alison. *How to Make It as a Woman: Collective Biographical History from Victoria to the Present.* Chicago: University of Chicago Press, 2004.
The Bouquet: Containing the Poetry and Language of Flowers. By a Lady. Boston: Oliver L. Perkins, 1845.
Bourne, Hermon. *Flores Poetici: The Florist's Manual.* Boston: Munroe and Francis, 1833.
Boyd, Melba Joyce. *Discarded Legacy: Politics and Poetics in the Life of Frances E. W. Harper, 1825–1911.* Detroit: Wayne State University Press, 1994.
Branch, Michael P. *"The Best Read Naturalist": Nature Writings of Ralph Waldo Emerson.* Charlottesville: University of Virginia Press, 2017.
———. *Reading the Roots: American Nature Writing Before Walden.* Athens: University of Georgia Press, 2004.
Brewer, William Henry, Sereno Watson, and Asa Gray. *Botany.* Vol. 1. Geological Survey of California. Cambridge, Mass.: Welch, Bigelow University Press, 1876.
Briggs, Cynthia K. "The Language of Flowers in *O Pioneers!*" *Willa Cather Pioneer Memorial Newsletter* (Summer 1986): 29–33.
Britton, N. L. "The Preservation of Native Plants." *Journal of the New York Botanical Garden* 3, no. 25 (January 1902): 1–3.
Brown, Marcella, owner. *The Scriptural Album with Floral Illustrations.* New York: J. C. Riker, n.d. American Antiquarian Society.
Brown, Thomas J. *Dorothea Dix: New England Reformer.* Cambridge, Mass.: Harvard University Press, 1998.
Bryan, Mary Lynn McCree, Barbara Bair, and Maree de Angury, eds. *The Selected Papers of Jane Addams.* Vol. 1, *Preparing to Lead, 1860–81.* Champaign: University of Illinois Press, 2003.
"By Telegraph: The Missing Italian Frigate Heard From." *Daily National Republican* (Washington, D.C.), April 28, 1864, 2.

"California Bound Passengers." *San Francisco Daily Evening Bulletin* 25, no. 90 (January 23, 1868): 2.
*Carter, Sarah C. *Lexicon of Ladies' Names with Their Floral Emblems*. Boston: J. Buffum, 1852.
Cary, Alice. "The Thistle Flower." In *The Poetical Works of Alice and Phoebe Cary*, 197. New York: Hurd and Houghton, 1876.
*Cary, Alice, and Phoebe Cary. *Flowers from Alice and Phoebe Cary*. Boston: DeWolfe and Fiske, 1906.
Casper, Scott E., Joanne D. Chaison, and Jeffrey D. Groves, eds. *Perspectives on American Book History: Artifacts and Commentary*. Amherst: University of Massachusetts Press, 2002.
Casper, Scott E., Jeffrey D. Groves, Stephen W. Nissenbaum, and Michael Winship, eds. *The Industrial Book, 1840–1880*. Vol. 3 of *A History of the Book in America*, edited by David D. Hall. Chapel Hill: University of North Carolina Press, 2007.
Cassey, Amy Matilda. "Original & Selected Poetry &c. Album." Library Company of Philadelphia. https://digital.librarycompany.org/islandora/object/Islandora%3A64815#page/5/mode/1up.
Caulkins, Frances M. *Eve and Her Daughters of Holy Writ, Or, Women Of The Bible*. New York: American Tract Society, 1850.
Chandler, Charles H., with Sarah Fiske Lee. *The History of New Ipswich, New Hampshire, 1735–1914, with Genealogical Records of the Principal Families*. Fitchburg, Mass.: Sentinel Printing, 1914.
"Changes to Washoe Life." "Washoe History" page, Washoe Tribe of California and Nevada website. https://washoetribe.us/aboutpage/4-Page-washoe-history.
Charlotte Perkins Gilman Papers, 1846–1961; Drawings; Advertising cards for Kendall Mfg. Co: Soapine French Laundry Soap. Collection 177, folder 321. Schlesinger Library, Radcliffe Institute, Harvard University, Cambridge, Mass.
*Chauncey, Mary, ed. *The Floral Gift, from Nature and the Heart*. New York: Leavitt and Allen, 1853.
*Clarkson, Lida. *The Gathering of the Lilies*. Philadelphia: J. L. Sibole, 1877.
Cobb, Daniel J. *The Family Adviser; Calculated to Teach the Principles of Botany*. Rochester: Marshall & Dean, 1828.
Cobb, Jasmine Nichole. *Picture Freedom: Remaking Black Visuality in the Early Nineteenth Century*. New York: New York University Press, 2015.
Cohen, Michael C. "Album Verse and the Poetics of Scribal Circulation." In *A History of Nineteenth-Century American Women's Poetry*, edited by Jennifer Putzi and Alexandra Socarides, 68–86. Cambridge: Cambridge University Press, 2017.
Colesworthy, Daniel Clement. "A Little Word." *Youth's Companion* 15, no. 29 (November 26, 1841): 116.
"The Collection and Preservation of Plants." *Godey's Lady's Book* 66 (April 1863): 350–352.
*Comstock, J. L. *An Introduction to the Study of Botany*. Hartford: D. F. Robinson, 1833.
———. *The Young Botanist: Being a Treatise on the Science Prepared for the Use of Persons Just Commencing the Study of Plants*. New York: Robinson, Pratt, 1836.
Conroy-Krutz, Emily. *Christian Imperialism: Converting the World in the Early American Republic*. Ithaca, N.Y.: Cornell University Press, 2015.
Coolbrith, Ina. "Among the Daisies." In *The Galaxy: An Illustrated Magazine of Entertaining Reading*, vol. 4, edited by William Conant Church, 151. New York: W. C. and F. P. Church, 1868.
Cordell, Ryan, and Abby Mullen. "'Fugitive Verses': The Circulation of Poems in Nineteenth-Century American Newspapers." *American Periodicals* 27, no. 1 (2017): 29–52.
*Cook, Harriet N. *The Trees, Fruits, and Flowers of the Bible*. New York: American Tract Society, 1846.
Cook, Sylvia Jenkins. "'Oh Dear! How the Factory Girls Do Rig Up!': Lowell's Self-Fashioning Workingwomen." *New England Quarterly* 83, no. 2 (June 2010): 219–249.
Coultas, Harland. "On the Formation of an Herbarium." *Arthur's Home Magazine* 6, no. 12 (December 1855): 317–318.
Cox, F. A. *Female Scripture Biography*. Boston: Lincoln and Edmands, 1831.
Coxe, Margaret. *Conversations on the Botany of the Scriptures*. New York: General Protestant Episcopal Sunday School Union, 1836.

———. *The Young Lady's Companion and Token of Affection.* Columbus, Ohio: Isaac N. Whiting, 1846.

*———. *The Young Lady's Companion and Token of Affection.* Columbus, Ohio: Isaac N. Whiting, 1851.

Crowdsourcing and Citizen Science Act of 2016 (15 U.S.C. 3724). https://uscode.house.gov/view.xhtml?req=granuleid:USC-prelim-title15-section3724&num=0&edition=prelim.

Crowe-Carraco, Carol. "'Perusin' the Pennyrile': Sadie Price, 19th-Century Botanist." *Filson Club History Quarterly* 65, no. 2 (April 1991): 252–267.

Cummins, Maria Susana. *The Lamplighter*. Edited by Nina Baym. New Brunswick, N.J.: Rutgers University Press, 1988.

Curtiss, A. H. "Hints on Herborizing." *American Naturalist* 6, no. 5 (May 1872): 257–260.

Dana, Henry Wadsworth Longfellow, and Christian Y. Dupont. "Longfellow and Dante." Special issue, *Dante Studies* 128 (2010): 221–278.

*Darby, John. *Botany of the Southern States*. New York: A. S. Barnes and H. L. Burr, 1859.

Davis, Donald G., Jr., and John Mark Tucker. "The Impact of the Christian Faith on Books, Publishing, and Libraries: American Organizations and Leaders in the Nineteenth and Twentieth Centuries." *Library and Information History* 32, nos. 1–2 (February/May 2016): 112–122.

Dean, Bradley P. "A Possible Source for Thoreau's Increased Interest in Natural Phenomena in 1851." *Thoreau Research Newsletter* 2, no. 2 (April 1991): 1–3.

———, ed. *Wild Fruits: Thoreau's Rediscovered Last Manuscript*. New York: Norton, 2000.

Deems, Charles Force. "Lines Written in the Album of a Little Stranger." *Godey's Lady's Book* 18 (March 1839): 107.

———. *The Triumph of Peace: And Other Poems*. New York: D. Fanshaw, 1840.

DeJong, Mary G. Introduction to *Sentimentalism in Nineteenth-Century America: Literary and Cultural Practices*, edited by Mary G. DeJong with Paula Bernat Bennett, 8–18. Madison, N.J.: Fairleigh Dickinson University Press, 2013.

DeSoto, Lindsey. "What is Black Cohosh?" Health.com, May 17, 2023. https://www.health.com/black-cohosh-7488797.

Dickerson, Mary Anne. Diary. Library Company of Philadelphia. https://digital.librarycompany.org/islandora/object/Islandora%3ADICM2.

Dickinson, Emily. *Emily Dickinson's Herbarium: A Facsimile Edition*. Cambridge, Mass.: Belknap Press of Harvard University Press, 2006.

———. *The Letters of Emily Dickinson*. Edited by Cristanne Miller and Domhnall Mitchell. Cambridge, Mass.: Belknap Press of Harvard University Press, 2024.

Dickinson, Mary Ann, owner. *The Scriptural Album with Floral Illustrations*. New York: J. C. Riker, n.d. American Antiquarian Society.

"Did You Know?" *Christian History and Biography* 90, no. 3 (Spring 2006): 2–4.

Diffenbaugh, Vanessa. *The Language of Flowers: A Novel*. New York: Ballantine, 2012.

*Dinnies, Anna Peyre. *The Floral Year*. Boston: Benjamin B. Mussey, 1847.

DiNoto, Andrea, and David Winter. *The Pressed Plant: The Art of Botanical Specimens, Nature Prints, and Sun Pictures*. New York: Stewart, Tabori and Chang, 1999.

Dix, Dorothea. *The Garland of Flora*. Boston: S. B. Goodrich, 1829.

"A Double Handful; or, The Heirs of Mariposa Joe." In *Frank Leslie's Pleasant Hours*, 15:449–459. New York: Frank Leslie, 1874.

Downing, Martha, owner. *The Scriptural Album with Floral Illustrations*. New York: J. C. Riker, n.d. John Hay Library, Brown University.

Downs, Cathy. "Ellen Glasgow's Darwinian Language of Flowers in *The Battleground* and *Barren Ground*." *Ellen Glasgow Journal of Southern Women Writers* 2 (2009): 38–54.

Drug Enforcement Administration, Diversion Control Division, Drug and Chemical Evaluation Section. "Jimson Weed (Datura stramonium)." September 2019. https://test.deadiversion.usdoj.gov/drug_chem_info/jimson_w.pdf.

"Drying and Preserving Flowers." *Godey's Lady's Book* 74 (October 1869): 354.

*Dumont, Henrietta. *The Floral Offering*. Philadelphia: H. C. Peck and Theo. Bliss, 1851.

———. *The Floral Offering*. Philadelphia: H. C. Peck and Theo. Bliss, 1853.
Dupree, A. Hunter. *Asa Gray: 1810–1888*. Cambridge: Belknap Press of Harvard University Press, 1959.
Dupuy, Eliza A. "The Father's Revenge." *Godey's Lady's Book* 23 (October 1841): 175–179.
Dustman, Karen. *Silver Mountain City: Ghost of the Sierra*. Markleeville, Calif.: Claritage Press, 2020.
Eaton, Amos. *Botanical Exercises*. Albany, N.Y.: Websters and Skinners, 1820.
*———. *A Manual of Botany, for the Northern and Middle States of America*. Albany, N.Y.: Websters and Skinners, 1824. Eliza Nye, owner.
———. *Manual of Botany, for North America: Containing Generic and Specific Descriptions of the Indigenous Plants and Common Cultivated Exotics, Growing North of the Gulf of Mexico*. Fifth edition. Albany, N.Y.: Websters and Skinners, 1829.
*Edgarton, Sarah Carter, ed. *Fables of Flora*. Lowell, Mass.: Powers and Bagley, 1844.
*———. *The Flower Vase, Containing the Language of Flowers and Their Poetic Sentiments*. Lowell, Mass.: Powers and Bagley, 1844.
Ellington, Brenda. "Sentiment and Satire: The Language of Flowers in Fanny Fern's *Ruth Hall*." *Explicator* 74, no. 4 (2016): 219–223.
*Embury, Emma C. *American Wild Flowers in Their Native Haunts*. Illustrated by Edwin Whitefield. New York: Appleton, 1845.
Emerson, Ralph Waldo. *Nature*. Boston: James Monroe, 1836.
Engelhardt, Molly. "Victorian Sélams and Talking Bouquets: Phallic Invasion of the Feminine/Floral Order." *Tulsa Studies in Women's Literature* 35, no. 2 (Fall 2016): 343–363.
"Evangelinos Apostolides Sophocles." American Hellenic Educational Progressive Association website. https://ahepahistory.org/biographies/Professor-Evangelinos-Apostolides-Sophocles-1804-1883.html.
Evans, Deborah M. "Annie Adams Fields." In *American Women Prose Writers, 1870–1920* (vol. 221 of *Dictionary of Literary Biography*), edited by Sharon M. Harris, Heidi L. M. Jacobs, and Jennifer Putzi. Detroit: Gale, 2000.
*Ewing, Francis A. *Bible Natural History; or a Description of the Animals, Plants, and Minerals, Mentioned in the Sacred Scriptures*. Philadelphia: American Sunday School Union, 1835.
Fay, Hattie S. "When the Autumn Comes Again." *Godey's Lady's Book* 54 (January 1857): 59.
Fee, Elizabeth, Theodore M. Brown, Jan Lazarus, and Paul Theerman. "Medical Education for Women, 1870." *American Journal of Public Health* 93, no. 3 (March 2002): 363.
Fern, Fanny. *Ruth Hall and Other Writings*. Edited by Joyce W. Warren. New Brunswick, N.J.: Rutgers University Press, 1986.
Fett, Sharla M. *Working Cures: Healing, Health, and Power on Southern Slave Plantations*. Chapel Hill: University of North Carolina Press, 2002.
Fields, Annie Adams. "Diary of Annie Adams Fields." *The Brownings' Correspondence: An Online Edition*. https://www.browningscorrespondence.com/fields-diary/.
———. "Fragrant Memories." Fields Addenda, Box 6 (1), Huntington Library, San Marino, Calif.
———. *James T. Fields: Biographical Notes and Personal Sketches*. Boston: Houghton Mifflin, 1881.
Flannery, Maura. *In the Herbarium: The Hidden World of Collecting and Preserving Plants*. New Haven, Conn.: Yale University Press, 2023.
Flannery, Michael A. *Civil War Pharmacy: A History of Drugs, Drug Supply and Provision, and Therapeutics for the Union and Confederacy*. New York: Informa Healthcare, 2009.
Flora's Pocket Dictionary. Philadelphia: Porter and Coates, 1870.
"Flowers." *American Monthly Magazine* 2, no. 9 (December 1830): 615.
**Flowers of the Holy Land*. Jerusalem, n.p., n.d.
**Flowers, Their Language, Poetry, and Sentiment*. Philadelphia: Porter and Coates, 1870.
"Forming and Preserving Herbariums or Dried Plants." *Farmer's Monthly Visitor*, May 31, 1844, 78.
Friedlander, Jay. "From Sustainability to Abundance." *Stanford Social Innovation Review*, Dec. 21, 2015. https://ssir.org/articles/entry/from_sustainability_to_abundance#.

*Frisbie, Fannie. *Songs of the Flowers*. Boston: G. W. Cottrell, 1855.

Fuller, Thomas. *Good Thoughts in Bad Times*. Boston: Ticknor and Fields, 1863.

Gaard, Greta. "Critical Ecofeminism: Interrogating 'Meat,' 'Species,' and 'Plant.'" In *Meat Culture*, edited by Annie Potts, 264–287. Leiden: Brill, 2016.

Gabler-Hover, Janet. *Dreaming Black/Writing White: The Hagar Myth in American Cultural History*. Lexington: University Press of Kentucky, 2000.

Gagliano, Monica, John C. Ryan, and Patrícia Vieira, eds. *The Language of Plants: Science, Philosophy, Literature*. Minneapolis: University of Minnesota Press, 2017.

Garvey, Ellen Gruber. *Writing with Scissors: American Scrapbooks from the Civil War to the Harlem Renaissance*. Oxford: Oxford University Press, 2013.

Genovese, Eugene D. "The Chivalric Tradition in the Old South." *Sewanee Review* 108, no. 2 (Spring 2000): 188–205.

Gianquitto, Tina. *"Good Observers of Nature": American Women and the Scientific Study of the Natural World, 1820–1885*. Athens: University of Georgia Press, 2007.

Gibson, Prudence, and Catriona Sandilands. "Introduction: Plant Performance." *Performance Philosophy* 6, no. 2 (2021): 1–23.

Giesburg, Judith Ann. *Emilie Davis' Civil War: The Diary of a Free Black Woman in Philadelphia, 1863–1865*. University Park: Pennsylvania State University Press, 2014.

Gilman, Charlotte Perkins. Soapine trade cards, Jay T. Last Collection of Horticultural Prints and Ephemera, 1840-1933, series 1, binder 1, Huntington Library, San Marino, California.

———. "Three Ages of Woman" cards. Charlotte Perkins Gilman Papers (D.513), Rare Books, Special Collections, and Preservation, University of Rochester. https://archives.lib.rochester.edu/repositories/2/archival_objects/314974.

Giraud, Jane Elizabeth. *The Flowers of Shakespeare*. London: Day and Haghe, 1845.

Gisel, Bonnie, with images by Stephen J. Joseph. *Nature's Beloved Son: Recovering John Muir's Botanical Legacy*. Berkeley, Calif.: Heyday Books, 2008.

Gniadek, Melissa. *Oceans at Home: Maritime Domestic Fictions in Nineteenth-Century American Women's Writing*. Amherst: University of Massachusetts Press, 2021.

Godfrey, Laura. Diary. Maine Women Writers Collection, University of New England, Portland, Maine.

Gollin, Rita K. *Annie Adams Fields: Woman of Letters*. Amherst: University of Massachusetts Press, 2002.

*Goodale, Elaine, and Dora Read Goodale. *In Berkshire with the Wild Flowers*. New York: G. P. Putnam's Sons, 1879–1880.

*Goodrich, Samuel. *The Child's Botany*. Boston: S. G. Goodrich, 1828.

Goodwin, Joan W. *The Remarkable Mrs. Ripley: The Life of Sarah Alden Bradford Ripley*. Boston: Northeastern University Press, 1998.

*Gray, Asa. *The Botanical Text-Book: An Introduction to Scientific Botany*. New York: George Putnam, 1850.

*———. *Botany for Young People*. Part 1, *How Plants Grow*. New York: American Book Company, 1858.

*———. *Botany for Young People*. Part 2, *How Plants Behave*. New York: Ivison, Blakeman, Taylor, 1872.

*———. *First Lessons in Botany and Vegetable Physiology*. New York: Ivison and Phinney, 1857.

———. *A Manual of the Botany of the Northern United States*. Boston: James Monroe, 1848.

———. *Manual of the Botany of the United States, Including Virginia, Kentucky, and All East of the Mississippi*. New York: George P. Putnam, 1856. Berg Collection, New York Public Library, with Henry David Thoreau's marginalia.

*Green, Frances, and Joseph Congdon. *Analytical Class-Book of Botany*. New York: Appleton, 1855.

*Greene, Harvey B. *Wild Flowers from the Holy Land*. New York: Christian Herald, 1895.

*———. *Wild Flowers from Palestine*. Lowell, Mass.: Dumas, n.d.

Greenwood, Laura. *The Rural Wreath; or Life Among the Flowers*. Boston: Wentworth, 1857.

Grier, Elizabeth J., owner. *The Scriptural Album with Floral Illustrations*. New York: J. C. Riker, n.d. American Antiquarian Society.

*Griffin, Mary M. *Drops from Flora's Cup*. Boston: Hill and Brodhead, 1846.

Griswold, Rufus. *The Poets and Poetry of America, with an Historical Introduction*. Philadelphia: Carey and Hart, 1842.

Gross, Robert A., and Mary Kelley, eds. *An Extensive Republic—Print, Culture, and Society in the New Nation, 1790–1840*. Vol. 2 of *A History of the Book in America*, edited by David D. Hall. Chapel Hill: University of North Carolina Press, 2010.

Gutierrez, Kathleen Cruz. "From Objects of Study to Worldmaking Beings: The History of Botany at the Corner of the Plant Turn." *History Compass* 21, no. 8 (2023): 1–11.

Haklay, Mordecai (Muki), Daniel Dörler, Florian Heigl, Marina Manzoni, Susanne Hecker, and Katrin Vohland. "What Is Citizen Science? The Challenges of Definition." In Vohland et al., *The Science of Citizen Science*, 15–21.

Hale, Sarah Josepha. *Flora's Interpreter, and Fortuna Flora*. Boston: Benjamin B. Mussey, 1849.

———. *Flora's Interpreter, and Fortuna Flora*. Boston: Benjamin B. Mussey, 1850.

———. *Flora's Interpreter, or, the American Book of Flowers and Sentiments*. Boston: Marsh, Capen and Lyon, 1832.

———. *Flora's Interpreter, or, the American Book of Flowers and Sentiments*. Boston: Marsh, Capen and Lyon, 1838.

*———. *Flora's Interpreter, or, the American Book of Flowers and Sentiments*. Boston: Marsh, Capen, Lyon, and Webb, 1842.

Hall, Matthew. *Plants as Persons: A Philosophical Botany*. Albany: State University of New York Press, 2011.

*Hanson, J. Wesley. *Flora's Dial*. Lowell: Jonathan Allen, 1846.

Harland, Marion. "Bred in the Bone." *Godey's Lady's Book* 85, no. 507–508 (September–October 1872): 213–219, 309–315.

———. "Rizpah's Idols." *Godey's Lady's Book* 72–73 (January–October 1866).

Harper, France Ellen Watkins. "The Mission of the Flowers." In *Moses: A Story of the Nile*, 44–47. Philadelphia: Merrihew and Sons, 1869.

Hasegawa, Guy R. "Quinine Substitutes in the Confederate Army." *Military Medicine* 172, no. 6 (June 2007): 650–655.

Hemans, Felicia. "The Dial of Flowers." In *The Amulet, or Christian and Literary Remembrancer*, 31–32. London: William Baynes and Son, 1828.

"The Herbarium." *Soil of the South*, April 1, 1852, 255.

*Herrick, Sophie Bledsoe. *The Wonders of Plant Life Under the Microscope*. New York: G. P. Putnam's Sons, 1883.

Heverly, C. F. *Annual* 11, Bradford County Historical Society, 1917–1918. Towanda, Pa.: Bradford Star, 1918.

Heyrman, Joy Peterson. "'Signature Drawings:' Social Networks and Collecting Practices in Antebellum Albums." PhD diss., University of Maryland, 2008. https://api.drum.lib.umd.edu/server/api/core/bitstreams/fa470749-634c-44a5-8f50-dd4f35f08ace/content.

Hooper, Lucy. *The Lady's Book of Flowers and Poetry*. New York: J. C. Riker, 1841.

———. *The Lady's Book of Flowers and Poetry*. New York: J. C. Riker, 1842.

*———. *The Lady's Book of Flowers and Poetry*. New York: Derby and Jackson, 1858.

[Ida]. "My Herbarium." *Country Gentleman*, October 1857, 290–291.

Iovino, Serenella. "Stories from the Thick of Things: Introducing Material Ecocriticism." *Interdisciplinary Studies in Literature and Environment* 19, no. 3 (Summer 2012): 449–460.

*Irving, C. *A Catechism of Botany*. New York: Collins and Hannay, 1826.

———. *A Catechism of Botany*. London: Charles H. Law, 1852.

Jasper, Kelli Towers. "Gathering Flowers: Romantic Era Botanico-Literary Production and the Transatlantic Mediation of Culture." PhD diss., University of Colorado, 2016. https://scholar.colorado.edu/concern/graduate_thesis_or_dissertations/jm214p25t?locale=fr.

Jefferson, Lynne. "The Emergence of a Pioneer: The Manipulation of Hagar in Nineteenth-Century American Women's Novels." PhD diss., Indiana University, 2008.

J. G. M. "To Emeline." In *Atkinson's Casket: Gems of Literature, Wit and Sentiment*, 11:200. Boston: Samuel C. Atkinson, 1836.

Johnson, Laura. *Botanical Teacher for North America*. Albany, N.Y.: Oliver Steele, 1834.

Johnson, Rochelle L. *Passions for Nature: Nineteenth-Century America's Aesthetics of Alienation*. Athens: University of Georgia Press, 2009.

Joseph, Stephen J. "Technical and Aesthetic Application." In Gisel, *Nature's Beloved Son*, 201–202.

Junior, Nyasha. *Reimagining Hagar: Blackness and Bible*. Oxford: Oxford University Press, 2019.

Keeney, Elizabeth. *The Botanizers: Amateur Scientists in Nineteenth-Century America*. Chapel Hill: University of North Carolina Press, 1992.

Keevak, Michael. *Becoming Yellow: A Short History of Racial Thinking*. Princeton: Princeton University Press, 2011.

Kelley, Mary. *Learning to Stand and Speak: Women, Education, and Public Life in America's Republic*. Chapel Hill: University of North Carolina Press, 2006.

Kemble, John Haskell. "The Panamá Route to the Pacific Coast, 1848–1869." *Pacific Historical Review* 7, no. 1 (March 1938): 1–13.

Keyser, Elizabeth. *The Portable Louisa May Alcott*. New York: Penguin, 2007.

Kilcup, Karen. *Fallen Forests: Emotion, Embodiment, and Ethics in American Women's Nature Writing, 1781–1924*. Athens: University of Georgia Press, 2013.

———. *Who Killed American Poetry? From National Obsession to Elite Possession*. Ann Arbor: University of Michigan Press, 2019.

Kimmerer, Robin Wall. *Braiding Sweetgrass: Indigenous Wisdom, Scientific Knowledge, and the Teachings of Plants*. Minneapolis: Milkweed, 2013.

*Kirtland, C. M. *Poetry of the Flowers*. New York: John W. Lovell, n.d.

*———. *Poetry of the Flowers*. New York: Thomas Y. Crowell, n.d.

Kohlstedt, Sally Gregory. "Parlors, Primers, and Public Schooling: Educating for Science in Nineteenth-Century America." *Isis* 81, no. 3 (September 1990): 424–445.

Kress, W. John, and Vicki A. Funk. "Herbarium Collections, Floras, and Checklists." In *Plant Conservation: A Natural History Approach*, edited by Gary A. Krupnick and A. John Kress, 209–218. Chicago: University of Chicago Press, 2005.

Kuhn, Mary. *The Garden Politic: Global Plants and Botanical Nationalism in Nineteenth-Century America*. New York: New York University Press, 2023.

LaFauci, Lauren E. "Peculiar Nature: Slavery, Environment, and Nationalism in the Antebellum South." PhD diss., University of Michigan, 2009. https://deepblue.lib.umich.edu/handle/2027.42/63797.

Lägerlof, Selma. *The Miracles of Antichrist*. Boston: Little Brown, 1910.

Laist, Randy, ed. *Plants and Literature: Essays in Critical Plant Studies*. Amsterdam: Rodopi, 2013.

LaMond, Annette. "A History Reclaimed: The Society for the Protection of Native Plants and the Cambridge Plant Club." Written for the Cambridge Plant and Garden Club, December 2020. https://www.masshort.org/hubfs/Lamond.Native%20Plants.pdf.

**The Language and Poetry of Flowers, and Poetic Handbook of Wedding Anniversary Pieces, Album Verses, and Valentines*. New York: Hurst, n.d.

Laughton, Mary W., owner. *The Scriptural Album with Floral Illustrations*. New York: J. C. Riker, n.d. John Hay Library, Brown University.

Lawrence, Anna M. "Listening to Plants: Conversations Between Critical Plant Studies and Vegetal Geography." *Progress in Human Geography* 46, no. 2 (2022): 629–651.

**Leaf and Flower Pictures, and How to Make Them*. New York: Anson D. F. Randolph, 1857.

Lehuu, Isabelle. *Carnival on the Page: Popular Print Media in Antebellum America*. Chapel Hill: University of North Carolina Press, 2000.

Lewis, Andrew J. *A Democracy of Facts: Natural History in the Early Republic*. Philadelphia: University of Pennsylvania Press, 2011.

Lightfoot, D. Tulla. *The Culture and Art of Death in 19th Century America*. Jefferson, N.C.: McFarland, 2019.

"A List of Books Published by Messrs. Ticknor and Fields." Advertisement from June 1864; back matter, 1–20, of Prescott, *Azarian*.

Little, Ellen. "A Brief History of the American Sunday School Union." Elizabeth Nesbitt Collection @ Pitt:

American Sunday School Union, University of Pittsburgh Library System. https://pitt.libguides.com/Nesbitt/ASSU.
Longfellow, Henry Wadsworth. *The Courtship of Miles Standish and Other Poems*. Boston: Ticknor and Fields, 1858.
———. "A Day of Sunshine." In *Tales of a Wayside Inn*, 220–221. Boston: Ticknor and Fields, 1863.
*Lorimer, Mary. *Among the Trees: A Journal of Walks in the Woods, and Flower-Hunting Through Field and by Brook*. New York: Hurd and Houghton, 1869.
MacDonnell, Kevin. "Trade Cards by Charlotte Perkins Gilman: The Young Artist's Soapine and Non-Soapine designs." *Advertising Trade Card Quarterly*, Fall 2001, 18–28.
*Manton, Walter P. *Field Botany: A Hand-Book for the Collector*. Boston: Lee and Shepard, 1882.
Marder, Michael. *The Philosopher's Plant: An Intellectual Herbarium*. New York: Columbia University Press, 2014.
———. "Vegetal Anti-Metaphysics: Learning from Plants." *Continental Philosophy Review* 44, no. 4 (2011): 469–489.
Marino, Elisabetta. "Powerful Emblems of Protest: Flower Language and Flower Imagery in the Writings of Margaret Fuller." *Gender Studies* 16, no. 1 (2017): 1–7.
Marsh, George Perkins. *Address Delivered Before the Agricultural Society of Rutland County, Sept. 30, 1847*. Rutland, Vt.: Printed at the Herald Office, 1848.
Marshall, Humphry. *Arbustrum Americanum: The American Grove, or, an Alphabetical Catalogue of Forest Trees and Shrubs, Natives of the American United States*. Philadelphia: Joseph Crukshank, 1785.
*Mason, Mary Ann. *A Wreath from the Woods of Carolina*. New York: General Protestant Sunday School Union, and Church Book Society, 1859.
Mathes, Valerie Sherer, ed. *Gender, Race, and Power in the Indian Reform Movement: Revisiting the History of the WNIA*. Albuquerque: University of New Mexico Press, 2020.
———, ed. *The Women's National Indian Association: A History*. Albuquerque: University of New Mexico Press, 2015.
Miller-Rushing, Abraham J., Richard B. Primack, Rick Bonney, and Emma Albee. "The History of Citizen Science in Ecology and Conservation." In *Handbook of Citizen Science in Conservation and Ecology*, edited by Christopher A. Lepczyk, Owen D. Boyle, and Timothy Boyle, 17–23. Oakland: University of California Press, 2020.
Mintz, Stephen. "Statistics: Education in America." Gilder Lehrman Institute of American History. https://www.gilderlehrman.org/history-resources/teacher-resources/statistics-education-america-1860-1950.
Mitchell, Clara, owner. *The Scriptural Album with Floral Illustrations*. New York: J. C. Riker, n.d. New York Public Library.
Moerman, Daniel E. *Native American Ethnobotany*. Portland, Oreg.: Timber Press, 1998.
Mohapatra, Sradhanjali, Ashif Iqubal, Mohammed Javed Ansari, Bisma Jan, Sultan Zahiruddin, Mohd Aamir Mirza, Sayeed Ahmad, and Zeenat Iqbal. "Benefits of Black Cohosh (*Cimicifuga racemosa*) for Women Health: An Up-Close and In-Depth Review." *Pharmaceuticals* (Basel) 13, no. 3 (February 2022): 278. https://doi.org/10.3390/ph15030278.
Montgomery, Beronda. *Lessons from Plants*. Cambridge, Mass.: Harvard University Press, 2021.
"Morrill Online: Ancestry of the Morrill and Morrell Families in America." https://morrillonline.com.
Moore, Gerry, James Macklin, and Lisa DeCesare. "A Brief History of Asa Gray's 'Manual of Botany.'" *Harvard Papers in Botany* 15, no. 2 (December 2010): 277–286.
*Morley, Margaret Warner. *A Few Familiar Flowers: How to Love Them at Home or in School*. Boston: Ginn, 1897.
Morrill, Martha B., owner. *The Scriptural Album with Floral Illustrations*. New York: J. C. Riker, n.d. John Hay Library, Brown University.
Morris, Ellen K., and Edward S. Levin. Introduction to *The Art of Publishers' Bookbindings, 1815–1915*, edited by Ellen K. Morris and Edward S. Levin, 13–17. Los Angeles: William Dailey Rare Books, 2000.
Moulton, Gary E., ed. *Journals of Lewis and Clark*. Vol. 12, *Herbarium of the Lewis and Clark Expedition*. Lincoln: University of Nebraska Press, 1999.
Moylan, Michele, and Lane Stiles, eds. *Reading Books: Essays on the Material Text and Literature in America*. Amherst: University of Massachusetts Press, 1996.

The Multigraph Collective. *Interacting with Print: Elements of Reading in the Era of Print Saturation*. Chicago: University of Chicago Press, 2018.

Musselman, Lytton John. *A Dictionary of Bible Plants*. Cambridge: Cambridge University Press, 2011.

*Nelson, Edward T. *Herbarium and Plant Descriptions*. Boston: Allyn and Bacon, n.d. Edna Smith, owner.

*———. *Herbarium and Plant Descriptions*. Boston: Allyn and Bacon, n.d.

New Ipswich Academy record books. Collection MS-1445, Rauner Library Collections, Dartmouth University, Hanover, N.H.

"The Newbury Street School: A Circular." Boston: Newbury Street School, 1874. https://books.google.com/books?id=pFcePKYLXBcC&printsec=frontcover&source=gbs_ge_summary_r&cad=0#v=onepage&q&f=false.

*Newman, John B. *Boudoir Botany, or, the Parlor Book of Flowers*. New York: Harper and Brothers, 1847.

———. *Illustrated Botany*. New York: J. K. Welman, 1846.

*———. *Illustrated Botany*. New York: Fowler and Wells, 1850.

———. *Illustrated Botany*. New York: Fowler and Wells, 1850. M. S. Snowden, owner. GEN P. L. N46, Arnold Arboretum Archives, Harvard University.

*———. *The Ladies' Flora*. New York: T. L. Magagnos, n.d.

Northrop, B. G. "Arbor Day." *Chautauquan* 5 (April 1885): 409–411.

Noyes, Penelope Barker. "From Lover's Lane to Sparks Street." *Proceedings of the Cambridge Historical Society* 41 (1967–1969): 156–169. https://historycambridge.org/wp-content/uploads/2017/08/Proceedings-Volume-41-1967-1969.pdf.

Nye, George Hyatt, and Frank E. Best, eds. *A Genealogy of the Nye Family*. Nye Family of America Association, 1907.

"Oakland Female Institute." *Ballou's Pictorial Drawing-Room Companion* 11, no. 2 (July 12, 1856): 24.

Oliver, Daniel. *Lessons in Elementary Botany*. London: Macmillan, 1864.

Osgood, Frances Sargent. *The Flower Alphabet, in Gold and Colors*. Boston: S. Colman, 1845.

*———, ed. *The Poetry of Flowers and Flowers of Poetry*. New York: J. C. Riker, 1841.

"Our Italian Visitors: The Italian Line-of-Battle Ship Re Galantuomo, The Object of her Visit, The Re d'Italia to be Convoyed to Italy, The New American Built Italian Iron-Clad." *New York Herald*, November 6, 1863.

Page, Louisa Jane, owner. *The Scriptural Album with Floral Illustrations*. New York: J. C. Riker, n.d. John Hay Library, Brown University.

Palleau-Papin, Françoise. "The Subversive Language of Flowers in *Sapphira and the Slave Girl*." *Willa Cather Pioneer Memorial Newsletter* 39 (1995): 41–44.

*Parrish, Edward. *The Phantom Bouquet: A Popular Treatise on the Art of Skeletonizing Leaves and Seed-Vessels and Adapting Them to Embellish the Home of Taste*. Philadelphia: Lippincott, 1862.

Pennisi, Elizabeth. "'A Tragic Mistake': Decision to Close Duke University's Herbarium Triggers Furor." *Science*, February 16, 2024. https://www.science.org/content/article/tragic-mistake-decision-close-duke-university-s-herbarium-triggers-furor.

Pennsylvania College Alumni Association. *General Report Made to the Alumni Association of Pennsylvania College*. Gettysburg: H. C. Neinstedt, 1860.

**Phantom Flowers: A Treatise on the Art of Producing Skeleton Leaves*. Boston: J. E. Tilton, 1864.

Phelps, Almira Hart Lincoln. *Botany for Beginners*. Hartford: F. J. Huntington, 1833.

*———. *Botany for Beginners*. Philadelphia: J. B. Lippincott, 1867.

———. *Familiar Lectures on Botany*. Hartford: F. J. Huntington, 1829.

———. *Familiar Lectures on Botany*. Hartford: F. J. Huntington, 1832.

*———. *Familiar Lectures on Botany*. Hartford: F. J. Huntington, 1836.

———. *Familiar Lectures on Botany*. New York: Huntington and Savage, 1849.

**The Philopena; or, Cousin Lill's Stories for her Pets*. New York: J. Q. Preble, n.d.

Pinkham, Lucy, owner. *The Scriptural Album with Floral Illustrations*. New York: J. C. Riker, n.d. John Hay Library, Brown University.

"The Plant Humanities Initiative." Dumbarton Oaks website. https://www.doaks.org/research/mellon-initiatives/plant-humanities-initiative.

Pollitt, Phoebe. *African American Hospitals in North Carolina: 39 Institutional Histories, 1880–1967*. Jefferson, N.C.: McFarland, 2017.

Portrait and Biographical Record of the State of Colorado. Chicago: Chapman, 1899.

Pratt, Anne. *Flora and Thalia, or, Gems of Flowers and Poetry*. Philadelphia: Carey, Lea and Blanchard, 1836.

———. *The Flowering Plants, Grasses, Sedges, and Ferns of Great Britain*. Vol. 3. London: Frederick Warne, n.d.

Prescott, Harriet Elizabeth. *Azarian: An Episode*. Boston: Ticknor and Fields, 1864.

"The Preservation of Wild Flowers." *Godey's Lady's Book* 23 (July 1841): 24.

"Preserving Plants, and Method of Making an Herbarium." *New-England Farmer*, July 31, 1844, 40.

Primack, Richard. *Walden Warming: Climate Change Comes to Thoreau's Woods*. Chicago: University of Chicago Press, 2014.

"A Proclamation, for a Day of National Humiliation, Fasting and Prayer!" The Alfred Whital Stern Collection of Lincolniana, Library of Congress, Broadside Portfolio 2, no. 1; Stern catalog 4721.

"Ralston Family Cased Photograph Collection." Catalog description, Digital Collections, Library Company of America, call number P.2012.5.1–22. https://digital.librarycompany.org/islandora/object/Islandora%3A66946?solr_nav%5Bid%5D=2bf02f5f42c901510ef8&solr_nav%5Bpage%5D=0&solr_nav%5Boffset%5D=5.

*Rattan, Volney. *A Popular California Flora, or Manual of Botany for Beginners*. San Francisco: A. L. Bancroft, 1879.

RBG Kew. *The State of the World's Plants Report—2016*. Kew, England: Royal Botanic Gardens, Kew, 2016. https://kew.iro.bl.uk/concern/reports/f931f1de-72c7-46b4-b57c-28eb417c53ec?locale=en.

"Reception and Departure of the Fifty-Fourth Regiment." *Boston Evening Transcript*, May 28, 1863, 2.

Reese, William S. *Stamped With a National Character: Nineteenth Century American Color Plate Books; An Exhibition*. New York: Grolier Club, 1999.

**The Religious Album*. New York: J. C. Riker, n.d.

Rhea, John M. *A Field of Their Own: Women and American Indian History, 1830–1941*. Norman: University of Oklahoma Press, 2016.

Rhyner, Corinne Kopcik. "Flowers of Rhetoric: The Evolving Use of the Language of Flowers in Margaret Fuller's Dial Sketches and Poetry, Elizabeth Stoddard's *The Morgesons*, Edith Wharton's *Summer*, Mary Austin's *Santa Lucia* and *Cactus Thorn*, and Susan Glaspell's *The Verge*." PhD diss., Georgia State University, 2012. https://scholarworks.gsu.edu/english_diss/88/.

Rice, Edwin Wilbur. *The Sunday-School Movement, 1780–1917 and the American Sunday-School Union, 1817–1917*. Philadelphia: American Sunday-School Union, 1917.

Richards, Thomas Addison. *The American Artist; or, Young Ladies' Instructor, in the Art of Flower Painting, in Water Colours*. Baltimore: Bayly & Burns, 1838.

Ringland, Holly. *The Lost Flowers of Alice Hart*. Toronto: Anansi International, 2018.

*Roberts, Jane, owner. *The Scriptural Album with Floral Illustrations*. New York: J. C. Riker, n.d.

Roman, Judith A. *Annie Adams Fields: The Spirit of Charles Street*. Bloomington: Indiana University Press, 1990.

"The Rose Without a Thorn." *Telescope* 1–2 (1824): 32.

Rosenthal, Debra. "Floral Counterdiscourse: Miscegenation, Ecofeminism, and Hybridity in Lydia Maria Child's *A Romance of the Republic*." *Women's Studies* 31 (2002): 221–245.

Roylance, Patricia. "Longfellow's Dante: Literary Achievement in a Transatlantic Culture of Print." Special issue, *Dante Studies* 128 (2010): 135–148.

"The Ruralist: The Herbarium." *Moore's Rural New Yorker* 9, nos. 21 and 26 (May 22 and June 26, 1858): 169 and 209.

Rusert, Britt. *Fugitive Science: Empiricism and Freedom in Early African American Culture*. New York: New York University Press, 2017.

Rutkowski, Alice. "Leaving the Good Mother: Frances E. W. Harper, Lydia Maria Child, and the Literary Politics of Reconstruction." *Legacy* 25, no. 1 (2008): 83–104.

"Sacred Trees and Flowers." *Quarterly Review* 114 (July 1863): 210–250.

Salmón, Enrique. *Iwígara: The Kinship of People and Plants.* Portland: Timber Press, 2020.

———. "Kincentric Ecology: Indigenous Perceptions of the Human-Nature Relationship." *Ecological Applications* 10, no. 5 (October 2000): 1327–1332.

Sandilands, Catriona. "Plants." In *The Cambridge Companion to Environmental Humanities,* edited by Jeffrey Cohen and Stephanie Foote, 156–169. Cambridge: Cambridge University Press, 2021.

Sattelmeyer, Robert. *Thoreau's Reading: A Study in Intellectual History, with Bibliographical Catalogue.* Princeton: Princeton University Press, 1988.

Scharnhorst, Gary. *Julian Hawthorne: The Life of a Prodigal Son.* Urbana: University of Illinois Press, 2014.

Schiebinger, Londa. *Secret Cures of Slaves: People, Plants, and Medicine in the Eighteenth-Century Atlantic World.* Stanford, Calif.: Stanford University Press, 2017.

Schneider, Robert A. "Royal G. Wilder: New School Missionary in the ABCFM, 1846–1871." *American Presbyterians* 64, no. 2 (Summer 1986): 73–82.

Seaton, Beverly. *The Language of Flowers: A History.* Charlottesville: University Press of Virginia, 1995.

Senchyne, Jonathan. *The Intimacy of Paper in Early Nineteenth-Century American Literature.* Amherst: University of Massachusetts Press, 2020.

———. "Vibrant Material Textuality: New Materialism, Book History, and the Archive in Paper." *Studies in Romanticism* 57, no. 1 (Spring 2018): 67–85.

A Series of Progressive Lessons, Intended to Elucidate the Art of Flower Painting in Water Colours. Philadelphia: Thomas Desilver, 1820.

"79 Marlborough." Genealogies of Back Bay Houses website. https://backbayhouses.org/79-marlborough/.

Shealy, Daniel, ed. *Louisa May Alcott's Fairy Tales and Fantasy Stories.* Knoxville: University of Tennessee Press, 1992.

Sheley, Nancy Strow. "The Language of Flowers in Harriet Beecher Stowe's *Uncle Tom's Cabin* and Other Nineteenth-Century American Works." *Resources for American Literary Study* 30 (2005): 76–103.

Sherman, Susan E., owner. *The Scriptural Album with Floral Illustrations.* New York: J. C. Riker, n.d. John Hay Library, Brown University.

Shinn, Kathryn K. "Gift Books." In *Edgar Allan Poe in Context,* edited by Kevin J. Hayes, 179–187. Cambridge: Cambridge University Press, 2012.

Smith, Catherine, owner. *The Scriptural Album with Floral Illustrations.* New York: J. C. Riker, n.d. John Hay Library, Brown University.

Smith, Corinne Hosfeld. *Westward I Go Free: Tracing Thoreau's Last Journey.* Winnipeg: Green Frigate Books, 2012.

Smithsonian Institution. "Spare the Flowers." Archives of American Gardens, The Garden Club of America collection.

*Soule, John. Abraham Lincoln "Skeleton Leaves" series stereoscope card. Boston: I. L. Rogers, 1874.

Southey, Robert. "The Holly Tree." *Penny Magazine* 11, no. 27 (September 1, 1832): 224.

Sprague, Charles J. "To My Herbarium." *Putnam's Monthly Magazine* 4 (December 1854): 632.

*Sprague, Isaac, illustrator, and A. B. Hervey, text. *Beautiful Wild Flowers of America.* Boston: Lothrop, 1882.

*———. *Flowers of the Field and Forest.* Boston: Cassino, 1882.

"Spring Violets." *Small Fruit Recorder and Cottage Gardener* 3, no. 4 (April 1, 1871): 49.

Steele, Eliza R. *Heroines of Sacred History.* New York: John S. Taylor and M. W. Dodd, 1841.

Stefaniak, Lisa. "Botanical Gleanings: Susan Fenimore Cooper, Catharine Parr Traill, and the Representation of Flora." In *Susan Fenimore Cooper: New Essays on Rural Hours and Other Works,* edited by Rochelle L. Johnson and Daniel Patterson, 232–248. Athens: University of Georgia Press, 2001.

Storey, Henry Wilson. *History of Cambria County, Pennsylvania.* Vol. 2. New York: Lewis Publishing, 1907.

Stowe, Harriet Beecher. *Women in Sacred History.* New York: J. B. Ford, 1873.

Student Name Index Project, Bradford Academy. Morgan Library of Ohio Imprints, 1796–1850. https://www.morganohiolibrary.com/BradfordAcademy.html.

*Sumner, George. *A Compendium of Physiological and Systematic Botany.* Hartford, Conn.: Oliver D. Cooke, 1820.

Sundberg, Marshall D. "Botanical Education in the United States, Part 2: The Nineteenth Century—Botany for the Masses vs. the Professionalization of Botany." *Plant Science Bulletin* 58, no. 3 (Fall 2012): 101–131.

Taber, Sarah Ann, owner. *The Scriptural Album with Floral Illustrations*. New York: J. C. Riker, n.d. Winterthur Library, Winterthur, Del.

"Tarbel Bancroft." Find a Grave. https://www.findagrave.com/memorial/68407374/tarbell-bancroft.

Thaxter, Celia. "Before Sunrise." *The Atlantic Monthly* 29, no. 171 (January 1872): 47–48.

———. *An Island Garden, with Pictures and Illuminations by Childe Hassam*. Boston: Houghton Mifflin, 1894.

———. *An Island Garden, with Pictures and Illuminations by Childe Hassam*. Boston: Houghton Mifflin, 2001.

*Thayer, Emma Homan. *Wild Flowers of the Pacific Coast*. New York: Cassell, 1887.

Thiers, Barbara M. *Herbarium: The Quest to Preserve & Classify the World's Plants*. Portland, Oreg.: Timber Press, 2020.

Thiselton-Dyer, Thomas Firminger. *The Folk-Lore of Plants*. London: Chatto & Windus, 1889.

Thoreau, Henry David. "The Flowering of Plants, Accidentally Observed in '51, with Considerable Care in '52; The Spring of '51 Being 10 Days and More Earlier than That of '52. The Names Those Used by Gray." Berg Collection, New York Public Library.

**The Token Album*. New York: J. C. Riker, n.d.

Tomisetta, Marie. *Poetical Selections from Celebrated Authors, Suitable for Inscription in Autograph Albums, Comprising a Choice Collection of Humorous, Friendly, Affectionate And Miscellaneous Verses*. New York: Sharps, 1880.

Tompkins, Jane. *Sensational Designs: The Cultural Work of American Fiction, 1790–1860*. Oxford: Oxford University Press, 1986.

Treat, Mary. "Botany for Invalids." *Herald of Health and Journal of Physical Culture* 8, no. 1 (July 1866): 39–40.

*[Treat, William]. *Angel Voices, or, Words of Counsel for Overcoming the World*. Boston: Ticknor and Fields, 1855.

*Turner, Cordelia Harris. *The Floral Kingdom, Its History, Sentiment and Poetry*. Chicago: Moses Warr, 1877.

Turner, Frederick Jackson. "The Significance of the Frontier in American History." *Annual Report of the American Historical Association*, 1893: 199–227.

Tuttle, Edward Mowbray. *The Book of Rural Life*. Vol. 4. Chicago: Bellows-Durham, 1925.

Ulrich, Laurel Thatcher, Ivan Gaskell, Sara Schechner, and Sarah Anne Carter. *Tangible Things: Making History Through Objects*. Oxford: Oxford University Press, 2015.

United States Census, 1870. https://www2.census.gov/library/publications/decennial/1870/population/1870a-05.pdf.

Urakova, Alexandra. "'The Ghost Book' and the Gift Book: Editorial and Marketing Strategies in Eliza Leslie's *The Gift*." *American Periodicals: A Journal of History and Criticism* 30, no. 1 (April 2020): 43–59.

U.S. Fish and Wildlife Service. "Price's Potato-Bean (*Apios priceana*) 5-Year Review: Summary and Evaluation." https://ecos.fws.gov/docs/five_year_review/doc4819.pdf.

Verberk, W. "Explaining General Patterns in Species Abundance and Distributions." *Nature Education Knowledge* 3, no. 10 (2011): 38.

*Vick, James. *Vick's Flower and Vegetable Garden*. Rochester: James Vick, n.d.

Vieira, Patrícia, Monica Gagliano, and John Ryan, eds. *The Green Thread: Dialogues with the Vegetal World*. Lanham, Md.: Lexington, 2016.

Vohland, Katrin, Anne Land-Zandstra, Luigi Ceccaroni, Rob Lemmens, Josep Perelló, Marisa Ponti, Roeland Samson, and Katherin Wagenknecht. "Editorial: The Science of Citizen Science Evolves." In Vohland et al., *The Science of Citizen Science*, 1–12.

Vohland, Katrin, Anne Land-Zandstra, Luigi Ceccaroni, Rob Lemmens, Josep Perelló, Marisa Ponti, Roeland Samson, and Katherin Wagenknecht, eds. *The Science of Citizen Science*. Cham, Switzerland: Springer, 2021.

Wall, L. L. "The Medical Ethics of Dr. J. Marion Sims: A Fresh Look at the Historical Record." *Journal of Medical Ethics* 32, no. 6 (June 2006): 346–350.

Walker, Tacy Anna, owner. *The Scriptural Album with Floral Illustrations*. New York: J. C. Riker, n.d. John Hay Library, Brown University.

Warner, Deborah Jane. "Science Education for Women in Antebellum America." *Isis* 69, no. 1 (March 1978): 58–67.
Warren, James Perrin. *Thoreau's Botany: Thinking and Writing with Plants*. Charlottesville: University of Virginia Press, 2023.
Waterman, Catharine H. *Flora's Lexicon*. Philadelphia: Hooker and Claxton, 1839.
*———. *Flora's Lexicon*. Philadelphia: Herman Hooker, 1840.
Webb, George E. "The Chemist as Consultant in Gilded Age America: Benjamin Silliman, Jr. and Western Mining." *Bulletin of the History of Chemistry* 15/16 (1994): 9–14.
Welch, Margaret. *The Book of Nature: Natural History in the United States, 1825–1875*. Boston: Northeastern University Press, 1998.
Weld, H. Hastings. *The Women of the Scriptures*. Philadelphia: Lindsay and Blakiston, 1848.
Welsh, Jane. *A Botanical Catechism*. Northampton, Mass.: T. W. Shepard, 1819.
———. *The Pastime of Learning*. Boston: Cottons and Barnard, 1831.
White, James Terry. *Captive Memories*. New York: James T. White, 1897.
*Whittier, John Greenleaf, ed. *Jack in the Pulpit*. New York: R. Worthington, 1884.
**Wild Flowers of America*. New York: G. H. Buek, 1894.
"The Wild Flowers of Early Spring-Time." *Godey's Lady's Book* 48 (April 1854): 343–345.
"Wild Geranium." In *Spring Wild Flowers*, 23. Leaflet 7, Popular Series: Botany. Chicago: Field Museum of Natural History, 1924.
Wilkinson, Lady Caroline Catherine. *Weeds and Wild Flowers: Their Uses, Legends, and Literature*. London: John Van Voorst, 1858.
Winship, Michael. *American Literary Publishing in the Mid-Nineteenth Century: The Business of Ticknor and Fields*. Cambridge: Cambridge University Press, 1995.
———. "Manufacturing and Book Production." In *A History of the Book in America*, vol. 3: *The Industrial Book 1840–1880*, edited by Scott Casper et al., 40–69. Chapel Hill: University of North Carolina Press, 2007.
Wirt, Elizabeth W. *Flora's Dictionary*. Baltimore: Fielding Lucas Junior, 1832.
*———. *Flora's Dictionary*. Baltimore: Fielding Lucas Junior, [ca. 1837].
Wolfe, Charles. "A Song—To Mary." In *English Literature of the Nineteenth Century*, edited by Charles Dexter Cleveland, 133–134. Philadelphia: Biddle, 1857.
"Women's Medical College of Pennsylvania Student Theses: 1851–1895, 1926–1942." Drexel University. https://drexel.edu/~/media/Files/legacy-center/women-in-medicine/a072_wmcp-theses.pdf.
**The Wonders of Vegetation: The Seed*. Philadelphia: American Sunday-School Union, 1845.
Wood, Alphonso. *A Class-Book of Botany*. Boston: Crocker and Brewster, 1845.
*———. *A Class-Book of Botany*. Boston: Crocker and Brewster, 1855. Hezekiah W. Marbourg, owner.
———. *A Class-Book of Botany*. New York: A. S. Barnes and Burr, 1861.
———. *First Lessons in Botany*. Boston: Crocker and Brewster, 1851.
———. *How to Study Plants*. New York: A. S. Barnes, 1879.
*———. *Leaves and Flowers; or, Object Lessons in Botany*. New York: A. S. Barnes, 1863.
Woods, Harriet F. *Historical Sketches of Brookline, Mass.* Boston: Robert S. Davis, 1874.
Wordsworth, William. *Peter Bell: A Tale in Verse*. London: Longman, Hurst, Reese, Orme, and Brown, 1819.
Youmans, Eliza A. *The First Book of Botany. Designed to Cultivate the Observing Powers of Children*. New York: D. Appleton and Company, 1870.
*———. *The First Book of Botany*. New York: D. Appleton and Company, 1873.
Zebuhr, Laura. "The Work of Friendship in Nineteenth-Century American Friendship Album Verses." *American Literature* 87, no. 3 (September 2015): 433–454.
Zimmer, Carl. "Duke Shuts Down Huge Plant Collection, Causing Scientific Uproar." *New York Times*, February 21, 2024. https://www.nytimes.com/2024/02/21/science/duke-herbarium.html.
Zimmerman, Erin. *Unrooted: Botany, Motherhood, and the Fight to Save an Old Science*. Brooklyn, N.Y.: Melville House, 2024.

Index

Page numbers in italic refer to images.